Optical Constants of Inorganic Glasses

The CRC Press
Laser and Optical Science and Technology Series

Editor-in-Chief: Marvin J. Weber

Handbooks of Laser Science and Technology

Volume I: Lasers and Masers, Edited by Marvin J. Weber

Volume II: Gas Lasers, Edited by Marvin J. Weber

Volume III: Optical Materials, Part 1, Edited by Marvin J. Weber

Volume IV: Optical Materials, Part 2, Edited by Marvin J. Weber

Volume V: Optical Materials, Part 3, Edited by Marvin J. Weber

Supplement I: Lasers, Edited by Marvin J. Weber

Supplement II: Optical Materials, Edited by Marvin J. Weber

Thermodynamic and Kinetic Aspects of the Vitreous State

Sergei V. Nemilov

Forthcoming Handbook Titles

Handbook of Laser Wavelengths, Edited by Marvin J. Weber

Optical Constants of Inorganic Glasses

Andrei M. Efimov
Research and Technological Institute
of Optical Materials Science
State Scientific Center
Vavilov State Optical Institute
St. Petersburg, Russia

CRC Press
Boca Raton New York London Tokyo

Library of Congress Cataloging-in-Publication Data

Efimov, Andrei M.
Optical constants of inorganic glasses / by Andrei M. Efimov
p. cm. — (Laser science & technology series)
Includes bibliographical references and index.
ISBN 0-8493-3783-6
1. Glass—Optical properties. 2. Physical constants. I. Title. II. Series.
TA450.E47 1995
620.1′4495—dc20 94-47106
CIP

International Standard Book Number 0-8493-3783-6
Library of Congress Card Number 94-47106
Printed in the United States of America 1 2 3 4 5 6 7 8 9 0
Printed on acid-free paper

PREFACE

Optical and spectroscopic properties of inorganic glasses have much in common with those of other inorganic solids; however, there is also a set of specific features distinguishing the vitreous state from the crystalline state. Here I mean specific features concerning the basic properties of a vitreous matrix as such, rather than those of various optical centers differing in nature from the "perfect" matrix, such as intrinsic point defects, impurity or dopant ions, various microcrystals, etc. Because of their applied prospects, it is the properties of such centers that have attracted the attention of the majority of investigators for the past decades. The physical optics of the vitreous matrix, however, remains to be of principal importance from the viewpoint of obtaining further insights into both glass structure and the nature of fundamental optical excitations in glass.

In spite of a long history of studies of the physical optics of inorganic glasses, such studies are far from completion. This is due mainly to difficulties originating from the lack of translational symmetry in the glass structure. Indeed, such key notions of solid state physics as the unit cell and the Brillouin zone are not applicable to vitreous solids. Hence, there are (1) substantial obstacles for developing rigorous theories of the optical spectra of glasses; (2) an abundance of prior crude approaches (such as those involving the molecular-like, crystallite-based, or small cluster models) to the interpretation of glass spectra; and (3) serious doubts concerning the expediency of applying to the spectroscopy of glasses powerful methods of quantitative data treatment or computer modeling borrowed straightforwardly from the spectroscopy of crystals. As a result, quantitative approaches have not been developed in glass spectroscopy until recently; moreover, a tendency arose to divide the unified totality of physical phenomena related to the interaction of a material with an electromagnetic field into topics of glass spectroscopy and of glass optics, these topics being treated separately. Thus, this field is not a simple one to discuss.

There is no doubt that the problem of the frequency dispersion of optical constants is of primary interest for both the spectroscopy of fundamental excitations in glass and the optics of glass in the transparency range. Moreover, it is this problem that connects the two topics together, thus revealing a broad unified field as it actually is. In particular, the pattern of frequency dispersion of the refractive index in the transparency range is only a small part of the totality of frequency dispersion of optical constants over the entire optical frequency range. This interconnection, though evident qualitatively, is taken into account on a quantitative level (if any) only approximately in many cases. Some recent findings contributing to the field have been published solely in Russian literature and have remained practically unnoticed by American, European, or Asian readers; therefore, a book devoted to the totality of problems related to the quantitative analytical presentation of the frequency dispersion of optical constants seems to be appropriate.

In the case of a broad field, this book, naturally, can pretend to be neither a review covering all available literature in the field nor a handbook of all computational methods developed up to the present; otherwise, the total volume of the book would exceed all reasonable limits. Rather, this book should present a self-consistent general treatment intended to orient readers in the scope of problems involved in the field. In particular, the book should equip readers with basic information concerning (1) the nature of the underlying physical phenomena and the various analytical presentations of the frequency dispersion of optical constants arising from these phenomena; (2) the most appropriate computational methods applicable for reconstructing the spectra of optical constants of glasses in the ranges of fundamental excitations; (3) available data on the spectra of optical constants of glasses; and (4) the interpretation of empirical regularities observed for the frequency dispersion of optical characteristics of glasses in the transparency range.

The choice of the most appropriate structure for this book was not a simple task. At first glance, it seemed natural to present each version of analytical presentation for the frequency dispersion of optical constants together with results obtained using this version. In this case, however, various presentations would be too separated, thus depriving readers of the opportunity to grasp all of the totality of presentations available and making the discussion of interrelations between them difficult. Therefore, I preferred to systemize the material in another way: namely, to present successively (1) the totality of analytical models available, (2) the totality of computational methods, and (3) the totality of results. Of course, this led immediately to the necessity of providing a lot of cross-references.

Historically, the numerical presentation of the frequency dispersion of optical constants of inorganic glasses was based on notions, analytical models, and computational methods borrowed from the physical optics of crystals. The expediency of using many of these models and computational methods (though not all of them) for glasses remains to a lesser or greater extent, up to the present. Therefore, a brief review of notions and analytical models for the dielectric constant assumed to be partially applicable for all solids, glasses included, is presented in Chapter 1. The specific features of optical spectrum formation and the resultant behavior of optical constants of glasses in the frequency ranges of fundamental excitations are described, based on recent findings, in Chapter 2. Methods for reconstructing the spectra of optical constants, including the one intended specifically for glasses, are reviewed briefly in Chapter 3. Experimental data on the optical constants of various inorganic glasses in the ranges of fundamental excitations obtained with these methods are presented in Chapter 4. These data concern mostly the binary oxide glasses because it is the latter that have been studied predominantly until now. In addition to oxide glasses, the little data available for halide and chalcogenide glasses are reviewed. The optical characteristics of interest in applied optics are reviewed and the interpretation of these characteristics of glasses in terms of the parameters of the fundamental excitations are presented in Chapter 5. Finally, the numerical values of the optical constants vs. frequency for various glasses are given in the Appendix for the broad frequency range of the fundamental excitations.

When writing this book, I naturally could not help but assume that the reader has some preliminary acquaintance with the field. In this aspect, various chapters of the book differ from each other to some extent. Chapter 1, being introductory in character, requires only limited preliminary information because most of the notions touched upon are explained in the text. This also applies, to some extent, for Chapter 5. At the same time, to keep the volume of the book within reason, Chapter 2 requires that basic notions of theories involved in the problem — such as phonon, exciton, wavevector, density of states, etc. — be known beforehand.

I hope that this book will be helpful to scientists and students interested in studies of both the spectroscopy of fundamental excitations in glasses and the optics of glass in the transparency range. The proof of the pudding is in the eating; let the readers judge if my attempt to write such a book is a success.

I would like to take this opportunity to acknowledge, first of all, my indebtedness to Dr. M. J. Weber, who presented me with the opportunity to write this book. I wish to acknowledge also my deepest respect to the memory of Prof. A. N. Lazarev, who for years encouraged me in my studies that are the basis for this book, and my gratitude to Dr. E. G. Makarova for her fruitful collaboration in these studies. Last, but not least, my thanks are addressed to many of my colleagues who helped me overcome numerous problems with preparing the manuscript.

Andrei M. Efimov

THE AUTHOR

Andrei Markovich Efimov was born in St. Petersburg (formerly Leningrad), Russia, in 1937. He graduated from the Leningrad State University (Department of Chemistry) in 1960. Since then, he has been associated with the S. I. Vavilov State Optical Institute in St. Petersburg. He earned his "Kandidat" degree in 1966 in Chemistry and his Doctor of Science degree in 1991 in Physics and Mathematics from the Vavilov State Optical Institute.

Dr. Efimov has extensive experience in various fields of glass science. During the last sixteen years he has developed quantitative models and approaches to understanding the optical and spectroscopic properties of inorganic glasses as related to their structure. Dr. Efimov has more than one hundred publications in scientific journals and conference proceedings. He has been a member of the program committees of many all-union conferences and of the Fifteenth International Congress on Glass. Since1985, he has also served as a member of the Editorial Board of the journal *Fizika i Khimiya Stekla* (Glass Physics and Chemistry). Recently, Dr. Efimov was appointed a professor at St. Petersburg's Institute of Fine Mechanics and Optics.

CONTENTS

Chapter 4
Spectra of Optical Constants in the Ranges of Fundamental Excitations for Particular Glass Types

Chapter 5
Refractive Index Dispersion of Glasses in the Transparency Range: Applied Aspects

LIST OF STANDARD SYMBOLS

c	light velocity in vacuum
e	electron charge
E	amplitude of electric field
E_g	band gap for direct transitions
f	oscillator strength (transition probability)
g	density of states
h	Planck's constant
m	mass of a particle (of electron, unless otherwise indicated)
N	number (of particles, oscillators, etc.) per unit volume
$\hat{n}$	complex refractive index
n	real part of refractive index
P (i)	polarization
P (ii)	relative partial dispersion
Q	error function
R	reflectivity
$\hat{r}$	complex amplitude of reflection
S	oscillator intensity
x	variable oscillator frequency
$\hat{x}$	complex displacement (of particle under field)
Z	effective charge
α	absorption coefficient
$\hat{\alpha}$	complex polarizability
γ	damping coefficient
$\hat{\varepsilon}$	complex dielectric constant
ε'	real part of complex dielectric constant
ε''	imaginary part of complex dielectric constant
ε_∞	dispersionless high-frequency part of dielectric constant
ε_0	"static" dielectric constant
κ	absorption index (imaginary part of complex refractive index)
λ	wavelength
υ	Abbe number (constringence)
ρ	density
σ	standard deviation
ϕ	phase angle of reflection
ω	angular frequency

Chapter 1

Frequency Dispersion of Optical Constants of a Solid: General Relationships

I. BASIC PHENOMENA AND TERMINOLOGY

A. OPTICAL CONSTANTS AND POLARIZATION PHENOMENA

Prior to discussing the problem of how to present analytically the frequency dependence of optical and spectroscopic properties of a solid, it is necessary to remember that the dielectric constant and refractive index of the condensed media with nonvanishing conductivity are, according to the solutions of Maxwell's equations, the complex quantities as follows:

$$\hat{\varepsilon}(\omega) = \varepsilon'(\omega) + i\varepsilon''(\omega) \tag{1}$$

and

$$\hat{n}(\omega) = n(\omega) + i\kappa(\omega). \tag{2}$$

Here $v(\omega)$ is the real refractive index, $\kappa(\omega)$ is the absorption index, and w is the current circular frequency; $\varepsilon' = n^2 - \kappa^2$ and $\varepsilon'' = 2n\kappa$. Thus both the characteristics of an electromagnetic wave in a medium and, as will be seen below, the response of the medium to the effect of an electromagnetic field can be fully described on a macroscopic level in terms of a pair of *optical constants,* namely, either n and κ or ε' and ε''. Of course, optical constants are constants only for a particular value of current frequency ω; with varying ω, they are the functions of the latter. Maxwell's theory as such does not presuppose any interpretation of the frequency dependence of optical constants; the analytical form of this dependence should follow from the particular physical model for the interaction between the field and the substance (see below).

It is the spectroscopic meaning of the term optical constants that is specified above. There also exists a practice in applied optics (especially in Russian literature) of using this term to mean something quite different, namely, that comprising refractive index values at some fixed wavelengths from the transparency range and various simple functions of these values, such as *Abbe number, main dispersion, and partial dispersions,* etc. (see Chapter 5, Section I.A). In this book, the term optical constants will refer solely to its spectroscopic meaning; to avoid misunderstanding, the quantities used in applied optics will be referred to as "optical characteristics".

The response of the medium to the effect of an electromagnetic field manifests itself in the phenomenon of *polarization,* i.e., shifting the electrical charges initially present in the medium and/or generating new charges under the effect of the field. The subject of the response of the medium to the effect of an electromagnetic field is discussed in detail in many review papers and monographs, such as those of Korff and Breit,[1] Seitz,[2] Moss,[3] Born and Wolf,[4] and Garbuny.[5]

The magnitude of the polarization *P*, being the total electric dipole moment of unit volume that is generated by the effect of an external electromagnetic field, is given by

$$P = N\hat{\alpha}E. \tag{3}$$

Here *E* is the external electric field, *N* is the number of charged particles (in the standard approach, bound electrons) per unit volume, and $\hat{\alpha}$ is the so-called *polarizability* of particles that plays the role of a time-invariant proportionality coefficient between the magnitudes of the applied field and the polarization of the material. It follows from the relations of electrostatic field theory that *P* is also connected to *E* via the dielectric constant as follows:

$$P = E \cdot \frac{\hat{\varepsilon} - 1}{4\pi}. \tag{4}$$

Combining (3) and (4) yields for the complex dielectric constant

$$\hat{\varepsilon} = 1 + 4\pi N\hat{\alpha}. \tag{5}$$

Thus a macroscopic quantity such as the dielectric constant is a measure of the polarizability; the latter, in a general case, is also the complex quantity. Hence, an interconnection between optical constants and microcharacteristics of a material is established by the particular analytical model assumed for the frequency dependence, or *frequency dispersion,* of the polarizability $\hat{\alpha}(\omega)$ and of the complex dielectric constant $\hat{\varepsilon}(\omega)$.

Formulas (3) through (5) and also those derived directly from the latter are related to the case when the internal electric field in a substance is taken to be the same as the external one. For condensed media, the problem of the so-called local field generated by the polarization of the medium and causing some difference between the internal and external fields has also been the subject of theoretical investigation and widely discussed. This problem is touched upon briefly in Section II.B.

B. KRAMERS-KRONIG RELATIONS

The pair of optical constants are not independent of one another but are related. An interconnection between them is the well-known Kramers-Kronig relations described elsewhere and also various sum rules.[6] For $\varepsilon'(\omega)$ and $\varepsilon''(\omega)$, the Kramers-Kronig relations are as follows:

$$\varepsilon'(\omega) = 1 + \frac{2}{\pi} \mathrm{P} \int_0^\infty \frac{\omega^* \varepsilon''(\omega^*)}{\omega^{*2} - \omega^2} d\omega^* \tag{6}$$

and

$$\varepsilon''(\omega) = -\frac{2\omega}{\pi} \mathrm{P} \int_0^\infty \frac{[\varepsilon'(\omega^*) - 1]}{\omega^{*2} - \omega^2} d\omega^*, \tag{7}$$

the frequency ω^* being the variable of integrating. Here and in Formulas (8) through (10) to follow, P denotes the Cauchy principal value of the integral.

Similar relations hold for an interconnection between $n(\omega)$ and $\kappa(\omega)$:

$$n(\omega) = 1 + \frac{2}{\pi} P \int_0^\infty \frac{\omega^* \kappa(\omega^*)}{\omega^{*2} - \omega^2} d\omega^* \tag{8}$$

and

$$\kappa(\omega) = -\frac{2\omega}{\pi} P \int_0^\infty \frac{[n(\omega^*) - 1]}{\omega^{*2} - \omega^2} d\omega^*. \tag{9}$$

Another Kramers-Kronig relation is used when considering the phase of the reflected beam (see Chapter 3, Section II.B).

Along with the interconnection between the optical constants, the Kramers-Kronig relations implicitly determine their frequency dispersion. As is seen from Formulas (6) through (9), if an optical constant is known for all frequencies then its counterpart can be calculated for any particular frequency of interest. In particular, the magnitude of $\varepsilon'(\omega)$ or of refractive index at any fixed frequency is wholly determined by the full spectrum of $\varepsilon''(\omega)$ or $\kappa(\omega)$, respectively. Therefore, it is the spectra of optical constants in the frequency ranges of fundamental excitations (both the vibrational and electronic ones) that substantially determine the absolute values and frequency dispersion of the optical constants in any other ranges, the transparency range included, hence the importance of studying the spectra of fundamental excitations for applied optics. This fact is more evident in the modification of (8) given by Efimov and Makarova[7] as

$$n(\omega) = 1 + \frac{2}{\pi} \left[P \int_0^a \frac{\omega^* \kappa(\omega^*)}{\omega^{*2} - \omega^2} d\omega^* + P \int_b^\infty \frac{\omega^* \kappa(\omega^*)}{\omega^{*2} - \omega^2} d\omega^* \right] + \eta(\omega), \tag{10}$$

where a and b are the frequency limits of the transparency range and $\eta(\omega)$ is the small contribution resulting from $\kappa(\omega^*)$ "tails" in this range.

Being restricted by no particular physical considerations, the Kramers-Kronig relations are quite universal, i.e., equally applicable to materials of any nature. However, Kramers-Kronig relations say nothing about the actual form of the analytical function describing the frequency dispersion of the optical constants. It is the analytical model of the dielectric constant that, as was already mentioned, determines this function explicitly.

C. GENERAL PATTERN OF FREQUENCY DISPERSION OF OPTICAL CONSTANTS FOR A SOLID

To illustrate the principal character of the frequency dispersion of optical constants in the neighborhood of a single strong excitation in a solid, the spectra of the optical constants of KBr are plotted (with minor irregularities smoothed) in Figure 1 using the experimental data of Reference 8. When plotting the IR spectra and denoting band locations, it is customary, for avoiding numbers that are too large, to present frequencies not in the units of the circular frequency itself (i.e., in s^{-1}) but in the

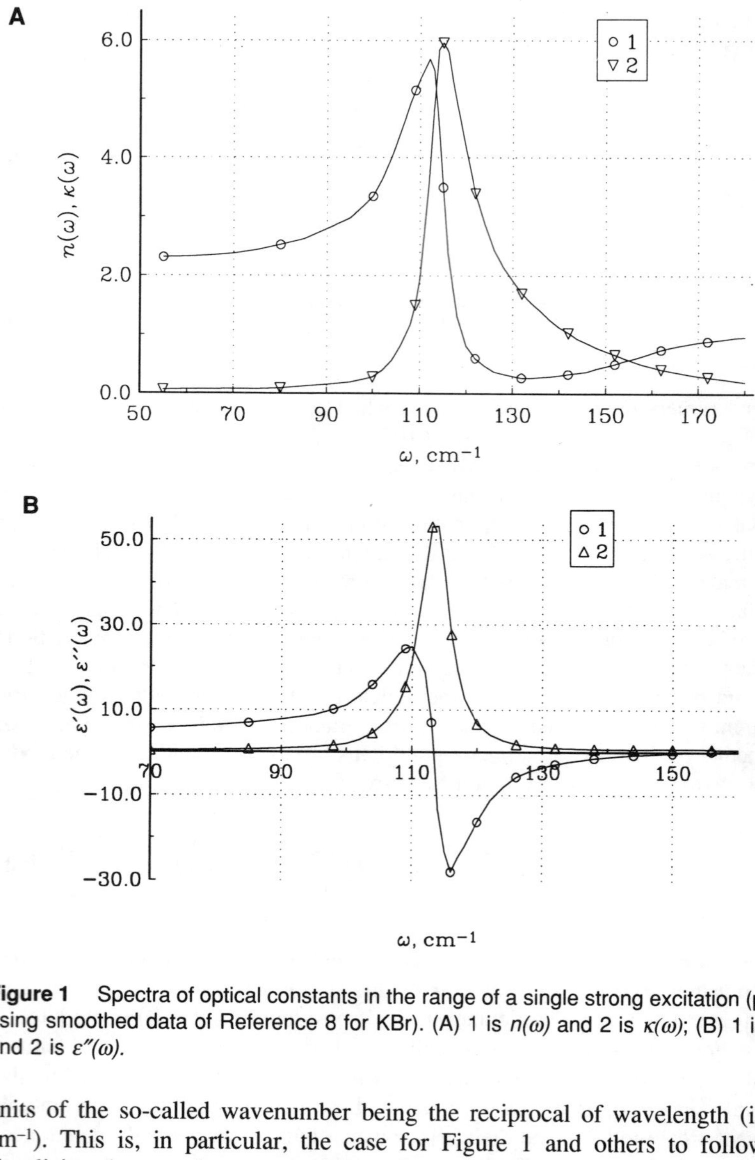

Figure 1 Spectra of optical constants in the range of a single strong excitation (plotted using smoothed data of Reference 8 for KBr). (A) 1 is $n(\omega)$ and 2 is $\kappa(\omega)$; (B) 1 is $\varepsilon'(\omega)$ and 2 is $\varepsilon''(\omega)$.

units of the so-called wavenumber being the reciprocal of wavelength (i.e., in cm^{-1}). This is, in particular, the case for Figure 1 and others to follow; for simplicity, the quantity presented in cm^{-1} remains to be referred to as frequency. The characteristic asymmetric contour of $n(\omega)$ or $\varepsilon'(\omega)$ and the symmetric one of $\varepsilon''(\omega)$ are evident; the phenomenon of decreasing refractive index with an increase in frequency in the vicinity of $\varepsilon''(\omega)$ maximum known as the so-called *anomalous dispersion* is observed explicitly. The physical nature of the excitations forming the fundamental spectra of various solids will be touched upon in Chapter 2; here, it seems expedient to outline phenomenologically the general pattern of these spectra as related to corresponding frequency ranges.

The optical spectrum related to the fundamental vibrational excitations in a solid is located in the IR range and usually comprises a limited number of bands. The shapes of the respective $\kappa(\omega)$ or $\varepsilon''(\omega)$ features are approximately similar for crystals to those shown in Figure 1. Thus, the IR spectrum of fundamental excitations in a crystal has, in principle, a discrete character.[a] The number of bands in the spectrum may range from 1 (for simple cubic lattices) to 20 to 30 (for very complicated structures). Some bands overlap in most cases, thus making the total pattern of optical constants much more complicated as compared to that shown in Figure 1. The fundamental vibrational spectra of glasses reveal many specific features that are discussed in detail in Chapter 2.

The optical spectrum related to the fundamental electronic transitions in dielectrics and ionic conductors is located mostly in the VUV and near UV ranges; for semiconductors, it extends into the visible or even near IR range.[b] For this spectrum, two kinds of excitations need to be distinguished, namely, the *interband* transitions and *excitons* (see Chapter 2, Section I.C). The spectrum of interband transitions is usually considered to be continuous and inseparable into individual components, with the shapes of singularities differing from those shown in Figure 1. Conversely, strong excitonic transitions are characterized by narrow lines, the variety of possible lineshapes comprising those illustrated by Figure 1.

The transparency range of solids, which may include the near IR, visible, and near UV, lies inbetween two regions of strong absorption related to two kinds of fundamental excitations mentioned above. The "normal" dispersion of the refractive index (i.e., that characterized by an increase in the refractive index with an increase in frequency) in the transparency range may be understood qualitatively as that occurring in the range between two strong excitations similar to those shown in Figure 1.

Due to substantial applied interest, the first attempts to describe quantitatively the frequency dispersion of optical constants dealt with the behavior of the refractive index in the transparency range. Early approaches gave the explicit analytical presentation of the refractive index dispersion, such as that of Cauchy and others (these are reviewed, for example, by Korff and Breit[1]). However, all of these approaches lacked a theoretical basis. Later, the problem of how to establish a correspondence between the fundamental spectra and the behavior of $n(\omega)$ in the transparency range was key when developing the dispersion theory.

In the course of developing the dispersion theory, various analytical models for the complex dielectric constant were put forth aimed at either solving the problem of interconnection between the fundamental spectra and the behavior of $n(\omega)$ in the transparency range or at describing the fundamental spectra as such.

It should be stressed that the problem of the theoretical approach to the frequency dispersion of the dielectric constant for a solid is complicated. Indeed, the analytical models pretending to provide a rigorous description of the electronic contribution to the dielectric constant should take into account the actual band structure of a material under consideration. Along with principal shortcomings inherent in the field (the discussion of which lies outside the scope of this book), one should also take into account that any rigorous analytical model of this kind would inevitably not be a general one, i.e., it would be applicable, at best, to a

[a] Generally, the pronounced discrete character, due to much lower linewidths and, hence, to negligibly small overlapping, is typical of spectra of gaseous substances; however, the particular physical nature of excitations in solids differs from that of the latter (see Chapter 2, Section I).

[b] The cases of narrowband and doped semiconductors and metals, for which the presence of free carriers (electrons and holes) determines substantially all optical properties, lies outside the topic of this book.

narrow set of solids with similar crystal lattices (see, for example, the studies of Pantelides[9] and Chelikowsky[10]). On the contrary, only universal models may form a basis for developing methods applicable for treating the experimental spectroscopic and optical data of a broad variety of materials. As will be seen later, such models are necessarily simplified; it is these models that are considered in this book. Restrictions imposed on the complexity of models that are intended for the treatment of experimental data are specified in Chapter 3, Section III.

II. THE CLASSICAL ANALYTICAL MODEL FOR THE COMPLEX DIELECTRIC CONSTANT

The early and simple analytical model for the complex dielectric constant, and one of the most widely used for treating experimental data, is borrowed from classical dispersion theory. This theory was derived mostly from the works of Drude, Voigt, and Lorentz and is summarized in many review papers and monographs, including those cited above.[1-5]

A. THE PRINCIPAL (DRUDE) MODEL

The classical dispersion theory, in its initial form, considered the polarization resulting from the displacement of bound electrons from their equilibrium positions under the effect of an external electromagnetic field. The Newtonian differential equation for an uncoupled damped oscillator was taken as the starting point:

$$\frac{m\mathrm{d}^2 x}{\mathrm{d}t^2} + m\gamma \frac{\mathrm{d}x}{\mathrm{d}t} + kx = eE_0 \exp(i\omega t), \tag{11}$$

where x is the displacement, m is the electron mass, e is the electron charge, γ is the damping coefficient of the oscillator, and k is the force constant determining the restoring force. As is seen from (11), the left-hand side is similar to that of the equation for an oscillator exhibiting the mechanical vibrations; the driving force on the right-hand part is the one created by the applied field. The solution of (11) gives for the displacement $\hat{x}(\omega)$:

$$\hat{x}(\omega) = \frac{eE_0}{m} \cdot \frac{1}{\omega_0^2 - \omega^2 - i\gamma\omega}. \tag{12}$$

For an undamped harmonic oscillator, $\hat{x}(\omega)$ is real and the resonant frequency, ω_0, is given simply by $\omega_0 = \sqrt{k/m}$.

The polarizability, $\hat{\alpha}(\omega)$, is connected to the displacement $\hat{x}(\omega)$ by $\hat{\alpha} = e\,\hat{x}/E$; thus one can express the frequency dispersion of $\hat{\varepsilon}(\omega)$ explicitly via the combination of (5) and (12). According to the later findings of the quantum mechanical dispersion theory, a dimensionless oscillator strength, f, characterizing the transition probability ($0 \le f \le 1$) should be introduced in addition to the parameters used in (12). Taking this into account, the combination of (5) and (12) yields for the complex dielectric constant of a material containing J kinds of oscillators:

$$\hat{\varepsilon}(\omega) = 1 + \sum_{j=1}^{J} \frac{4\pi N_j e^2 f_j}{m} \cdot \frac{1}{\omega_{0j}^2 - \omega^2 - i\gamma_j\omega}. \tag{13}$$

Initially, as already mentioned, the classical dispersion theory dealt with oscillators related solely to electronic excitations (taken in the classical approach); therefore, in (13), m was taken to be the free electron mass and e to be the electron charge. Later, when considering the electronic excitations, the free electron mass m was replaced, naturally, by the effective electron masses, m^*. Moreover, the notion of an oscillator acquired a more generalized meaning, namely, that of any microscopic system (a particle or some combination of particles) capable of transferring from the ground state to the excited state under the effect of optical excitation; thus, it also became applicable to vibrational excitations. In this case, m should be replaced by the reduced masses, m_j, of atomic pairs or groups[c] participating in the vibrations and e should be replaced by Z_je, Z_j being the effective charge of the atoms or ions. Many monographs omit this consideration and keep treating (13) solely in terms of electronic excitations.

The most generalized version of analytical model (13), taking into account the above considerations and assuming that $\omega_{0j} = \omega_{0j}(\omega)$, $f_j = f_j(\omega)$, and $\gamma_j = \gamma_j(\omega)$ (or, equivalently, $J \to \infty$), turns out to be valid from the viewpoint of any rigorous quantum mechanical dispersion theory.[11,12] However, classical dispersion theory in its historical form and modern applications of (13) presuppose that the J magnitude is limited (in fact, rather small) and the parameters of every kind of oscillator are constants, thus providing the simplified character of the model (see, also, Chapter 3, Section III).

Introducing for brevity the intensities, $S_j = 4\pi N_j e^2 f_j/m_j$, (13) takes the more compact form,

$$\hat{\varepsilon}(\omega) = 1 + \sum_{j=1}^{J} \frac{S_j}{\omega_{0j}^2 - \omega^2 - i\gamma_j\omega}. \quad \textbf{(14)}$$

Thus, in this model, every kind of oscillator is characterized by three free parameters, ω_{0j}, S_j, and γ_j that are frequency independent. The real and imaginary parts of (14) are:

$$\varepsilon'(\omega) = 1 + \sum_{j=1}^{J} \frac{S_j\left(\omega_{0j}^2 - \omega^2\right)}{\left(\omega_{0j}^2 - \omega^2\right)^2 + \gamma_j^2\omega^2} \quad \textbf{(15)}$$

and

$$\varepsilon''(\omega) = \sum_{j=1}^{J} \frac{S_j\gamma_j\omega}{\left(\omega_{0j}^2 - \omega^2\right)^2 + \gamma_j^2\omega^2}. \quad \textbf{(16)}$$

In (13) through (16), the summation is made over all optical excitations contributing to $\varepsilon''(\omega)$. Most experimental studies consider only a rather limited frequency range that does not comprise the highest oscillator frequencies. Denoting the latter by ω_{0k}, one may assume $\omega^2_{0k} - \omega^2 - i\gamma_j\omega \approx \omega^2_{0k}$ and rewrite (14) replacing the contributions of such oscillators by a single high-frequency dielectric constant, ε_∞, and defining $J = J - K$:

---

[c] For a pair of atoms with mass m_1 and m_2, the reduced mass m is given by $m_1m_2/(m_1 + m_2)$.

$$\hat{\varepsilon}(\omega) = \varepsilon_\infty + \sum_{j=1}^{J} \frac{S_j}{\omega_{0j}^2 - \omega^2 - i\gamma_j\omega}. \quad \textbf{(17)}$$

Here, $\varepsilon_\infty = \sum S_k/\omega^2_{0k}$, $\omega_{0k} > \omega_{0(k+1)}$. It is this form of the classical $\hat{\varepsilon}(\omega)$ model, after Spitzer and Kleinman,[13] that is usually used in IR spectroscopy for treating experimental data with the method of *dispersion analysis* (see Chapter 3, Section III). Moreover, the intensities S_j may be replaced by the contributions of oscillators, $\Delta\varepsilon_{0j} = S_j/\omega^2_{0j}$, to the so-called static dielectric constant, ε_0, that comprises the contributions of all optical excitations at zero frequency:

$$\varepsilon_0 = \varepsilon_\infty + \sum_{j=1}^{J} \frac{S_j}{\omega_{0j}^2} = \varepsilon_\infty + \sum_{j=1}^{J} \Delta\varepsilon_{0j}. \quad \textbf{(18)}$$

In this case, (14) takes the following form:

$$\hat{\varepsilon}(\omega) = \varepsilon_\infty + \sum_{j=1}^{J} \frac{\Delta\varepsilon_{0j}\,\omega_{0j}^2}{\omega_{0j}^2 - \omega^2 - i\gamma_j\omega}. \quad \textbf{(19)}$$

This expression, or its simpler form for the case $J = 1$,

$$\varepsilon(\omega) = \varepsilon_\infty + \frac{(\varepsilon_0 - \varepsilon_\infty)\omega_0^2}{\omega_0^2 - \omega^2 - i\gamma\omega}, \quad \textbf{(20)}$$

is also used in some works.

Some remarks of a more general character also should be made. First, it should be noted that the band shape (16) assumed by the classical dispersion theory is Lorentzian. In absorption spectroscopy, this type of contour is usually presented in a somewhat different form that was derived earlier irrespective of dispersion theory:

$$I(\omega) = I_0 \frac{\Gamma^2/4}{(\omega - \omega_0)^2 + \Gamma^2/4}. \quad \textbf{(21)}$$

Here $I(\omega)$ is the intensity at the current frequency ω, I_0 is the peak intensity of spectrum line, and Γ is the halfwidth of the latter. It is easy to demonstrate that (21) is the simplified form of (16) valid for a single weak band and for frequencies lying near the absorption maximum. Indeed, with the assumptions $n(\omega) \approx 1$, $\omega_{0j} + \omega \approx 2\omega_{0j}$ and, hence, $\omega_{0j}^2 - \omega^2 \approx 2\omega_{0j}(\omega_{0j} - \omega)$, a single term from (16) is reduced immediately to the form of (21).

Second, some aspects of applying the classical analytical model for the complex dielectric constant to various media should be touched upon. As was already mentioned, the oscillator parameters ω_{0j}, S_j, and γ_j are assumed to be frequency independent. This may be an adequate approximation when considering the properties of gases (which were the first type of substances studied with classical dispersion theory). In this case, oscillator parameters may be close to the actual frequencies, intensities, and widths of spectrum features related to the particular electronic transitions or normal vibrations of a molecule. However, this is not so for solids for several reasons. First of all, a certain degree of oscillator coupling should always take place. Due to oscillator coupling, the parameters of the classical analytical model for the complex dielectric constant (and of other simplified models either) are some effective quantities providing a generalized representation for the properties of a certain set of elementary excitations rather than true microscopic characteristics of these excitations. In particular, more rigorous theoretical approaches to the vibrational excitations in solids reveal that, for $J > 1$, oscillator coupling results in both ω_{0j} and γ_j becoming functions of the frequency.[11,12]

The classical analytical model has been widely used in the spectroscopy of crystals for treating the IR reflection spectra (see Chapter 3, Section II.B). Hence it follows that the effect of the dependences $\omega_{0j} = \omega_{0j}(\omega)$ and $\gamma_j = \gamma_j(\omega)$ on IR band contours is negligible for most cases and is hidden by experimental errors. There are, however, a few examples of appreciable deviations of IR band shapes from those of the classical dispersion theory caused by extraordinarily strong oscillator coupling. For such IR spectra, some modifications of the classical analytical model for the complex dielectric constant were proposed to take into account the effect of oscillator coupling. In particular, Barker and Hopfield[14] used a model involving both uncoupled and coupled oscillators. For two coupled oscillators, the analytical model for the complex dielectric constant becomes the sum of two components, $\hat{\varepsilon}_1(\omega)$ and $\hat{\varepsilon}_2(\omega)$, the first of which is given by

$$\varepsilon_1 = 4\pi Ne^2 \frac{Z_1^2 - iZ_1Z_2\omega\gamma_{12}/\left[\omega_2^2 - \omega^2 - i\omega(\gamma_2 + \gamma_{12})\right]}{\omega_1^2 - \omega^2 - i\omega(\gamma_1 + \gamma_{12}) + \omega^2\gamma_{12}^2/\left[\omega_2^2 - \omega^2 - i\omega(\gamma_2 + \gamma_{12})\right]}, \quad \textbf{(22)}$$

where γ_{12} is the damping coefficient characterizing the effect of coupling between the oscillators denoted by 1 and 2. The expression for $\hat{\varepsilon}_2(\omega)$ has the same form as (22) with indices 1 and 2 interchanged. This model allowed Barker and Hopfield to improve the description of the IR spectra of some titanates and tantalates (see Chapter 3, Section III.B). In most cases, however, such spectra are treated with the so-called *factorized* analytical model for the complex dielectric constant discussed below in Section III.

A special case is the use of the classical analytical model for treating the optical spectra related to the fundamental electronic excitations. As noted above, the continuous nature of the interband constituent of these spectra prevents one from separating these spectra into the contributions of particular oscillators (see Chapter 2, Section I and also Chapter 3, Section IV.B). Therefore, when dealing with solid dielectrics, the application of the classical model (and other models listed in Chapter 1) for treating quantitative data is now confined mostly to IR spectra. There are, however, examples of successful usage of the classical $\hat{\varepsilon}(\omega)$ model for treating the optical spectra of electronic excitations for typical semiconductors (see, for example, Reference 15 and also Chapter 3, Section III.B). It should be noted, moreover, that the behavior of $\varepsilon'(\omega)$ for solids at frequencies much greater than those corresponding to any interband transitions of valence electrons is often described in terms of the simple formula (see, for example Reference 16),

$$\varepsilon'(\omega) = 1 - \omega_p^2/\omega^2, \quad \textbf{(23)}$$

where ω_p is the so-called plasma frequency determined by relation $\omega_p^2 = 4\pi N_v e^2/m$, and N_v is the total number of valence electrons per unit volume. This formula was derived theoretically in the single-electron approximation; it also follows directly from (13) when putting $\omega_{0j}^2 - \omega^2 \approx -\omega^2$, neglecting $\gamma_j\omega$, and assuming $\Sigma N_j f_j = N_v$. Some authors (such as Martin and Wallis[15]) introduce particular plasma frequencies, ω_{pj}, for each kind of electronic oscillator, thus obtaining, by substituting ω_{pj}^2 for S_j in (14) through (17), the following expression for the complex dielectric constant

$$\hat{\varepsilon}(\omega) = \varepsilon_\infty + \sum_{j=1}^{J} \frac{\omega_{pj}^2}{\omega_{0j}^2 - \omega^2 - i\gamma_j\omega}, \tag{24}$$

which is another presentation of the classical analytical model.

B. THE PROBLEM OF LOCAL FIELD CORRECTION, THE LORENTZ-LORENZ MODEL

The approaches discussed above take the internal electric field, E_{int}, in a condensed medium to be equal to the external field, E. In general, however, the internal field may be influenced by the effect of polarization which generates an additional *local field*, E_{loc}, at a particle in the medium. Various approaches were proposed to take this effect into account. The earliest was the Lorentz-Lorenz approach proposed in the course of developing the classical dispersion theory.[1-5] This approach is based on a purely electrostatic model involving the effect of charges induced on the inner surface of a sphere around an ion. In this approach, the local field $E_{loc} = 4\pi P/3$ and, hence, the internal field E_{int} is

$$E_{int} = E + \frac{4\pi P}{3}. \tag{25}$$

Therefore, (3) should be replaced by

$$P = N\alpha\left(E + \frac{4\pi P}{3}\right). \tag{26}$$

Hence, instead of the Drude model (13), the Lorentz-Lorenz version of classical analytical model for the complex dielectric constant is

$$\frac{\hat{\varepsilon}(\omega) - 1}{\hat{\varepsilon}(\omega) + 2} = \frac{1}{3}\sum_{j=1}^{J} \frac{4\pi N e^2 f_j}{\omega_{0j}^2 - \omega^2 - i\gamma_j\omega}. \tag{27}$$

The relative advantages and shortcomings of the Drude and Lorentz-Lorenz versions, it seems, might become clear when treating a given set of experimental data with both models. However, this is not so, because the Lorentz-Lorenz model may be reduced to the same analytical expression as that of the Drude model, with dispersion frequencies, ω'_{0j}, and intensities, S'_j, differing from the initial ω_{0j} and S_j values. For a simplified case with no damping (see Section II.C.1), such ω'_{0j} values, according to Born and Wolf,[4] are the roots of the following equation:

$$3-\sum_{j=1}^{J}\frac{S'_j}{\omega_{0j}'^2-\omega^2}=0. \tag{28}$$

The ω'_{0j} values should be lower than the ω_{0j} values in the initial Lorentz-Lorenz model, the relation between them, for a case of single undamped oscillator, being[4,5]

$$\omega_0'^2=\omega_0^2-\frac{4\pi Ne^2f}{3m}=\omega_0^2-\frac{1}{3}S. \tag{29}$$

Earlier, in the monographs cited above,[1,2,4,5] the Lorentz-Lorenz model was considered to be more justified theoretically. Special studies into the problem of local field correction showed, however, that the validity of the Lorentz-Lorenz model is greatly overestimated, at least for materials other than ionic crystals. As was pointed out by Moss,[3] some experimental data can be represented better by a Drude-type formula (that, as is seen from above, comprises no local field correction) than by the Lorentz-Lorenz model. Generally, materials with high atomic polarizabilities and a low degree of bond ionicity (such as typical semiconductors) do not follow the Lorentz-Lorenz model. In particular, Nozieres and Pines[17] found, on the basis of quantum mechanical considerations, that the local field correction for these substances could be omitted. The conclusions[17] were cited by Moss[3] who supported the views of the authors. Furthermore, Burckhardt, Feltz, and Bussemer[18] interpreted their data on the composition dependence of the parameters of two-term Sellmeier-type and Lorentz-Lorenz-type dispersion formulas (see Section II.C.1) as an indirect proof that the Lorentz-Lorenz local field correction was not valid for chalcogenide glasses. An elaborate approach to calculating the effect of the local field for molecular liquids and solids that differed from the Lorentz-Lorenz approach was developed by Onsager, Bottcher, and their successors (see, for example, the comprehensive monograph of Bakhshiev[19]). This approach is based on the fact that the local field around an individual molecule is determined mainly by van der Waals and like forces. However, the Onsager method is not applicable to substances lacking molecular structure, because in this case no individual particle can be separated unambiguously from the rest of structure, and the forces influencing a particle are not simply van der Waals forces.

For non-molecular solids with strong electron localization, according to Nozieres and Pines,[17] some local field correction does exist; however, its magnitude is probably substantially lower than that predicted by the Lorentz-Lorenz model and it does not alter perceptibly the magnitude of dielectric constant. This conclusion should evidently hold for oxides, both crystals and glasses. The determination of whether a local field correction is required to account for the spectra of optical constants of a material could be sought on the basis of a comparison between the experimental spectroscopic data (potentially influenced by local field effects) and some independent data (i.e., neither spectroscopic nor optical data). Lazarev and his co-workers theoretically calculated, involving no local field corrections, the values of phonon frequencies and intensities for a set of simple and complex oxides (see, for example, Reference 20); data obtained were in excellent agreement with the experiment. This agreement supports the notion that the local field has a negligibly small influence on the spectroscopic properties of the oxides in the IR range. On the other hand, there are studies taking the Lorentz-Lorenz local field correction into account when treating the IR and Raman spectroscopy data for some solids, including glasses.[21]

C. SIMPLIFIED APPROACHES RELATED TO THE CLASSICAL ANALYTICAL MODEL FOR THE DIELECTRIC CONSTANT

There exists a number of simplified approaches to treating optical properties of solids that may be related to the classical analytical model for dielectric constant. Among these are the totality of approaches to the refractive index dispersion, the concept of refractivity, and the Wemple-DiDomenico approach.

1. Simplified Models for Approximating the Refractive Index Dispersion in the Transparency Range

Early studies of the dispersion of the refractive index dealt solely with the transparency range and therefore involved simpler analytical models for this dispersion than those discussed above. Neglecting the simplest Cauchy model, the first of the models worth mentioning is given by the expression

$$n^2(\lambda) - 1 = \sum_{j=1}^{J} \frac{A_j \lambda^2}{\lambda^2 - \lambda_{0j}^2}, \tag{30}$$

where λ is the current wavelength, λ_{0j} is the resonance wavelength of j-th oscillator, and A_j is a constant. This formula is referred to as the Sellmeier dispersion formula. Formula (30) was initially derived in the 1770s based on concepts that differed substantially from modern views; however, it remains in use even now for various purposes, such as approximating the frequency dispersion of refractive index of optical materials (see Chapter 5, Section II.A.1). In fact, the Sellmeier formula is a simplified version of the classical analytical model discussed above. This follows either from (11), when assuming the damping term to be zero, or directly from (15), when taking into account that, for transparency region, either $\omega_{0j} \gg \omega$ (the electronic excitations) or $\omega \gg \omega_{0j}$ (the vibrational ones). In both the cases, $|\omega^2_{0j} - \omega^2| \gg \gamma_j\omega$, thus damping may be neglected and the denominator in (15) may be reduced to $(\omega^2_{0j} - \omega^2)^2$. When the absorption index is negligibly small, the real dielectric constant may be replaced by the square of the refractive index, yielding the formula

$$n^2(\omega) - 1 = \sum_{j=1}^{J} \frac{S_j}{\omega_{0j}^2 - \omega^2}. \tag{31}$$

Using wavelengths instead of frequencies, (31) may be transformed into (30) with $A_j = S_j \lambda^2_{0j}/4\pi^2c^2$. Thus, Equation (31) is the Sellmeier dispersion formula rewritten for frequencies.

The major application of the Sellmeier dispersion formula is the interpolation of refractive index values for various optical materials (see Chapter 5, Section II.A.1). For applied optics, the problem of refractive index interpolation is important because the refractive indices of commercial optical materials are usually measured only for particular frequencies corresponding to the brightest spectral lines of standard light sources. Since the frequency ranges between these lines may be rather large, an interpolation procedure is necessary to obtain refractive index values for intermediate frequencies. Various analytical forms of the Sellmeier formula including some differing appreciably from the one in (30) have been developed (see, for example, a review of such forms in Reference 22). Some of these are listed here briefly. For example, in early publications, the so-called Ketelaar-Helmholtz formula appeared; it was derived from (30) by introducing the static dielectric constant ε_0 (18):

$$n^2(\lambda) = \varepsilon_0 + \sum_{j=1}^{J} \frac{M_j}{\lambda^2 - \lambda_{0j}^2}, \tag{32}$$

where $M_j = S_j \lambda_{0j}^4 / 16\pi^4 c^4$. Other modifications of (30) took into account the contribution of the IR excitations only approximately. This was because it was well known that the refractive index in the transparency range is determined predominantly by the contributions of electronic excitations. Denoting the dispersion wavelengths of the electronic and vibrational excitations by λ_{0l} and λ_{0m}, respectively, the sum of terms in (30) can be presented as

$$\sum_{j=1}^{J} \frac{A_j \lambda^2}{\lambda^2 - \lambda_{0j}^2} = \sum_{l=1}^{L} \frac{A_l \lambda^2}{\lambda^2 - \lambda_{0l}^2} + \sum_{m=L+1}^{M} \frac{A_m \lambda^2}{\lambda^2 - \lambda_{0m}^2}. \tag{33}$$

Because $\lambda_{0m} \gg \lambda$, one may put $\lambda^2 - \lambda_{0m}^2 \cong -\lambda_{0m}^2$, and (30) becomes

$$n^2(\lambda) - 1 \approx \sum_{l=1}^{L} \frac{A_l \lambda^2}{\lambda^2 - \lambda_{0l}^2} - \sum_{m=L+1}^{M} \frac{A_m \lambda^2}{\lambda_{0m}^2} = \sum_{l=1}^{L} \frac{A_l \lambda^2}{\lambda^2 - \lambda_{0l}^2} - K\lambda^2, \tag{34}$$

where K is a constant. Moreover, when applying dispersion formulas to treat data solely in the visible range, no IR term was taken into account at all. For certain purposes, greatly simplified formulas involving only a single term were used. The single λ_0 value entering into such formulas played the role of *center of gravity* for the entire range of fundamental electronic excitations.

With approximations similar to those used when deriving Formula (31), the Lorentz-Lorenz version of the classical analytical model for the complex dielectric constant (27) is reduced to the simplified Lorentz-Lorenz dispersion formula for the refractive index:

$$\frac{n^2(\omega) - 1}{n^2(\omega) + 2} = \frac{1}{3} \sum_{j=1}^{J} \frac{S_j}{\omega_{0j}^2 - \omega^2}. \tag{35}$$

This formula, like those of the Sellmeier type, was used in some cases to approximate the refractive index dispersion of optical materials. The main field in which the Lorentz-Lorenz approach remains in wide use is the one related to the concept of *refraction,* or *refractivity.*

2. Refractive Index vs. Density, the Concept of Refractivity

The idea of refraction, or refractivity, as the quantity characterizing the refractive power of a substance was initially introduced by Newton long before the development of classical dispersion theory. The specific refractivity, r, was considered to be approximately a constant that connects the refractive index in the transparency range to the density, ρ, of a substance regardless of its particular state (gaseous, liquid, or solid) and was defined as:

$$r = \frac{1}{\rho}\left(n^2 - 1\right). \tag{36}$$

Further interest in the concept of refractivity arose because its magnitude could be expressed as a linear function of the chemical composition of a substance (see below). In the course of further development of the refractivity concept, Drude combined (36) with the above Sellmeier-Drude formula for $n(\omega)$ dispersion (31), thus introducing the frequency dependence of refractivity. Therefore, (36) will be referred to as the Newton-Drude (ND) specific refractivity, r_{ND}.

Experimental studies showed that r_{ND} varies with temperature and with changes in the state of a substance. This stimulated attempts to modify the analytical form of refractivity to stabilize its magnitude. As a result, many formulas for refractivity containing the factor $1/\rho$ were introduced. The Lorentz-Lorenz refractivity, r_{LL}, based on the Lorentz-Lorenz formula (35) for the refractive index dispersion became widely used. Along with r_{LL}, the other formulas, as a rule, involved no theoretical basis. The main difference between these empirical formulas and those for R_{ND} or R_{LL} was in the power of n in both the numerator and denominator. A popular example of the former is the Gladstone-Dale formula:

$$R_{GD} = \frac{1}{\rho}\left(n - 1\right). \tag{37}$$

In addition to the specific refractivity, the notion of molar refractivity, R, was also introduced. For example, the Lorentz-Lorenz molar refractivity, R_{LL}, is given by

$$R_{LL} = \frac{M}{\rho} \cdot \frac{n^2 - 1}{n^2 + 2}, \tag{38}$$

where M is the molar mass of the compound. The comprehensive review of various formulas for refractivity and of related problems may be found, for example, in the monograph by Batsanov.[23]

It became customary to believe that R_{LL} values, being admittedly more justified, vary with changes in temperature and other conditions to a lesser degree than R_{ND}. Moreover, the R_{LL} value was treated as representing the actual magnitude of volume occupied by atoms or molecules in the bulk of a solid. The relative variations depending on chemical composition, $\Delta R_{LL}/R_{LL}$ and $\Delta R_{ND}/R_{ND}$, however, may turn out to be comparable in some cases. Also, the above ambiguity of introducing the Lorentz-Lorenz local field correction for many substances (see Section II.B) should evidently be taken into account when discussing data on refractivities. In other words, there is no reason to treat the Lorentz-Lorenz refractivity as being the only preferable presentation of refractivity. One should not overestimate the physical significance of the notion of refractivity, bearing in mind that it lacks rigorous physical meaning. Indeed, the single $n(\omega)$ value for a particular frequency (the latter being either some finite one or zero) used to calculate the refractivity is a macroscopic quantity that clearly integrates the contributions of various oscillators of different origin. Therefore, refractivity can, at best, be correlated with some principal physical characteristics of particles making up a solid rather than serve as direct measure of these characteristics. This, of course, by no means denies the practical usefulness of the notion of refractivity for many purposes.

It was shown experimentally that the molar refractivity of a substance may be treated approximately as the additive sum of contributions of constituent structural fragments (atoms, ions, radicals, molecules, etc.). Therefore, many further studies involving the concept of refractivity were directed to obtaining the individual molar refractivity increments for the constituents and applying the increments found to the calculation of the molar refractivity of a substance from its chemical composition. In these studies, the refractivities were calculated initially using some particular $n(\omega)$ value for the visible range. Later, to avoid problems connected to the frequency dispersion, the refractivities related to infinite wavelength, R_∞, i.e., those calculated from ε_0 values found with the help of single-term dispersion formula of (31) type, were used. However, the magnitude of refractivity, as expected, depends greatly on the degree of ionicity of chemical bonds. It increases with a decrease in the effective electrostatic charge of electropositive atoms and with an increase in the charge of electronegative ones. Therefore, an atom is necessarily characterized by a variable refractivity magnitude depending on the nature of the chemical bond in a compound under consideration. Numerous studies of this subject were conducted from the end of the nineteenth century to the 1970s. Various systems of molar refractivity increments were developed for free ions (such as that of Fajans[24]), covalently bonded atoms, and atoms with intermediate bond ionicity. These systems are reviewed by Batsanov.[23] The refractivity calculated for a substance was used further either for reconstructing the refractive index or for judging, based on a comparison between calculated and experimental values, whether an assumed structural model was valid. Such calculations were widely applied in glass science and glass technology. First, refractivities were involved in a set of methods for calculating the refractive index of glass via its chemical composition (see Chapter 5, Section III.A). Attempts were also made to correlate R_{LL} values to principal characteristics of the glass network. The evident shortcomings of such attempts resulted from both the approximate nature of the notion of refractivity and, especially, the use of an ionic system of refractivities which, naturally, is not appropriate for typical glass-forming oxides, the latter being characterized by a bond ionicity of no more than 50%. The purely ionic (i.e., greatly underestimated) R_{LL} increments for Si (0.08 cm^3/mol), B (<0.01 cm^3/mol), and like elements were used for calculating the R_{LL} of oxygen in glasses by subtracting the contributions of these elements from the experimental R_{LL} values, thus leading to substantially overestimating the oxygen increments. Because neither the absolute values of bond ionicity degree nor, especially, the variations in the latter with glass composition can be estimated very accurately, the idea of dividing the total glass refractivity into the contributions of individual chemical elements is not practical. It is more reasonable to separate the total oxide glass refractivity into only two main contributions, namely, the contribution of glass network as a whole and the contribution of modifier cations, such as alkali and alkaline earth metal cations.[25]

Recently one more version of the relationship between refractive index and density of a solid based on the so-called extended point dipole theory was tested:[26]

$$n^2 - 1 = \frac{4\pi\rho}{M} \sum_{j=1}^{J} \frac{A_j B_j \lambda_{0j}}{\lambda^2 - B_j \lambda_{0j}^2}, \tag{39}$$

where A_j is the same constant as that in the Sellmeier formula (30), $B = 1/(1 - b_j A_j \rho/M)$, and b_j is the constant characterizing the electron overlapping between adjacent dipoles. As can be seen, the refractive index dispersion in (39) is indistinguishable from that in the usual Sellmeier-type formula (30), because (39) and (30) differ only by the introduction of additional frequency-independent parameters. Using (39), Kitamura, Toguchi, Funo, Yamashita, and Kinoshita[26] described (with errors in the fifth decimal place) the refractive indices in the 238 to 1710 nm range for vitreous silica subjected to pressures up to 200 MPa.

3. The Wemple-DiDomenico Approach

The Wemple-DiDomenico approach[27-31] is essentially one more example of approaches that involve approximating the refractive index dispersion in the transparency range with simplified Sellmeier-type dispersion formulas. In contrast to other approaches using single- or two-term Sellmeier-type formulas, this approach seeks to reveal an interconnection between the effective parameters of such a dispersion formula and the principal characteristics of the chemical bonding in a material. The refractive index dispersion is fitted to a single-term formula,[27-30] rewritten for energies as

$$n^2 - 1 = \frac{E_d E_0}{E_0^2 - E^2}, \tag{40}$$

where E_0 is the energy of an effective oscillator, E is the photon energy, and E_d is the so-called dispersion energy. Later, Wemple[31] added an asymptotic term, E_L^2/E^2, to (40) to account for the vibrational contributions, thus providing a two-term dispersion formula instead of a single-term. E_d and E_0 values were obtained for a large number of semiconductors,[27,29,30] halides,[27,30,31] simple and complex oxides,[27,30,31] and glasses[28,29,31] based on experimental data for the refractive index dispersion.

Wemple and DiDomenico view Formula (40) as a physically meaningful relationship that enables one to derive rather broad conclusions concerning various characteristics of the structure and chemical bonding in solids. In particular, Wemple and DiDomenico assume the E_0 value to be connected straightforwardly to the bond length, d, $E_0 \propto d^{-s}$, with s varying in the $2 < d < 3$ range, and the dispersion energy E_d value to be determined by either

$$\left.\begin{array}{ll} & E_d = \beta N_c Z_a N_e \quad \text{[Reference 27]} \\ \text{or} & \\ & E_d = f^* N d^3 Z_a \quad \text{[Reference 30]} \end{array}\right\}, \tag{41}$$

where N_c is the coordination number of the cation, Z_a is the formal chemical valency of the anion, N_e is the effective number of valence electrons, N is the volume density of valence electrons, and β and f^* are assumed to be constants (here the symbol for the second of the two quantities referred to in Reference 30 as the "normalized oscillator strength" has an added asterisk to distinguish it from the commonly used dimensionless oscillator strength f). The magnitudes of β and f^* were found to be either of 0.37 ± 0.004 eV and 4.8 ± 0.3 eV for the majority of covalently bonded solids, or of 0.26 ± 0.004 eV and 3.0 ± 0.2 eV for the majority of ionic ones.

Numerical values for E_d and E_0 for a broad set of simple silicate and commercial glasses and some general assumptions made by Wemple[29] are discussed later in Chapter 5, Section III together with data on the refractive index dispersion found with other simplified dispersion formulas.

The physical significance of the parameters involved in the Wemple-DiDomenico approach warrant comment. It should be noted, first of all, that merely substituting energies for frequencies does not alter the simplified nature of a single-term formula such as (40) or a two-term formula. Hence it follows that E_0 and E_d (and, as a result, β and f^*) cannot have a clearly defined physical meaning. Indeed, the optical spectrum of fundamental electronic excitations for solids such as oxides and halides covers a very broad frequency range, extending into the vacuum and extreme ultraviolet as far as 20 to 21 eV and has a complicated structure with features of different origin (see Chapter 2, Section II.C). Therefore, an attempt to describe such a spectrum in terms of so-called constant conductivity model[27,30] seems to be an extreme oversimplification. All of the transitions in this complicated optical spectrum participate in determining the frequency dispersion of refractive index in the visible range; hence both the ω_0 value calculated with the single-term version of (31) and E_0 value calculated with (40) are no more than the center of gravity for the entire spectrum. The location of such a center of gravity is not connected straightforwardly to any particular spectral feature, including the absorption edge. For SiO_2, for example, the E_0 value found by Wemple[30] is 13.4 eV, which falls into the lower energy wing of the third peak of the spectrum, whereas the band gap estimates for the direct transitions given by different authors vary from 8.9 eV to 11.5 to 12.0 eV (see Chapter 2, Section II.C) and the position of the most intense peak is 10.2 to 10.4 eV. This is also clearly illustrated by the data of Reference 18 for chalcogenide glasses. Thus, there are no physical grounds to relate the E_0 value obtained from (40) either to the optical band gap E_g[d] or to the position of an intense low-frequency spectrum feature.

The empirical quantities E_d or f^* are treated in References 27 through 31 as measures of oscillator strength; however, there are two other quantities known and

---

[d] Wemple[27,30] matches E_0 values resulting from (40) with the E_g values calculated theoretically by Penn[32] (4.8 and 4.2 eV for Si and Ge, respectively) and by Van Vechten.[33] These have little in common with the standard band gap E_g (in Wemple's notation, E_t) values derived from the absorption edge of the direct interband transitions (such as 2.5 and 0.8 eV for Si and Ge, respectively[3]). However, it is the standard physical meaning of band gap E_g (namely, that of energy interval between the top of the valence band and the bottom of the conductivity band) that is presupposed by the studies of Penn[32] and Van Vechten[33] cited in References 27 and 30. Therefore the difference between E_g values calculated in References 32 and 33 and the standard ones[3] results solely from the crudely simplified nature of these calculations. Thus, there is no reason to refer to Penn's band gap E_g as a quantity having a special physical meaning differing from the standard one. Hence, a similarity between E_g values calculated in References 32 and 33 and E_0 values obtained by Wemple with Equation 40 is not an argument for the physical significance of these E_0 values.

commonly referred to in optics and spectroscopy as the oscillator strength: the normalized probability of a particular transition, f_j (in Wemple's notation,[30] F_{ij}), and the contribution of the transition to the static dielectric constant, $\Delta\varepsilon_{0j}$.[e] In terms of classical dispersion parameters (see Section II.A), f_j is independent of ω_{0j}, whereas $\Delta\varepsilon_{0j} = 4\pi N_j e^2 f_j / m_j \omega_{0j}^2$. Thus, when considering these two quantities in relation to ω_0, they may be treated as differing in the power of ω_0, the power being zero and –2, respectively. With E_0 replaced by $(h/2\pi)\omega_0$ and E by $(h/2\pi)\omega$, h being the Planck's constant, matching (40) with the usual single-term Sellmeier-dispersion formula (31) yields for E_d:

$$E_d = \frac{\hbar S}{\omega_0} = \frac{\hbar 4\pi N e^2 f}{m\omega_0}. \tag{42}$$

As is seen from (42), ω_0 enters to the –1 power in E_d and hence in f^*. Thus f^* with respect to ω_0 lies in between f and $\Delta\varepsilon_0$ and therefore loses the distinct physical meaning of both. No other distinct physical meaning for E_d or f^* is presented in References 27 through 31. As for representing E_d via different moments of the $\varepsilon''(\omega)$ spectrum,[27,30] this should be considered as another way to characterize the center of gravity rather than as a means of adding anything of physical significance. The assumed invariability of E_d and f^* for a given type of solid should probably be taken as being of an algebraic nature. Indeed, all the physical characteristics entering into (42) can vary with chemical composition only within rather narrow limits for a given type of solids, thus accounting for the seeming constancy of any voluntarily constructed product or ratio of these physical characteristics, including E_d.

Last, but not least, another oversimplification was introduced by Wemple and DiDomenico[27-31] when they used sharp distinctions between cations and anions, covalent and ionic solids, etc. In particular, Wemple and DiDomenico assumed, for the empirical quantities β and f^* entering into (41), two different fixed magnitudes for covalent and ionic solids, reminiscent of the tendency practiced when developing early systems of refractivities. This does not seem too convincing because, as is well known, the transition from the purely covalent solid to a predominantly ionic one is, in fact, continuous.

Although the Wemple-DiDomenico approach may lack physical significance, this does not rule out the practical usefulness of this approach for obtaining approximate estimates for quantities such as the so-called material dispersion,[31] etc. (see Chapter 5, Section III.B).

III. FACTORIZED ANALYTICAL MODEL FOR THE COMPLEX DIELECTRIC CONSTANT

Along with the various versions of the classical analytical model for the complex dielectric constant noted above, Berreman and Unterwald[34] introduced the so-called "factorized" analytical model,

---

[e] In the case of the single- or two-term dispersion formula, the physical meaning of f is not rigorously determined; crudely, the single f value characterizes, somehow, the effective total number of valence electrons per atom. The second quantity, $\Delta\varepsilon_0$, retains a definite physical meaning irrespective of the number of terms. For the visible frequency range and the single-term dispersion formula, $\Delta\varepsilon_0$ is the total contribution of all electronic excitations to the static dielectric constant ε_0.

$$\hat{\varepsilon}(\omega) = \varepsilon_\infty \prod_{j=1}^{J} \frac{\omega_{\mathrm{L}0j}^2 - \omega^2 - i\gamma_{\mathrm{L}j}\omega}{\omega_{\mathrm{T}0j}^2 - \omega^2 - i\gamma_{\mathrm{T}j}\omega}, \quad (43)$$

where the letters L or T denote the frequencies and damping coefficients for longitudinal or transverse optical modes, respectively. The properties and specific features of Model (43) were described in detail by Berreman and Unterwald.[34] The band contour in (43) is similar to that in the classical analytical model for the complex dielectric constant and is also referred to as Lorentzian. For $\omega = 0$, (43) has the same form as the well-known Lyddane-Sachs-Teller formula[35] derived earlier for the case $J = 1$:

$$\varepsilon_0/\varepsilon_\infty = \prod_{j=1}^{J} \left(\frac{\omega_{\mathrm{L}0j}}{\omega_{\mathrm{T}0j}} \right)^2. \quad (44)$$

In a general case, Model (43) assumes that $\gamma_{\mathrm{L}j} \geq \gamma_{\mathrm{T}j}$; hence, the model has four free parameters per oscillator instead of three as in the classical model. The opportunity to vary $\gamma_{\mathrm{T}j}$ and $\gamma_{\mathrm{L}j}$ independently makes (43) much more flexible than any version of the classical analytical model, thus allowing one to successfully describe band contours differing from those determined by the classical model due to strong oscillator coupling (see Section II.A). Because of this advantage, the factorized Model (43) soon became widely used in the spectroscopy of crystals (see Chapter 3, II.B).

Berreman and Unterwald's Model (43) was proposed from the general mathematical considerations that were not based on any physical concept. Later, however, an attempt was made to provide a physical basis for this model. As already mentioned, rigorous quantum-mechanical considerations reveal that dispersion frequencies ω_{0j} and damping coefficients γ_j for a combination of oscillators should be treated as functions of the frequency. Based on ideas borrowed from such a rigorous approach,[12] Gervais and Piriou[36] showed that an analytical model for the complex dielectric constant taking into account the $\gamma_j = \gamma_j(\omega)$ dependence may be reduced to the form of (43) (with $\omega_{0j} = \omega_{0j}(\omega)$ dependence neglected). Therefore, Gervais and Piriou consider Model (43) to be more justified as compared to the classical one and refer to it as the "four-parameter-semi-quantum" (FPSQ) model. According to their approach,[36] the inequalities $\gamma_{\mathrm{L}j} - \gamma_{\mathrm{T}j} \geq 0$ and $\gamma_{\mathrm{T}j}/\gamma_{\mathrm{L}j} \geq (\omega_{\mathrm{T}0j}/\omega_{\mathrm{L}0j})^2$ should always hold in order to keep the physical meaning of damping intact. The analysis shows that in Gervais and Piriou's approach, $\gamma_j(\omega)$ values are treated as increasing monotonically with frequency, the $\gamma_{\mathrm{T}j}$ and $\gamma_{\mathrm{L}j}$ constants in (43) being the $\gamma_j(\omega)$ magnitudes at ω equal to $\omega_{\mathrm{T}0j}$ and $\omega_{\mathrm{L}0j}$, respectively. As was shown by Kopilevich and Makarova,[37] Model (43) may be considered as the generalization of other analytical models for complex dielectric constant, the classical one and the model of coupled oscillators (22) included. This explains the great flexibility of (43).

It should also be noted that there were attempts to show the actual character of the $\gamma = \gamma(\omega)$ dependence directly from the experimental IR spectrum of a crystal. This has been done, for example, for LiF[38] and CaF_2, SrF_2, and BaF_2.[39]

Model (43) is also used assuming that $\gamma_{\mathrm{L}j} \equiv \gamma_{\mathrm{T}j}$. In this case, (43) becomes the equivalent of the classical model (17).

As is seen from (43), this model does not involve the oscillator intensities S_j. However, in the approximation of small γ_j, the intensities can be calculated from $\omega_{\mathrm{TO}j}$ and $\omega_{\mathrm{LO}j}$ using the following expression:

$$S_j = \varepsilon_\infty\left(\omega^2_{\mathrm{LO}j} - \omega^2_{\mathrm{TO}j}\right)\prod_{i\neq j}^{l}\frac{\omega^2_{\mathrm{LO}i} - \omega^2_{\mathrm{TO}j}}{\omega^2_{\mathrm{TO}i} - \omega^2_{\mathrm{TO}j}}. \qquad (45)$$

Note that calculating the oscillator intensities with (45) presupposes that the complex dielectric constant in (43) can be divided into the contributions of individual oscillators. In a general case, this possibility is justified from the physical viewpoint and may be confirmed, in many cases, by comparison between S_j values found with dispersion analysis based on (17) and on (43) (see, for example, Chapter 3, Section III.B.2). However, as was stressed by Pavinich and Lazarev[40] and also by Kopilevich and Makarova,[37] such a division is not the unequivocal one in the mathematical sense. In particular, a weak j-th oscillator having an $\omega_{\mathrm{TO}j}$ value that lies within the range of LO-TO splitting for the neighboring stronger (j-1)-th oscillator so that $\omega_{\mathrm{TO}\,(j-1)} < \omega_{\mathrm{TO}j} < \omega_{\mathrm{LO}\,(j-1)}$, should be characterized by the inequality $\omega_{\mathrm{LO}j} < \omega_{\mathrm{TO}j}$, otherwise the regular sequence of zeros and poles determined by (43) would be violated. At the same time, a choice of physically nonexistent "oscillators" with intensities determined by the pair combinations $\omega_{\mathrm{TO}\,(j-1)}$ and $\omega_{\mathrm{LO}j}$ or $\omega_{\mathrm{TO}j}$ and $\omega_{\mathrm{LO}\,(j-1)}$, instead of actual pair of oscillators, becomes formally possible. Hence some independent physical information is necessary in such cases to choose the correct combination of oscillator frequencies.

Chapter 2

Frequency Dispersion of Optical Constants for a Glass: Specific Features

I. GENERAL REMARKS

Prior to discussing the specific features inherent in the optical and spectroscopic properties of a glass, it should be stressed that all of the analytical models for the complex dielectric constant in Chapter 1 are based on concepts involving the presence of translational symmetry in the structure of a solid. Though the full number of vibrational degrees of freedom for a three-dimensional lattice is $3N - 3$, N being the number of atoms per unit volume, the fundamental IR spectrum of a crystal, as was mentioned in Chapter 1, Section I.C, is of discrete character and comprises a rather limited number of bands. According to the dynamical theory of solids (see, for example, Reference 41), this is due to the selection rule by translational symmetry (see also Section II.B). Only states characterized by near-zero values of a so-called wavevector can be active in the IR and/or Raman spectra of regular crystals, so bands in these spectra are related solely to respective near-zero-wavevector phonons. Thus, when applying a particular analytical model for the complex dielectric constant to the IR spectrum of a crystal, every kind of oscillator is connected to a particular type of near-zero-wavevector phonon. Bearing this in mind, one may conclude that, when applying to IR spectra, both the standard classical model and the rest of the models for the complex dielectric constant in Chapter 1 should be treated as involving, implicitly, the occurrence of translational symmetry in the structure of a solid.

An even more complicated situation, in relation to the models in Chapter 1, is characteristic of the optical spectra of fundamental electronic excitations. As was mentioned in Chapter 1, Sections I.C and II.A, the optical spectrum related to interband electronic transitions in a crystal is usually treated as a unified totality inseparable into individual components. In particular, the singularities in the spectrum are considered to be the thresholds or edges related directly to the structure of the first Brilloin zone[42] and are referred to as so-called Van Hove features rather than δ-functions or symmetric peaks (see also Section II.C). Such a concept of optical spectrum, naturally, is also based wholly on the translational symmetry in the structure of a solid. Moreover, this concept is evidently in apparent contradiction with the discrete oscillator distribution and symmetric band contours presupposed by both the classical analytical model (17) and the factorized Model (43).

Thus, the validity of analytical models for the complex dielectric constant discussed in Chapter 1, when applied to the disordered solids and, in particular, to glasses, is questionable (for more details, see Section III.A). As a result, the majority of investigators consider the models for the complex dielectric constant discussed above to be inappropriate for the spectra of fundamental excitations in glasses (see, for example Reference 43). Therefore, attempts to use the methods of thorough data treatment based on these models for glass spectra are rather scarce (see Section III.A; Chapter 3, Section IV.A; and Chapter 4, Section II). On the other hand, the simplified versions of the classical analytical model for the dielectric constant and related notions mentioned above in Chapter 1, Section II.C are used rather widely for approximating experimental data on the refractive index of glasses in the transparency range (see Chapter 5, Section II.A.1).

In view of the above, it seems expedient to start with a brief discussion of the specific features of optical spectrum formation under the effect of structural disorder.

II. STRUCTURAL DISORDER AND THE FORMATION OF OPTICAL SPECTRA

A. KINDS OF STRUCTURAL DISORDER IN A GLASS NETWORK

The notion of structural disorder comprises a broad variety of particular realizations of the latter; for example, Bonch-Bruevich[44] specifies eight different types of materials, the properties of which are determined substantially by respective specific kinds of disorder. For polymeric solids, including glasses, two kinds of disorder should be taken primarily into account. Following the notation introduced by Zakis and Klyava,[45] these two kinds of disorders will be further referred to as *continuous disorder* and *discrete disorder.*

The continuous disorder has a fluctuational nature and consists in the random distribution of some short-range- and/or intermediate-range-order characteristics, such as Si-O-Si bond angles, dihedral, or torsion, angles between the adjacent polyhedra, and also, for highly polymerized networks, ring statistics. As follows from most modern studies (see, for example, reviews in References 46 through 48), the continuous disorder is the acknowledged principal feature of the structure of inorganic glasses. In the case of continuous disorder, the standard deviations for the distributions of characteristics such as bond and torsion angles are perceptible, though not too large,[46,47,49-51] whereas bond lengths and angles within the SiO_4 and other kinds of polyhedra (including polyhedra formed by some typical cations[48]) vary negligibly because the symmetry of the polyhedra is determined by regularities similar to those for crystals. It is this kind of disorder that is responsible for the principal spectroscopic manifestations of the disordered structure of glasses.

The discrete disorder is one caused by strong local deviations from the short-range order typical of the "perfect" matrix of a material, these deviations being considered as the defects of the matrix. Naturally, this kind of disorder can take place to some extent for all solids, including real crystals. For inorganic glasses synthesized in the standard way, i.e., with no methods of quite rapid quenching and subjected to no special treatment (for example, X-ray or gamma exposure), the discrete disorder does not play an appreciable role in the formation of their structure though some of their properties may depend greatly on the nature and concentration of defects.

Naturally, the principal general effect of appreciable structural disorder of any kind on the formation of optical spectra is the violation of the selection rule by translational symmetry. Consequences and specific features of the effect depend on the kind of excitations involved.

The continuous disorder predetermines the random nature of all microcharacteristics and, hence, the statistical character of magnitudes of macroscopic properties of glasses. In the spectroscopy of any excitations greatly localized in space (both vibrational and electronic), such as those of point defects,[52] impurity or dopant ions,[52,53] and small molecules,[54] this situation manifests itself in the well-known phenomenon of inhomogeneous, or fluctuational, broadening. In the case of inhomogeneous broadening, the contour of a feature in the experimental spectrum results from overlapping the quasi-infinite number of individual components related to the random realizations of force field and/or electric field around the optical center under excitation. In some cases, inhomogeneous broadening of narrow resonance lines may also be caused by the discrete disorder due to the random character of the electric or force field generated by defects present in high concentrations. The shape of the inhomogeneously broadened line or band can vary substantially depending on the character of frequency distribution of individual components determined by the particular cause of broadening. Sometimes the Lorentzian contour (with halfwidth increased as compared to that of initial line) can be observed.[52] As the simplest approximation, the purely Gaussian contour is used for many cases (see, for example, Reference 54); it is described with the probability density function,

$$\varphi(x) = \frac{1}{\sqrt{2\pi}\sigma} \exp\left[-(x-\xi)^2/2\sigma^2\right], \tag{46}$$

where ξ is the frequency location of the distribution center and σ is the standard deviation. Band shapes intermediate between Lorentzian and Gaussian are described mostly with either the Holtsmark distribution[52] that includes the factor $\exp(-t^{3/2})$, or the well-known Voigt function[55] (derived earlier for a principally different case of lineshapes in the absorption spectra of gases). The Voigt function is the convolution of Lorentzian and Gaussian contours of the following form:

$$I(\omega) = I_0 \frac{a}{\pi} \int_{-\infty}^{+\infty} \frac{\exp(-y^2)}{a^2 + (\omega - y)^2} dy, \tag{47}$$

where I_0 is the peak intensity and a is a constant determined by the relationship between the widths of two contours to be convoluted. Also, various empirical asymmetric contours were used for approximating the band shapes in the absorption spectra of rare-earth ions in glasses.[53]

Naturally, the concept of inhomogeneous broadening is not valid for fully delocalized excitations such as the fundamental vibrational excitations in perfect crystal lattice[41] (see also Section II.B) or the electronic excitations of typical intrinsic semiconductors. When considering spectra of crystals with some traces of discrete disorder generated by defects, the question of whether a particular excitation is localized or not and hence if the notion of inhomogeneous broadening is applicable to these spectra is usually clear beforehand. On the contrary, when considering the fundamental excitations in glasses, the situation becomes complicated, because the problem arises as to how the continuous disorder influences the degree of localization of excitations.

Discussion of this problem will be confined initially to the fundamental vibrational excitations, because the behavior of these excitations under disordering is better understood now (at least, in its principal aspects) than that of the fundamental electronic excitations (especially those lying far from the absorption edge).

B. DISORDER AND THE FUNDAMENTAL VIBRATIONAL SPECTRA

As is known, the spectrum of vibrational density of states, $g(\omega)$, of any solid is continuous (at least, within the limits of a particular vibrational subband), whereas the optical IR or Raman spectra of a crystal are discrete because, as was mentioned above, they correspond solely to respective near-zero-wavevector phonons. With damping effects neglected, these phonons are commonly considered to be fully delocalized, i.e., to be flat waves propagating throughout a crystal. As a result of violation of the selection rule by translational symmetry due to the structural disorder regardless of a particular kind of the latter, all the states of the entire $g(\omega)$ spectrum become active in the IR and/or Raman spectrum.[a] This phenomenon should manifest itself in asymmetric broadening of the fundamental IR and Raman bands, because the frequencies of near-zero-wavevector phonons are located at the frequency limits of $g(\omega)$ subbands and phonon intensities decrease rapidly with an increase in their wavevector. Such asymmetric broadening was considered to be the only theoretically acknowledged broadening mechanism for the fundamental IR and Raman bands; it was calculated theoretically and observed experimentally for the IR spectra of disordered organic polymers.[57]

However, disordered organic polymers, according to Reference 57, are related to the case of discrete disorder caused by the defects of perfect crystal lattice. In this case, the local modes of defects can present in the vibrational spectrum of a material. These rather intense local defect modes, related mostly to the stereoisomers of monomeric groups, were indeed observed, along with asymmetrically broadened fundamental bands, in the IR spectra of disordered organic polymers.[57]

It is commonly assumed that the $g(\omega)$ spectra of inorganic glasses are rather similar to their experimental spectra of neutron scattering (see, for example, References 47 and 58). The $g(\omega)$ spectrum of a glass can also be calculated with computer modeling. The pioneering and most well-known examples of such calculations are those of Bell and Dean[59-63] conducted with the numerical modeling (initially mechanical and then computer) of structure and dynamics of big clusters having the structure of the continuous random network. As follows from Bell and Dean's data, the $g(\omega)$ spectra of glasses such as SiO_2, GeO_2, and BeF_2 comprise several vibrational subbands related to the different types of atomic motion. Later, the major conclusions of Bell and Dean concerning the nature of $g(\omega)$ spectrum of glasses were confirmed with simpler structural models (see, for example, References 47, 64, and 65).

It is of importance that, according to Bell and Dean's data, the vibrational excitations in glasses are characterized by some intermediate degree of localization, i.e., the excitations are neither flat phonon waves (as is the case for perfect crystals) nor the local modes of defects. The relative number of network-forming atoms, such as silicon, participating in a certain normal mode, N_{eff}/N, depends on mode frequency and, moreover, on the particular form of vibration. Some high-frequency vibrations near band edges may involve the displacements of just two or three atoms of silicon (or germanium, etc.).[62]

Unfortunately, there is no method yet for rigorously calculating the IR or Raman spectrum of a disordered solid from its $g(\omega)$ spectrum. Hence, it is impossible to check, by calculation, if the phenomenon of asymmetric band broadening due to $g(\omega)$ spectrum activation determines substantially the pattern of IR or Raman spectrum of a glass. Therefore, a crude assumption is used frequently that one may neglect the frequency dependence of matrix element, M, entering the expression for the intensity,

[a] In the case of discrete disorder, this effect can only arise, naturally, when the defect concentration is high enough for treating an initial lattice to be separated into spatially limited fragments. (See, for example, Reference 57.)

$$I(\omega) \propto g(\omega) \cdot M(\omega). \tag{48}$$

When following this assumption, the $g(\omega)$ spectrum computed for some structural and dynamical model is treated as mimicking the $\varepsilon''(\omega)$ spectrum; therefore, the $g(\omega)$ spectrum is compared straightforwardly to the experimental one. For the locations of high-frequency $g(\omega)$ maxima and $\varepsilon''(\omega)$ peaks, a rough agreement within the ± 100 cm^{-1} error usually holds.[59-66] However, subtler features such as relative peak intensities and widths in the $g(\omega)$ and $\varepsilon''(\omega)$ spectra differ substantially, thus showing that neglecting the frequency dependence of matrix element $M(\omega)$ is too crude an assumption. Moreover, the $g(\omega)$ and optical spectra of a material in the low-frequency range usually have little in common.

An attempt to calculate the IR absorption spectra of vitreous SiO_2, GeO_2, and BeF_2 based on the $g(\omega)$ spectrum with the simplest point charge model was made later by Bell and Hibbins-Butler.[67] To obtain band shapes compatible with the experimental ones, the absorption lines calculated were replaced by Lorentzian contours with halfwidths of about 30 cm^{-1}. The approximate similarity between the calculated and experimental spectra was indeed observed; however, a satisfactory agreement between relative band intensities and band shapes was not obtained in this case, too.

As is known, the experimental IR and Raman spectra of glasses are substantially broadened as compared to those of respective crystalline phases. However, the IR spectra of glasses do not show, unlike the spectra of polymers, definite signs of asymmetry of strong bands that could be indicative, in accordance with the above, of $g(\omega)$ spectrum activation. At the same time, there is no doubt that strong bands in the IR spectra of glasses are indeed fundamental. First, the positions and relative intensities of strong bands in the IR spectra of glasses are approximately similar, in most cases, to those in the spectra of corresponding crystals. Moreover, it seems improbable that these bands might be formed by the totality of local modes of structural defects because, as was already mentioned, the continuous disorder implies rather small differences in the spatial locations of neighboring structural fragments and in the force constants of interatomic bonds. Therefore, the continuous disorder in glass structure should be, as a rule, insufficient for giving rise to local mode formation. Thus, local modes in glass spectra (if any) are probably related to some strong local deviations from the "perfect" glass matrix (known mostly as the intrinsic point defects) rather than to variations in structure resulting from the continuous disorder.[b]

Unlike the IR spectra of glasses, the asymmetric bands were observed in the Raman spectra of vitreous silica, germania, and some binary and ternary glasses.[66,70,72] The origin of this asymmetry, however, is related commonly to overlapping with weaker neighboring bands rather than to $g(\omega)$ spectrum activation.[70,72]

The above lack of appreciable signs of asymmetry of broadened band shapes in the experimental IR spectra of glasses accompanied by the great magnitudes of band broadening is in contradiction to the theoretical justification of $g(\omega)$ spectrum activation as the major disorder-induced mechanism of band broadening for the

---

[b] Indeed, the Raman and IR spectra of inorganic glasses comprise just a few very weak bands (such as those at 495 and 606 cm^{-1} in the Raman spectrum of vitreous silica) that might be tentatively attributed to local modes of either standard intrinsic point defects[68,69] (such as the broken bonds and peroxide bridges) or some unusual structural groups (such as the "breathing" modes of regular four- and threefold rings[70,71]).

fundamental IR spectra. Qualitatively, it has been stressed by many authors (see, for example, Reference 43) that a certain unappreciated broadening mechanism related directly to the disorder in glass structure should manifest itself in their IR spectra. However, attempts to develop a corresponding approach were scarce for a rather long period. Among such attempts, that of Mitchell, Bishop, and Taylor[73,74] should be mentioned. They interpreted band shapes in the spectra of the IR optical conductivity, $\rho(\omega) = \omega\varepsilon''(\omega)$, reconstructed from the experimental reflection spectra of some chalcogenide glasses with the Kramers-Kronig transform in terms of the Gaussian distribution of the Lorentzian oscillators, and estimated the band parameter values. The data of References 73 and 74 are discussed in detail in Chapter 4, Section II.B.2. The standard deviations for this distribution were calculated from band shapes in the vicinity of band maxima and damping coefficients were estimated from the shapes of band wings. For chalcogenide glasses with chain- and layer-type structures, band shapes were found to be predominantly Gaussian, as will be shown in Chapter 4, Section II.B.2 for the case of vitreous arsenic selenide. This conclusion overestimates the actual Gaussian contribution to bandwidths. For vitreous SiO_2 taken as an example of three-dimensional glass structures, band shapes were found in Reference 73 to be predominantly Lorentzian. The authors stated that the Gaussian distribution somehow reflects the random disorder in glass, but no assumption regarding the physical meaning of Gaussian standard deviation was proposed. The practice of resolving the Raman spectra of glasses into the components with purely Gaussian contours (see, for example, Reference 72) became rather widely used. Such a choice of band shapes is validated in Reference 72 in terms of a qualitative assumption that band contours should be determined by a function describing the bond angle distribution rather than by that describing the standard Raman contour.

There were also a few attempts to treat the IR glass spectra with the same $\hat{\varepsilon}(\omega)$ models as those used for crystals, thus ignoring any principal differences between the formation of the glass spectrum and crystal spectrum. In view of the above, the physical meaning of results obtained with such an approach is rather dubious; the data of such studies are touched upon below in Chapter 3, Section IV.A; Chapter 4, Sections II.A.2, II.B.1, and II.B.2.

Recently, the mechanism of symmetric band broadening in the fundamental IR spectra of glasses that allows for overcoming the above contradiction was proposed when developing a specific analytical model for the complex dielectric constant of glasses[75-80] (see Section III.B). This mechanism involves the notion of inhomogeneous-like broadening specified in References 76 and 79, this notion being derived from considerations that take into account the above data of Bell and Dean[59-63] on the localization of vibrational excitations in glasses. A characteristic of degree of mode localization such as the size of localization region (the latter being the microscopic volume in which a particular excitation generated by a single IR quantum is localized) is chosen as a basis. According to References 59 through 62, the degree of spatial localization of higher frequency vibrational modes in disordered networks of glasses is rather substantial; therefore, the size of the localization region of a single IR quantum should be rather small. As a result, there cannot be full structural similarity between different localization regions picked up by IR quanta at random in the bulk of a glass. The parameters of short- and/or medium-range order influenced by continuous disorder, such as the accurate values of bond angles, dihedral (torsion) angles, etc., should vary at random from one localization region to another, thus determining the random realizations of structure and of force constants. Hence it follows that various microscopic unit volumes, v_i (the v_i value is determined by the relationship $v_i = v_{loc}^{max}$, v_{loc}^{max} being the maximum size of localization region for the frequency range under consideration) making up the bulk of a glass should exhibit random variations in their individual vibrational densities of states, $g_i(\omega)$, including the frequencies of near-zero-wavevector phonons. Thus the notion of inhomogeneous-like broadening presupposes that, for a glass, the total spectrum of vibrational density of states is formed statistically from individual $g_i(\omega)$ spectra for various virtual v_i volumes. This means that a certain frequency distribution should take place for every kind of near-zero-wavevector phonon. Therefore band shapes in the IR $\varepsilon''(\omega)$ or Raman spectrum of a glass are determined mostly by overlapping the quasi-infinite number of individual components (each of them being related to a particular realization of a near-zero-wavevector phonon) distributed over frequency, hence band broadening.

The term inhomogeneous-like broadening is chosen because of evident similarity between this type of broadening and the well-known inhomogeneous broadening in the spectra of greatly localized excitations (see Section II.A). Indeed, a comparison shows that the kind of broadening specified in References 76 and 79 and referred to as inhomogeneous-like broadening differs from the standard notion of inhomogeneous broadening mainly in the size of the localization region of an excitation involved. For the acknowledged cases of standard inhomogeneous broadening mentioned above in Section II.A, a localization region of an excitation is of either atomic dimensions[52-53] or, at greatest, of dimensions of small molecules.[54] Conversely, as follows from the above, a localization region for the case referred to in References 76 and 79 as inhomogeneous-like broadening is no less than medium-range-order dimensions, i.e., substantially larger. From the viewpoint of a particular type of band shape, this difference is of importance (see Section III.B.1); at the same time, the essence of the phenomenon of broadening is similar enough in both cases compared. Therefore, the term inhomogeneous-like broadening is expedient enough.

C. DISORDER AND THE FUNDAMENTAL SPECTRA OF ELECTRONIC EXCITATIONS

As was mentioned in Section I, the principal features of the fundamental spectra of interband electronic excitations for crystalline solids are connected intimately to the concept of translational symmetry (see, for example, Reference 42). According to the electronic theory of disordered semiconductors described in terms of the single-electron approximation,[44] the following major consequences of continuous disorder for the spectra of the interband transitions can take place: (1) the appearance of the exponential absorption edge described by the empirical Urbach rule and (2) the loss of rigorous meaning of a notion such as the Brilloin zone. As a result, the Van Hove features in the spectrum of the combined density of states tend to vanish and, hence, respective features in the absorption spectra can disappear. In addition, discrete disorder can also result in the discrete defect levels in the band gap and in the appearance of additional spectral lines or bands near the fundamental absorption edge.

Unlike typical semiconductors, the spectra of wide band gap solids, such as oxides and including glasses, cannot be described satisfactorily in terms of the single-electron approximation. The effects of the electron-hole interaction (similar in nature to those forming the spectra of model insulators[42]) play a substantial role for such materials, thus providing the formation of strong exciton peaks and making the total pattern of excitations much more complicated. Factors that do not allow for assuming a simple interpretation of the optical spectrum of such substances, even in the crystalline state, are considered in detail by Laughlin[81] for the case of SiO_2. Along with the relative complexity of the structure of SiO_2, these factors include: (1) a large band gap making the single-electron approximation invalid for the interpretation of optical spectra and (2) the presence of lone-pair orbitals providing the existence of filled overlapped valence subbands that should be treated as a whole.

As a result of these difficulties, a comprehensive rigorous theory of optical spectra related to the fundamental electronic excitations in oxides and like substances is unavailable, even for the crystalline state. Therefore, as is seen from the literature on SiO_2, which is the major object of theoretical studies (see, for example, References 81 through 85), the calculations of its $\varepsilon''(\omega)$ spectrum in the VUV range on the basis of a variety of structural models ranging from the simplified ones (such as small cluster models[82]) to the actual structure of α-quartz[84] do not provide a satisfactory agreement with the experiment. Among such calculations, the best coincidence with experiment was obtained by Laughlin[81] with the idealized model of β-cristobalite. In view of this situation, there is no chance to rigorously specify the effect of disorder on the contours of principal features in the VUV electronic spectra of oxides and like solids. Luckily, judging by experimental data on SiO_2,[86-88] changes in these spectra under the effect of disorder are probably much less pronounced than is the case for the vibrational spectra.

All investigators of the VUV spectrum of vitreous silica mention that it is quite similar to that of α-quartz (see below). The entire fundamental spectrum of electronic excitations in SiO_2 extends into VUV range as far as 21 eV,[85-89] but it is the frequency range of the absorption edge that is studied most. Unlike typical semiconductors, silica and other oxides are characterized by the Urbach absorption edge in both an amorphous and crystalline state,[88] which is usually interpreted by analogy with the case of alkali halides[90] in terms of exciton autolocalization. The similarity of Urbach parameters for α-quartz and vitreous silica points out the common nature of their absorption edge; the influence of disorder is reduced to a slight long-wavelength shift of the edge interpreted in terms of inhomogeneous broadening caused by structural nonidentity of localization regions for excitons.[88] The estimates of the optical band gap for SiO_2 vary broadly from 8.9 eV[81] to 11.5 to 12.0 eV.[88]

As is known, the experimental VUV spectrum of α-quartz at energies exceeding those of the absorption edge comprises four peaks at 10.4 to 10.5 eV, 11.5 to 11.6 eV, about 14 eV, and about 17.5 eV.[86,87] There is no unified viewpoint concerning the problem of how this spectrum is formed. Griscom[83] distinguished five equally possible but substantially different versions of interpretation for this spectrum; later, additional versions emerged, including that of Laughlin.[81] The first of the peaks is the most sharp; from the beginning of studies, this has been attributed to the excitonic transition. This attribution was acknowledged later but the nature of the exciton remains disputable to some extent, especially for the case of vitreous silica (see below). As for the second peak at about 11.5 eV, it was assumed in Reference 89 that this peak has a triplet structure, its components differing in origin. A sharp minimum between the first and second peaks is usually considered, for the case of α-quartz, as resulting from the so-called antiresonance, or Fano resonance.[91] It is not entirely clear at the present whether the second peak at about 11.5 eV and two farther peaks at 14 and 17.5 eV should be attributed to the excitonic or interband transitions. Many investigators relate to the exciton-like transitions two of the four peaks, namely, those at 10.4 and 11.5 eV,[88] whereas others tend to extend this interpretation also to the third peak at 14 eV. However, the best correlation between the theoretically calculated and experimental VUV spectrum of α-quartz was obtained, as was already mentioned, by Laughlin[81] using an approach presupposing that all four maxima in the spectrum are the excitonic ones.

When passing from the VUV spectrum of α-quartz to that of vitreous silica, the peak contours, unlike the IR bands, are broadened only insignificantly; other spectrum variations consist of a very slight (of about 0.2 eV) low-energy shift and of a weak decrease in peak reflectivity values.[86-88] The pronounced similarity of these spectra usually is interpreted as speaking for the identical origin of excitations involved due to the predominant role of the short-range order in the formation of the spectra. The experimental studies cited in Reference 88 confirm that disorder does not impede the formation of the same types of excitons as those in α-quartz and results only in slightly increasing the exciton-phonon interaction and also in altering the migration ability of excitons due to their autolocalization. It was assumed, in particular, that the lowest energy peak at 10.2 eV can be an envelope of five individual Gaussian contours related to different kinds of autolocalized excitons.[88]

As to the theoretical calculations of the combined density of states and the VUV optical spectra of nonstoichiometric glasses, such as those conducted for sodium silicate[92] and lead silicate glasses,[93] the results of these calculations show rather poor agreement with experiment. Among the experimental VUV spectra of glasses other than SiO_2, only that of vitreous GeO_2 can be compared directly to the spectrum of respective crystal; as follows from the data of Pajasova,[94] the variations in the VUV spectrum of germania during the transition from the crystalline (hexagonal) to the vitreous state also are rather negligible though are somewhat larger than those for SiO_2 (see Chapter 4, Section III.B). The data on the VUV spectra of binary and more complex glasses[92,93,95-99] show, in general, that features in these spectra are broader than those in the spectrum of silica. It is difficult, however, to say anything more definitive about the influence of disorder on the VUV spectra of these glasses due both to the lack of data for the crystals of corresponding complex oxides and to the fact that some of the spectra do not seem to be reliable. As will be shown in Chapter 4, Section III.C, the VUV spectra of sodium silicate glasses obtained with all precautions have little in common with the spectrum of silica; conversely, the spectra of samples polished with standard procedure using water comprise strong peaks similar in location to those in the spectrum of vitreous silica. Hence, these peaks should be attributed to surface layers differing in chemical composition rather than to bulk glasses under study. Such strong silica-like peaks are seen clearly in the VUV spectra of silicate glasses presented in some publications, such as References 92 and 99, thus giving rise to serious doubts about the reliability of these spectra.

On the whole, changes in the qualitative pattern of the VUV spectra under the effect of the continuous disorder in the structure of oxide glasses probably may be considered to be less crucial than is the case for the vibrational spectra; however, the mechanism of influence of disorder on the shapes of spectrum features related to high-energy electronic excitations and also the very nature of these excitations in oxides as a whole are not understood well enough presently.

III. EFFECT OF DISORDER ON THE ANALYTICAL PRESENTATION OF THE COMPLEX DIELECTRIC CONSTANT FOR A GLASS

A. SHORTCOMINGS OF ANALYTICAL MODELS FOR THE COMPLEX DIELECTRIC CONSTANT DEVELOPED FOR CRYSTALS WHEN APPLIED TO GLASSES

As mentioned above in Section I, all analytical models for the complex dielectric constant that describe the spectra of the fundamental excitations of a solid in terms of a finite number of oscillators related to respective spectrum features, or bands, are connected to the concept of translational symmetry in the structure of the solid. In the IR spectrum, every oscillator is related, according to the pattern specified above in Section II.B, to a particular kind of delocalized, wave-like near-zero-wavevector phonon, the respective band contour being Lorentzian. Hence, applying these models to the IR spectra of glasses would mean that (1) the vibrational excitations are taken to be delocalized as they are in crystals (thus contradicting Bell and Dean's findings[59-63]) and (2) the band broadening that is characteristic of the spectra of glasses as compared to crystal spectra is ascribed solely to an increase in damping coefficients (thus ignoring other possible manifestations of disorder specified above in Section II.B). Such a substantial increase in damping coefficients in the vitreous state has no solid theoretical justification; moreover, there are experimental data speaking against this increase. For example, Risen[100] pointed out, for vitreous potassium arsenate, that an attempt to describe its IR spectrum as a result of an increase in damping coefficients of corresponding crystalline phases did not provide satisfactory agreement with experiment. These reasons enabled the majority of investigators to consider the analytical models for the complex dielectric constant developed for crystals to be inappropriate for the spectra of fundamental excitations of glasses (see, for example, Reference 43).

There are, however, studies involving the quantitative treatment of the fundamental glass spectra with these models. In accordance with the above, attempts to use the dispersion analysis version based on the classical analytical model for the complex dielectric constant usually do not provide a satisfactory agreement with experiment. As a rule, the best computer fits to the IR reflection spectra of glasses obtained with the classical dispersion analysis (such as those demonstrated, for example, in Reference 101) are characterized by appreciable errors, especially in spectral regions in between the reflectivity maxima (see also Chapter 3, Section IV.A). Particular data of this kind are discussed in detail in Chapter 4, Sections II.B.1 and II.B.2. Recently, Ljundahl, Fellman, and Ribbing[102] succeeded in obtaining rather good computer fits to the reflection spectra of several multicomponent silicate glasses with the classical dispersion analysis, thus revealing an exclusion from the usual situation outlined above. This probably is explained by extremely smoothed band contours in the experimental spectra studied in Reference 102 (see Chapter 3, Section IV.A and Chapter 4, Section II.A.2). Furthermore, the use of the dispersion analysis version based on the factorized $\hat{\varepsilon}(\omega)$ model (43) also allowed Gervais et al.[103,104] to describe rather satisfactorily the IR reflection spectra of various oxide glasses. On this basis, a conclusion assuming the substantial degree of coherence of vibrations was made. In view of the above in Section II.B, this is hardly the case. Probably the good quality of best fits obtained by Gervais et al.[103,104] is due to the above-mentioned great flexibility of the factorized model (43) rather than to the actual coherence of vibrations throughout the network of glasses studied.

B. SPECIFIC ANALYTICAL MODEL FOR THE COMPLEX DIELECTRIC CONSTANT OF GLASS

A specific analytical model for the complex dielectric constant of glasses that considers large bandwidths in the fundamental IR spectra of glasses in terms of symmetric inhomogeneous-like broadening (see Section II.B) was developed recently in References 75 and 76. This model is discussed below in further detail.

1. Validation of the Model

According to Efimov,[76,79] the predominant manifestation of continuous disorder in the vibrational spectra of glasses is the so-called inhomogeneous-like band broadening. The origin of this broadening, as was specified above in Section II.B, is traceable to the random distribution of particular realizations of structure of virtual microscopic unit volumes, in which vibrational excitations might localize over the bulk of a glass under study. This distribution leads, in turn, to the random frequency distribution of near-zero-wavevector phonons, which results in band broadening.

The frequency distribution of oscillators corresponding to the near-zero-wavevector phonons should be specified quantitatively in order to take into account the influence of inhomogeneous-like broadening on the analytical form of the $\hat{\varepsilon}(\omega)$ model. Generally, the particular kind of distribution observed for a quantity depends on the number of random factors influencing the quantity and on the degree of coupling between these factors. For acknowledged cases of standard inhomogeneous band broadening, the oscillator distribution varies substantially depending on the nature of optical center and the particular kind of disorder, thus resulting in a variety of band shapes (see Section II.A). For the case of inhomogeneous-like broadening, a situation seems to be more clear. Indeed, the number of random factors influencing near-zero-wavevector phonon frequencies should be rather large due to a much greater number of particles involved in a single excitation as compared to the standard cases of inhomogeneous broadening, such as those related to local modes of point defects and other greatly localized excitations.[52-54] At the first approximation, these random factors may probably be considered to be statistically independent. This provides a good reason to approximate the frequency distribution of oscillators of the j-th kind, $N_j(x)$, originated from inhomogeneous-like broadening by Gaussian probability density function of (46) type,

$$N_j(\omega) = \frac{N_{0j}}{\sqrt{2\pi}\,\sigma_j} \exp\left[-\left(x-\omega_j\right)^2 / 2\sigma_j^2\right], \tag{49}$$

N_{0j} being the total number of oscillators of a given kind, x being the variable oscillator frequency, ω_j being the central frequency of oscillator distribution, and σ_j being the standard deviation for this distribution.

For developing the analytical model for the complex dielectric constant of glasses using the oscillator distribution (49), the classical model (17) was taken in

References 75 and 76 as the initial one. The frequency-independent oscillator numbers N_j entering into the expression for S_j were replaced by the integral distribution functions corresponding to the oscillator densities of (49) type, and fixed oscillator frequencies ω_{0j} were replaced by the variable oscillator frequency, x, thus yielding the modified analytical presentation of $\hat{\varepsilon}(\omega)$:

$$\hat{\varepsilon}(\omega) = \varepsilon_\infty + \sum_{j=1}^{J} \frac{S_j}{\sqrt{2\pi}\,\sigma_j} \int_{-\infty}^{+\infty} \frac{\exp\left[-\left(x-\omega_j\right)^2 / 2\sigma_j^2\right]}{x^2 - \omega^2 - i\gamma_j \omega} dx. \qquad \textbf{(50)}$$

According to (50), the $\varepsilon'(\omega)$ and $\varepsilon''(\omega)$ functions are as follows:

$$\varepsilon'(\omega) = \varepsilon_\infty + \sum_{j=1}^{J} \frac{S_j}{\sqrt{2\pi}\,\sigma_j} \int_{-\infty}^{+\infty} \frac{\exp\left[-\left(x-\omega_j\right)^2 / 2\sigma_j^2\right] \cdot \left(x^2 - \omega^2\right)}{\left(x^2 - \omega^2\right)^2 + \gamma_j^2 \omega^2} dx \qquad \textbf{(51)}$$

and

$$\varepsilon''(\omega) = \sum_{j=1}^{J} \frac{S_j}{\sqrt{2\pi}\,\sigma_j} \int_{-\infty}^{+\infty} \frac{\exp\left[-\left(x-\omega_j\right)^2 / 2\sigma_j^2\right] \cdot \gamma_j \omega}{\left(x^2 - \omega^2\right)^2 + \gamma_j^2 \omega^2} dx. \qquad \textbf{(52)}$$

Naturally, the Gaussian distribution is only an approximation of actual oscillator distributions in the spectra of glasses; the actual oscillator distributions should be limited by some finite frequency ranges that do not include the point $\omega = 0$. However, when $\sigma_j/\omega_j \ll 1$, this approximation may be considered as satisfactory. The classical (17) rather than the factorized model (43) was chosen as the initial model to be modified because (17) comprises the minimum number of free parameters per oscillator type. This is of importance because an additional free parameter σ_j per oscillator kind is introduced when replacing constant N_j values by functions (49), thus substantially increasing the total number of free parameters for the case of multiband spectrum. When choosing the factorized model as a basis, the introduction of σ_j would result in the number of free parameters as large as five per oscillator kind, thus probably hindering the correctness of computational procedures based on the model (see Chapter 3, Section III.C).

Based on Model (50), a new version of the dispersion analysis equipped to cope with multiband, poorly resolved glass spectra was developed;[75,76,80,114] this version is discussed in detail, together with other methods for quantitative spectrum treatment, in Chapter 3 (see Section IV.B).

Lately, practically the same model as (50) — the only difference being the usage of squares of plasma frequencies instead of intensities like Equation (24) — was suggested independently by Brendel and Bormann,[105] without mentioning References 75 and 76. They gave an approximate analytical solution of the convolution integral entering into this model in terms of error functions. For this solution, the Fourier decomposition of the exponential term and then the residue theorem were applied. Using an algorithm based on this solution, Brendel and Bormann[105] successfully described the IR spectra of thin amorphous films of silicon monoxide, silicon nitride, amorphous silicon, and aluminium monoxide at frequencies above 400 to 500 cm^{-1}. Unlike References 75 and 76, no attempts to describe the total combination of the IR fundamental bands including those below 400 cm^{-1} were made in Reference 105.[c]

2. Comparison of the Model with Other Available Models

As is seen from (52), this specific analytical model for the complex dielectric constant presupposes that band shapes in the IR spectra of glasses are described by the convolution of Lorentzian and Gaussian contours rather than by any of these two contours as such or by empirical asymmetric contours, such as those used for the description of inhomogeneously broadened spectra of electronic transitions of rare-earth ions in glasses.[53] The analytical presentation of terms in Equation (52) determining the imaginary part of the complex dielectric constant (50) is rather similar to the well-known Voigt function (47); however, (52) and (47) are not identical to one another. The Voigt function, unlike (52), involves a simplified form of the Lorentzian contour (21) that, as was stressed above in Chapter 1, Section II.A, holds only for bands weak enough to keep the relation $n(\omega) \cong 1$ intact and for a narrow frequency range in the vicinity of the band frequency.

Kopilevich and Makarova[37] conducted the theoretical study of Efimov's analytical model (50) in comparison with other $\hat{\varepsilon}(\omega)$ models mentioned in Chapter 1. They showed that Model (50), unlike the classical one, cannot be reduced to the factorized model (43); at the same time, this model obeys all requirements following from general principles. With $\sigma_j \to 0$, Model (50) transforms, according to Reference 37, into the classical one; conversely, with $\gamma_j/\sigma_j \to 0$, the contour of a peak of the absorption coefficient $\alpha(\omega) = 2\omega\kappa(\omega)/c = 4\pi\kappa(\lambda)/\lambda$, c being the light velocity, tends to become purely Gaussian. Hence it follows that the analytical model for the complex dielectric constant (50) should be considered as a generalization of the classical model (17), this generalization being valid for rather a broad range of variations in the degree of disorder. It should be expected, therefore, that Model (50) might also be applicable to the spectra of some other types of disordered solids. Of course, reservations concerning the inevitably approximate nature of all analytical models of the type discussed in this book (see Chapter 3, Section III) hold for this model, too. However, Model (50), as is seen from the above, lacks major shortcomings inherent in other models already discussed when applied to the spectra of glasses; on the whole, Model (50) should be as valid for glasses as the classical model (17) is for most crystals.

3. Testing the Model

Model (50) was tested widely by applying it to the quantitative analysis of the IR spectra of various glasses. Such an analysis, discussed in detail in Chapter 3, Section IV.B and Chapter 4, presents many experimental proofs of the model, such as the high quality of the best computer fits and the reliability of the spectra of

---

[c] In the period of editing this book, another paper appeared (Kucirkova, H. and Navratil, K., Interpretation of infrared transmittance spectra of SiO_2 thin films, *Appl. Spectr.*, 48, 113, 1994.) that also used practically the same model as (50) borrowed from Naiman, M. L., Kirk, C. T., Aucoin, R. J., Terry, F. L., Wyatt, R. W., and Seinturia, S. D., Effect of nitridation of silicon dioxide on its infrared spectrum, *J. Electrochem. Soc.*, 131, 637, 1984.

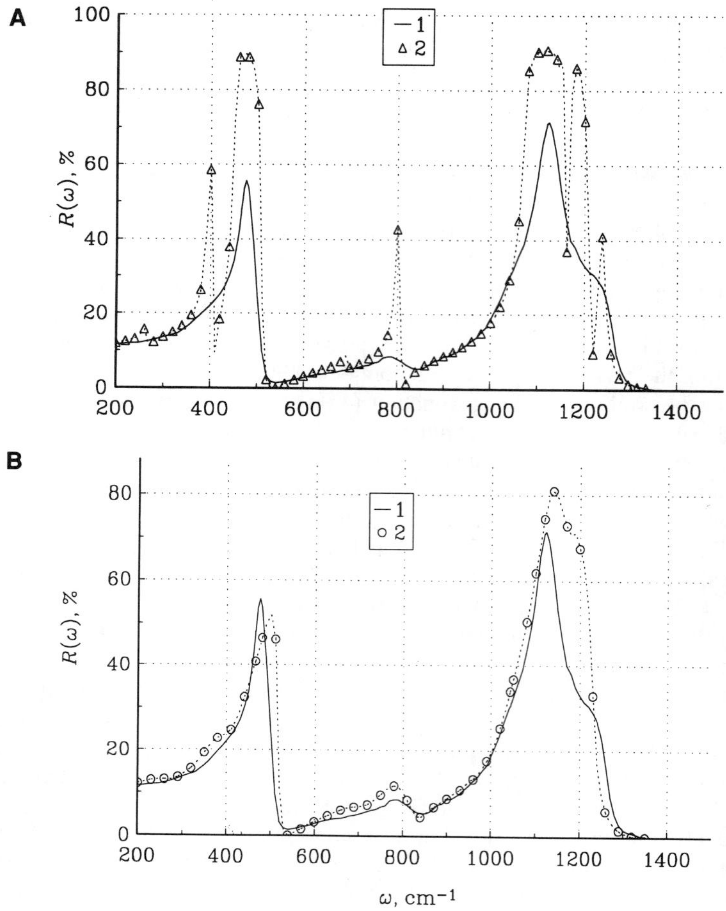

Figure 1 Comparison of the model IR reflection spectra calculated with the parameters of E modes of α-quartz and the experimental spectrum of vitreous silica. (A) 1 is silica glass spectrum and 2 is the preliminary model spectrum calculated with the parameters of E modes according to the data of Reference 13 for the case of unbroadened bands ($\sigma_j = 0$). (B) 1 is silica glass spectrum and 2 is the final model spectrum calculated with the parameters of E modes according to the data of Reference 13 for the case of broadened bands with $\sigma_j = 23$ cm^{-1}. (From Efimov, A. M., *Fiz. Khim. Stekla* (Sov. J. Glass Phys. Chem.) 15, 366, 1989. With permission.) (C) 1 is silica glass spectrum and 2 is the preliminary model spectrum calculated with the parameters of E modes according to the data of Reference 150 for the case of unbroadened bands ($\sigma_j = 0$). (D) 1 is silica glass spectrum and 2 is the final model spectrum calculated with the parameters of E modes according to the data of Reference 150 for the case of broadened bands with $\sigma_j = 23$ cm^{-1}. (From Efimov, A. M., *Dispersion of Optical Constants of Vitreous Solids,* D.Sc. Thesis, Vavilov State Optical Institute, Leningrad, 1990. With permission.)

optical constants obtained. Further, the physical significance of Model (50) was confirmed by the results of numerical experiments[77-79] in which changes in the IR spectra of some crystals that might be caused by random disordering were modeled. As starting data, the IR band frequencies, intensities, damping coefficients, and ε_∞

C

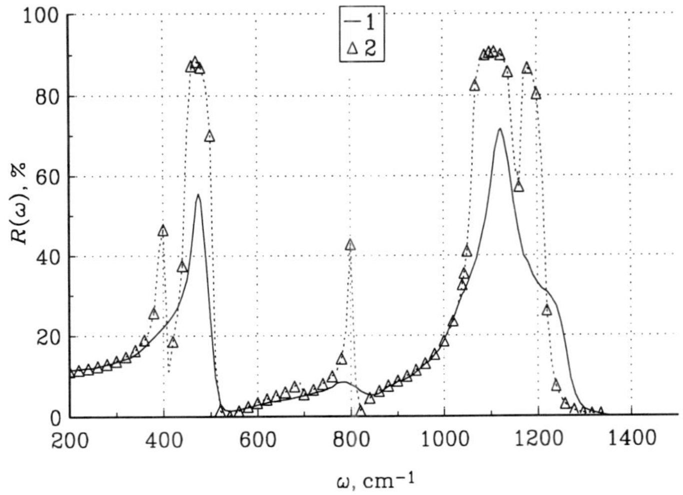

D

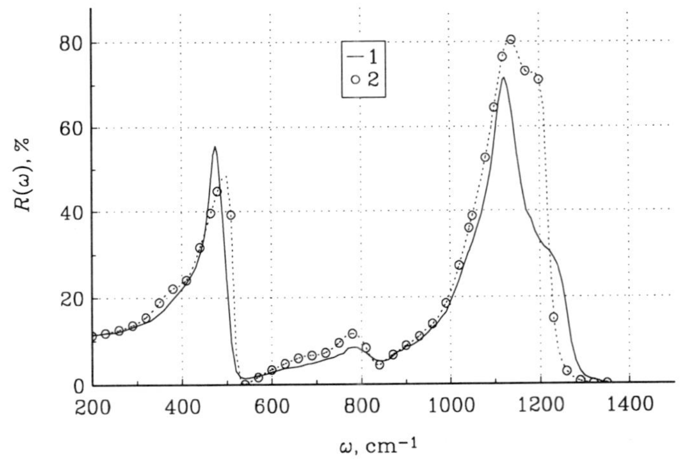

Figure 1 (continued)

values for α-quartz[d] and crystalline lithium disilicate borrowed from literature were used. Because the IR band intensities are density dependent, they were scaled by factors equal to ratios between the densities of crystals used and corresponding glasses. It was assumed that the effect of random disordering can be treated, to the first approximation, to result only in inhomogeneous-like band broadening. To simulate this broadening, the standard deviations σ_j similar to those typical of the IR spectra of oxide glasses (see Chapter 4, Section II.A) were introduced in addition to parameters borrowed from crystals, and resultant artificial spectra of disordered materials were calculated with (50). The pattern of the IR reflection spectra of such artificial disordered materials calculated using the IR active modes

[d] Concerning a comparison between the IR band parameters of vitreous silica and of quartz, see also Chapter 4, Section II.A.1.

A

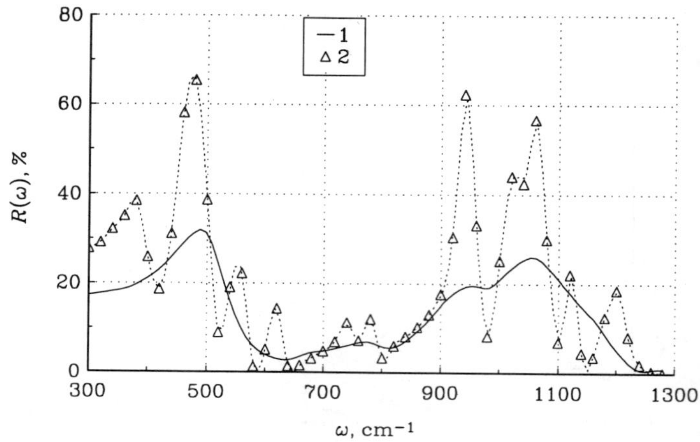

B

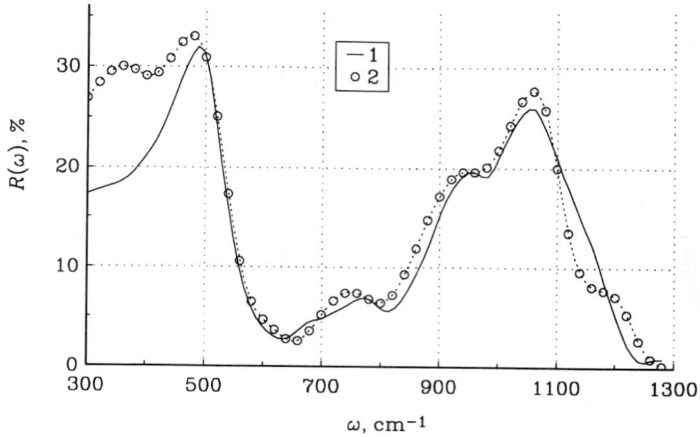

Figure 2 Comparison of the model IR reflection spectrum calculated with the parameters of all modes of $Li_2O \cdot 2SiO_2$ crystal according to the data of Reference 20 and the experimental spectrum of $Li_2O \cdot 2SiO_2$ glass (by synthesis). (A) 1 is glass spectrum and 2 is the preliminary model spectrum for the case of unbroadened bands ($\sigma_j = 0$). (From Efimov, A. M., *Fiz. Khim. Stekla* (Sov. J. Glass Phys. Chem.) 15, 366, 1989. With permission.) (B) 1 is glass spectrum and 2 is the final model spectrum for the case of broadened bands with $\sigma_j = 35$ cm^{-1}. (From Efimov, A. M., *Fiz. Khim. Stekla* (Sov. J. Glass Phys. Chem.) 15, 366, 1989. With permission.)

of α-quartz and lithium disilicate crystal are illustrated by Figures 1 and 2 in comparison with the spectra of respective glasses. The initial spectra combined from the totality of modes of a crystal (Figures 1A and C and 2A) drastically change under the effect of the introduction of standard deviations σ_j determining the Gaussian distribution of oscillators, thus becoming strikingly similar to the spectra of corresponding glasses. As is seen from Figure 1B and D and 2B, the spectra of a modeled disordered material and of glass agree rather well with one another in the region of main reflection peaks, in peak reflectivities, and even in their shapes, perceptible differences being observed only in certain rather narrow spectral regions. It should be stressed especially that, when constructing the model spectra of disordered materials shown in Figures 1 and 2, no parameters borrowed from data on crystals were adjusted, thus these spectra are by no means the computer fits to the experimental glass spectra. Therefore, the approximate similarity between the model spectra of disordered materials and the experimental glass spectra is of physical significance. Evidently, this similarity confirms that glass spectrum formation indeed follows the pattern specified above when deriving Model (50), thus demonstrating the predominant influence of inhomogeneous-like broadening on band shapes in the IR spectra of glasses. Minor deviations of model spectra from glass spectra observed at certain frequencies are discussed in detail in Reference 77 either in terms of a lack of certain modes of a crystal among those forming a glass spectrum (as is the case for A_2 modes of α-quartz) or in terms of some differences in the short-range order of a glass from that of a corresponding crystal (as is probably the case for vitreous lithium disilicate).

Chapter 3

Methods for Reconstructing the Spectra of the Optical Constants in the Ranges of Fundamental Excitations

I. METHODS AND LIMITATIONS IMPOSED BY STRONG ABSORPTION

In the ranges of low absorption, obtaining data on the optical constants meets no principal obstacles because both the refractive index $n(\omega)$ and the absorption coefficient $\alpha(\omega) = 4\pi\kappa(\lambda)/\lambda$ can be measured directly and separately from each other with well-known experimental methods. However, this is not the case for the ranges where absorption becomes strong, especially for those of the fundamental excitations. In these ranges, the maximum $\kappa(\omega)$ values can be of about unity or even greater (see, for example, Chapter 1, Figure 1), thus corresponding to quite high magnitudes of the absorption coefficient $\alpha(\omega)$ (of the order 10^4 cm^{-1} for the vibrational and 10^6 cm^{-1} for the electronic excitations). Such a strong absorption makes standard methods of absorption spectroscopy impractical because a probing beam is absorbed almost completely in the layer with thickness of about λ. Indeed, preparing samples as thin as that leads inevitably to substantial changes in chemical composition and/or structure of materials, especially for binary and more complex glasses. Therefore, reflectivity becomes the only measurable quantity, so it is the methods of the reflection spectroscopy that play the major role when studying the fundamental excitations. For the simplest case of near-normal incidence, the reflectivity $R(\omega)$ is expressed using the optical constants as follows:

$$R(\omega) = \frac{\hat{\varepsilon}(\omega) - 1}{\hat{\varepsilon}(\omega) + 1} = \frac{[n(\omega) - 1]^2 + \kappa^2(\omega)}{[n(\omega) + 1]^2 + \kappa^2(\omega)}. \tag{53}$$

The complex amplitude of reflection, $\hat{r}(\omega)$, is represented by

$$\hat{r}(\omega) = r(\omega)\exp[i\phi(\omega)], \tag{54}$$

where $r(\omega) = \sqrt{R(\omega)}$ and $\phi(\omega)$ is the phase angle of reflected beam. As is seen from (53), reflectivity is the function of two unknowns, namely, both optical constants; hence, neither $n(\omega)$ nor $\kappa(\omega)$ can be obtained directly from a single $R(\omega)$ value for a particular frequency ω. To obtain the optical constants, it is necessary to reconstruct the phase angles in addition to $R(\omega)$.

For oblique incidence, $R(\omega)$ depends greatly, in accordance with the well-known Fresnel equations described elsewhere, on both the incidence angle and the direction of polarization of the incident beam. Using the Fresnel equations, the phase angle of the reflected beam and, further, optical constants for a particular frequency can be calculated from reflectivity. To provide such calculations, two different values of reflectivity at a particular frequency should be measured either for two perpendicular directions of incident beam polarization at a single incidence angle or for two different incidence angles. For the case of limited ranges of variations in the magnitudes of optical constants and not too high absorption ($10^{-4} \leq \kappa \leq 10^{-1}$), methods using the Fresnel equations can provide, with all optimization conditions held, an absolute accuracy of optical constants as good as 10^{-4} to 10^{-5}.[106,107] For the broad ranges of the fundamental excitations characterized by variations in optical constants as high as about unity or even greater, these methods are not too reliable, because large errors can arise in such conditions.

Dispersive Fourier transform spectroscopy, which allows for direct measurement of the phase angle with a Michelson interferometer, was used for calculating the optical constants of some crystals[8] in the region of the IR fundamental excitations and also of vitreous silica[108] in the 20 to 120 cm^{-1} range. However, this technique did not become widely used yet.

The *Kramers-Kronig transform* and *dispersion analysis* methods are considered to be the major ones for reconstructing the spectra of optical constants in the ranges of fundamental excitations. Both methods are based on the mathematical treatment of experimental $R(\omega)$ spectrum measured in a broad frequency range for a single incidence angle and single direction of polarization, the case of near-normal incidence of natural light being most commonly used. According to the above, information on the phase angle of reflected beam and, hence, on the magnitudes of optical constants cannot be obtained directly from such data. Therefore, in this case, the problem of calculating the spectra of optical constants turns out to be an example of the so-called inverse spectroscopic problem, i.e., the problem of how to reconstruct, by means of some mathematical treatment, the principal physical characteristics of a material from "raw" experimental data unsuitable for direct calculation. From the mathematical viewpoint, the inverse spectroscopic problem is related to the type of incorrect problems that cannot be solved without involving some independent additional information. The Kramers-Kronig transform and dispersion analysis differ principally from each other in the kind of additional information used and also in the relative sensitivity to various kinds of experimental errors in the $R(\omega)$ spectrum under treatment. These methods are discussed below in Sections II through IV.

Recently, Vartiainen, Peiponen, and Asakura[109] proposed a different numerical method based on applying the maximum entropy model. This method, from its authors' viewpoint, is applicable for the $R(\omega)$ spectra with complicated structure, thus presenting an alternative to the Kramers-Kronig transform. Data presented in Reference 109 show that the spectra of optical constants of KCl in the 2 to 10 eV range reconstructed with this method from the $R(\omega)$ spectrum are in agreement with those taken from literature, the agreement being good indeed for $n(\omega)$ and a little less satisfactory for $\kappa(\omega)$. The method of Reference 109 is taken by its authors to be applicable to the case of purely Gaussian band shapes for which, as is thought in Reference 110, the Kramers-Kronig transform fails.

II. KRAMERS-KRONIG TRANSFORM

The Kramers-Kronig transform usually is considered to be an acknowledged method for reconstructing the optical constants of solids in the standard reflection spectroscopy using the data on the reflectivity at near-normal incidence; moreover, it also is used in a specific field of reflection spectroscopy, internal reflection spectroscopy.[111] The procedure of the Kramers-Kronig transform uses one more Kramers-Kronig relation for the phase angle of reflected beam, $\phi(\omega)$,

$$\phi(\omega) = -\frac{\omega}{\pi} \mathrm{P} \int_0^\infty \frac{\ln R(\omega^*)}{\omega^{*2} - \omega^2} d\omega^*, \tag{55}$$

here and in Formula (58) to follow P denotes the Cauchy principal value of the integral. After Robinson[112] this relationship was used widely, in various modifications, for reconstructing the phase of the reflected beam; the data obtained allow for calculating the spectra of optical constants $n(\omega)$ and $\kappa(\omega)$ via the Fresnel equations as

$$n(\omega) = \left[1 - R(\omega)\right] \Big/ \left[1 + R(\omega) - 2\sqrt{R(\omega)} \cos\phi(\omega)\right] \tag{56}$$

and

$$\kappa(\omega) = -2\sqrt{R(\omega)} \sin\phi(\omega) \Big/ \left[1 + R(\omega) - 2\sqrt{R(\omega)} \cos\phi(\omega)\right]. \tag{57}$$

A. SPECIFIC FEATURES OF THE METHOD

As was specified before (see Chapter 1, Section I), the Kramers-Kronig relations are not related to any microscopic physical model for the response of a medium to the effect of the electromagnetic field. Therefore the Kramers-Kronig transform method is commonly assumed not to be connected to any particular analytical model for the contour of components of a spectrum under treatment and, hence, is usually taken to be equally applicable to any optical spectra irrespective of frequency range and physical nature of a material.

Equation (55) presupposes integrating over the entire frequency range from zero to infinity, whereas any actual measurement, naturally, covers only a limited frequency range. An assumption of the behavior of $R(\omega)$ function in the ranges for which experimental data are unavailable serves as independent additional information used in the Kramers-Kronig transform method for solving the inverse spectroscopic problem.

It should be noted, however, that the relation (55) is not characterized by such an unambiguous validity as is the case for other Kramers-Kronig relations (7) through (10). It was shown theoretically[113] that the relation (55) can include, for some cases, additional unknown terms. In practice, such cases can occur when using the oblique incidence and/or the contact with optically dense media.[114] Therefore the use of the Kramers-Kronig transform for treating data recorded with the internal reflection technique[111] can sometimes meet certain difficulties.

B. AVAILABLE VERSIONS OF THE METHOD

In the course of the Kramers-Kronig transform, the integral in (55) is usually divided into three parts:

$$\phi(\omega) = -\frac{\omega}{\pi} \mathrm{P} \int_0^a \frac{\ln R_1^{\mathrm{mod}}(\omega^*)}{\omega^{*2} - \omega^2} d\omega^* - \frac{\omega}{\pi} \mathrm{P} \int_a^b \frac{\ln R(\omega^*)}{\omega^{*2} - \omega^2} d\omega^* -$$

$$\frac{\omega}{\pi} \mathrm{P} \int_b^\infty \frac{\ln R_2^{\mathrm{mod}}(\omega^*)}{\omega^{*2} - \omega^2} d\omega^* = -I_1 - I_2 - I_3, \tag{58}$$

where a and b are the limits of a frequency range for which the experimental $R(\omega^*)$ data are available and $R_1^{mod}(\omega^*)$ and $R_2^{mod}(\omega^*)$ are certain model functions chosen arbitrarily for simulating the actual $R(\omega^*)$ dependence in the ranges $0 < \omega < a$ and $b < \omega < \infty$, respectively. When treating the VUV spectra, the asymptotic behavior of R (ω^*) in the $b < \omega < \infty$ range usually is approximated by $R_2^{mod}(\omega^*) = C\omega^{-k}$, C being a constant and $3 \leq k \leq 4$ (see, for example, Reference 115). Moreover, a correcting function, $\psi(\omega)$, may be introduced in order to diminish the error of calculating the phase angle (see, for example, Reference 116):

$$\phi(\omega) = -I_1 - I_2 - I_3 + \psi(\omega). \quad \textbf{(59)}$$

The $\psi(\omega)$ function should be chosen from the condition of coincidence, for certain frequencies, between the phase angle values calculated and taken from some independent measurements.

Various algorithms were developed for conducting the Kramers-Kronig transform that differ from each other in the presentation of $R(\omega^*)$ for the ranges $0 < \omega < a$ and $b < \omega < \infty$ and also in the way of minimizing the errors of phase angle. The practice of using reference $R(\omega^*)$ values taken from independent measurements at $\omega < a$ and $\omega > b$ is rather widely applied. For example, the functions $R_1^{mod}(\omega^*)$ and $R_2^{mod}(\omega^*)$ in (58) may be assumed to be constants, and the correction $\psi(\omega)$ in (59) (that should take into account the effect of actual dispersion of optical constants) can be calculated using reference data for some frequencies.[117] Various methods of choosing such reference $R(\omega^*)$ values are discussed; in particular, reference $R(\omega^*)$ values calculated from $n(\omega^*)$ in the transparency range, where the phase angle value is zero, are frequently used. In general, the farther the frequencies of reference $R(\omega^*)$ values are from a and b, the smaller the errors. Along with literature mentioned before, the origin of some errors and ways to avoid them were mentioned briefly in Reference 43 in connection with treating the IR spectra of various SiO_2 modifications. A comprehensive review of the totality of error sources for various algorithms of the Kramers-Kronig transform is given in Reference 118.

As was stressed in References 114 and 118, even crude information on the locations and intensities of spectrum components lying far enough from the range under study can be vital for providing much better results than those obtained without such information. In particular, when calculating optical constants in the IR range, it is quite useful to take into account any data, whatever approximate, on the structure of the VUV fundamental spectrum.

Along with relations (55) or (58), the modified presentation of (55) with an integrand comprising the factor $\partial[\ln R(\omega^*)]/\partial\omega^*$ is used in some algorithms of the Kramers-Kronig transform. This presentation is more convenient for broad frequency ranges and provides, in some cases, a better accuracy in the neighborhood of the $\kappa(\omega^*)$ maxima. However, it imposes especially stringent requirements on the smoothness of the experimental $R(\omega^*)$ spectrum to be analyzed. Smoothness violated by some kink on the $R(\omega^*)$ spectrum can result in large errors of optical constants, including a false $\kappa(\omega^*)$ maximum.[118]

As was already mentioned, the Kramers-Kronig transform is used in the spectroscopy of various solids for obtaining data on the optical constants in both the IR and VUV ranges of fundamental excitations. As for glasses, the Kramers-Kronig transform was applied to the IR and VUV spectrum of vitreous SiO_2, the VUV spectrum of vitreous GeO_2, and also to the IR spectra of various simple, binary and ternary glasses and to the VUV spectra of some of the glasses. All available data are discussed in Chapter 4.

III. DISPERSION ANALYSIS: GENERAL ASPECTS

The method of the dispersion analysis is based, unlike the Kramers-Kronig transform, on a particular analytical model for the complex dielectric constant with which the contours of individual $\varepsilon'(\omega)$ and $\varepsilon''(\omega)$ components of a spectrum under treatment are to be approximated. It is this model that serves as an independent additional information used for solving the inverse spectroscopic problem. The term "dispersion analysis" was put into use by Spitzer and Kleinman[13] who described in detail the procedure of the dispersion analysis based on the classical analytical model (17). Along with the latter, other analytical models for the complex dielectric constant discussed above can be used in the dispersion analysis. Therefore, it is reasonable to describe the main features of the procedure in terms of a generalized idea of the dispersion analysis; the mathematical study of such a generalized form of the dispersion analysis for a variety of analytical models for the complex dielectric constant was conducted recently by Kopilevich and Makarova.[37] In the course of the dispersion analysis procedure, every observed (or assumed) component of a complicated experimental spectrum should be related to a particular kind of oscillator characterized by a set of dispersion parameters inherent in the analytical model chosen. As follows from Chapter 1, Sections II.A and III, the dispersion parameters for the classical model are ω_{0j}, S_j, and γ_j; those for the factorized model are $\omega_{\mathrm{T}0j}$, $\omega_{\mathrm{L}0j}$, $\gamma_{\mathrm{T}j}$, and $\gamma_{\mathrm{L}j}$. The procedure consists of computer simulation of an experimental spectrum by means of an analytical model for the complex dielectric constant chosen and Fresnel equations. In the beginning, a certain assumed number of oscillators, J, and some initial values of dispersion parameters should be arbitrarily chosen and introduced by the investigator. Using these data, the arrays of optical constants and, further, of $R(\omega)$ values are computed; then a minimization of differences between the computed and experimental $R(\omega)$ spectra is conducted by successively varying the dispersion parameter values. When the best fit to the spectrum is obtained, the corresponding values of dispersion parameters are assumed to be close enough to the actual spectrum component parameters. The particular algorithms used in the computational programs of dispersion analysis are discussed in many original publications cited below.

Available versions of the dispersion analysis are based on the analytical models for the complex dielectric constant that are discussed in Chapters 1 and 2. The reason why the simplified analytical models should be used in the dispersion analysis follows from the nature of the dispersion analysis as a method for the solution of the inverse spectroscopic problem. As was stressed by Kopilevich and Makarova,[119] the procedure of the dispersion analysis may be considered to be correct only in the case when the experimental data under study match certain requirements, such as the data being informative enough, i.e., the amount and accuracy of the data should allow for reliably calculating all free parameters for the model involved. Providing the correctness of dispersion analysis is more difficult the greater the number of free parameters, because opportunities to make experimental data more informative by means of broadening the frequency range of spectral measurements or of lowering experimental errors are usually rather limited due to methodological reasons. For the case when both the frequency range of measurement and the limits of experimental errors are fixed, there exists some reasonable limit for the total number of free parameters. When increasing the number of free parameters above this limit, the errors in calculating of both the optical constants and band parameters can rise instead of diminish, hence the expediency of choosing an analytical model for the complex dielectric constant that is capable of providing the required accuracy of the best fit to a spectrum of a material with the smallest number of free parameters. Thus, complicated many-parameter analytical models for the complex dielectric constant developed in rigorous theoretical approaches are of no use for the dispersion analysis.[119]

A. BASIC FEATURES OF THE METHOD COMPARED TO THOSE OF THE KRAMERS-KRONIG TRANSFORM

There are some specific features of the dispersion analysis as compared to those of the Kramers-Kronig transform. These specific features result mostly from the fact that the dispersion analysis and Kramers-Kronig transform use entirely different kinds of additional information for solving the inverse spectroscopic problem.

When solving the inverse spectroscopic problem for the entire IR or VUV range of fundamental excitations of a material, two stages of this task should be recognized. The first stage is the reconstruction of spectra of optical constants from "raw" experimental spectrum specified above when discussing the Kramers-Kronig transform. In this stage, the Kramers-Kronig transform has some well-known advantages over the dispersion analysis. The main advantage is the above-mentioned universal character of the underlying relationship which guarantees an applicability to any material irrespective of its physical nature. On the contrary, the applicability of any particular version of the dispersion analysis is reduced to materials for which the corresponding analytical model for the complex dielectric constant is valid. Moreover, the Kramers-Kronig transform uses a computational procedure that is simpler than that used in the dispersion analysis. On the other hand, the dispersion analysis is rather insensitive to some experimental errors that are crucial for the Kramers-Kronig transform (see Section II.B). Therefore, it is reasonable to consider these methods as complementing each other; Belousov and Pogarev[120] even proposed combining both methods into a unified procedure.

The second stage of solving the inverse spectroscopic problem for the range of fundamental excitations consists in describing the total $\varepsilon''(\omega)$ spectrum in terms of parameters of individual spectral components related to particular fundamental excitations. It is these parameters that are of primary interest for both the quantitative spectroscopy of fundamental excitations and applied optics. The parameters of individual spectral components play the role of dispersion parameters and therefore determine explicitly, through a $\hat{\varepsilon}(\omega)$ model chosen for a material, the frequency dispersion of $\varepsilon'(\omega)$ and $\varepsilon''(\omega)$. Thus, these parameters allow one to extrapolate, with no additional assumptions involved, the contribution of any spectrum component into the $\varepsilon'(\omega)$ values at any required frequency. The examples of such a use of band parameters for glasses can be found in References 77, 78, and 121. These examples are discussed in Chapter 2, Section III.B.3; Chapter 4, Sections II.A.2 and II.B; and Chapter 5, Section IV.B.

For the second stage of data treatment, the situation with the two methods under discussion becomes quite the opposite. The dispersion analysis acquires evident advantages against the Kramers-Kronig transform, especially when treating the spectra of glasses (see Section IV.A). As is seen from the above, the dispersion analysis combines, in one procedure, opportunities both to reconstruct the spectra of optical constants and to calculate the spectral component parameters; thus, no additional procedure for resolving the complicated spectrum into individual components is required. On the contrary, the opportunities provided by the Kramers-Kronig transform are exhausted, strictly speaking, at the first stage, because the Kramers-Kronig relations do not presuppose the very idea of individual spectral components. Therefore the Kramers-Kronig transform as such does not allow, in general, for obtaining the parameters of the components, so an additional procedure should be used for this purpose. The seriousness of problems arising in this connection depends on the character of the spectrum of a material under study. In the case of a single band or a few bands substantially separated from each other, the band frequency, band intensity, and bandwidth may be taken either directly from the plot of $\varepsilon''(\omega)$ vs. frequency or by means of a widespread practice to derive the transverse and longitudinal band frequencies from the locations of $\varepsilon''(\omega)$ and $Im[-\hat{\varepsilon}^{-1}(\omega)]$ maxima, respectively (see, for example, Reference 122). However, appreciable band overlapping such as that typical of the IR spectra of glasses, greatly impedes these possibilities; therefore it is necessary to supplement the Kramers-Kronig transform with some method of resolving a spectrum into individual components. This immediately involves a particular type of band contour (Lorentzian, Gaussian, intermediate,[55] or some other) and, hence, a particular analytical model for the complex dielectric constant, thus depriving the entire procedure of universal character and increasing the total error of results obtained. The above also holds, along with standard deconvolution methods such as those reviewed in Reference 55, for approaches that do not explicitly involve the analytical presentation of band contour, such as the Fock method.[123] This method assumes the invariability of both the shapes of spectral components and their locations on the frequency axis for the series of spectra under study; such an assumption implicitly presupposes that any other $\hat{\varepsilon}(\omega)$ models, except the particular one remaining unknown explicitly, are rejected.

B. DISPERSION ANALYSIS USED FOR TREATING THE SPECTRA OF CRYSTALS

The important advantage of the dispersion analysis specified above is often underestimated. The viewpoint that this method can be ignored on the grounds of the simplified character of analytical models used is sometimes expressed even in the spectroscopy of crystals. In spite of this, tasks involving the estimation of band frequencies, band intensities, and bandwidths in the spectra of inorganic crystals are solved mostly with the dispersion analysis rather than with the Kramers-Kronig transform. However, the total combination of data obtained has not been systemized. Therefore it is useful to cite briefly available publications on the dispersion analysis of the spectra of crystals; this should illustrate the potentialities of the dispersion analysis in general. The information including the specific features of the dispersion analysis version specially equipped to cope with glass spectra and on the data obtained with this version for various glasses is presented in detail in Section IV and Chapter 4, respectively.

The classical version of the dispersion analysis based on Model (17) was used for treating the IR spectra of crystals such as silicon carbide,[124] typical semiconductors,[125] various oxides (SiO_2,[13,126] GeO_2,[127] TiO_2,[128] Al_2O_3,[129] TeO_2,[130] ZnO,[131] and MgO[132,133]), halogenides (LiF,[132,133] $NaCl$,[134] MgF_2,[135] CaF_2, SrF_2, and BaF_2,[136]

BaFCl,[137] and ZnF_2[135]), nitrates and carbonates,[138] silicates,[139–141] phosphates,[142–144] some arsenates,[143] titanates,[128] vanadates,[143] niobates,[145] molybdates and tungstates,[146] etc. Special attention was paid to the details of computational procedure in some of the studies.[132,133,147] Except for rare occasions,[128,129,146] errors of the best fits to the experimental reflection spectra of crystals are close to the error of $R(\omega)$ measurement; this holds for crystals with both simple and quite complicated IR spectra, such as those of α-quartz,[13,126] silicates,[139–141] and niobates.[145] Hence, the effect of dependences $\omega_{0j} = \omega_{0j}(\omega)$ and $\gamma_j = \gamma_j(\omega)$ on the IR band shapes mentioned in Chapter 1, Section II.A is negligible in these spectra and therefore is fully concealed by experimental errors. Thus, the approximation of the complex dielectric constant with Model (17) is rather good for the IR spectra of most crystals.

As mentioned in Chapter 1, Section I.C, the optical spectra of interband electronic transitions are principally inseparable into independent spectral components, which is incompatible with the discrete oscillator distribution and with the symmetric band shapes presupposed by the classical analytical model (17) and the factorized model (43). Nonetheless, the classical dispersion analysis was used successfully for treating the fundamental spectra of electronic excitations of some typical semiconductors, such as elemental Si[148] and Ge.[15,148] Some objectives for this were given in Reference 148.

A few cases of rather poor results obtained for crystals with the classical dispersion analysis based on (17) were related, first of all, to the IR spectra of TiO_2 and titanates[128] for which the substantial deviations of computer fits from the experimental spectra in the regions of $R(\omega)$ maxima were observed. Weaker but perceptible deviations of computer fits were obtained when treating the IR spectra of Al_2O_3,[129] $CaMoO_4$, and $CaWO_4$,[146] and like crystals.

For crystals such as $SrTiO_3$, $BaTiO_3$, and $KTaO_3$, Barker and Hopfield[14] developed a more complicated version of the dispersion analysis based on Model (22) that explicitly involved coupling two classical oscillators. The spectra of these crystals comprise three bands. In Reference 14 coupling was assumed to take place between the most intense mode with resonance frequency in the 30 to 90 cm^{-1} range and the highest frequency mode ω_{0j} in the 510 to 550 cm^{-1} range, whereas the third mode with an intermediate ω_{0j} value (170 to 200 cm^{-1}) was taken to be uncoupled. Such a version allowed Barker and Hopfield to obtain a satisfactory quality of best fits to the spectra of the above crystals for which the usual classical model (17) failed.

Another widely used version of the dispersion analysis is the one based on the factorized analytical model for the complex dielectric constant (43). This version, in the modification for $\gamma_{Lj} \neq \gamma_{Tj}$, was used, with success, for treating the IR spectra of crystals such as some oxides (TiO_2,[36,149] Al_2O_3,[36] and SiO_2[150]), chlorides ($LaCl_3$ and $PrCl_3$[34]), silicates,[20] etc. As was mentioned in Chapter 1, Section III, Model (43) is much more flexible than the classical one due to the opportunity to independently vary γ_{Tj} and γ_{Lj} magnitudes. Therefore the version of dispersion analysis based on (43) is capable of providing better fits to the experimental spectra of crystals with an extraordinarily large $\omega_{L0j}/\omega_{T0j}$ (ratio (such as TiO_2 and titanates), for which the classical version fails. For α-quartz, on the contrary, the magnitudes of all band parameters found by Spitzer and Kleinman with the classical dispersion analysis[13] and by Gervais and Piriou[150] with the version based on Model (43) almost coincide. For example, the ω_{0j} and S_j/ω_{0j}^2 magnitudes for the E mode related to the asymmetric Si-O-Si stretch are found by Gervais and Piriou to be 1065 cm^{-1} and 0.65, whereas the magnitudes obtained by Spitzer and Kleinman are 1072 cm^{-1} and 0.67, respectively. Such a coincidence is explained by much lower band intensities in the spectrum of α-quartz ($\omega_{L0j}/\omega_{T0j} \leq 1.15$) as compared to that of a band with ω_{T0} = 189 cm^{-1} in the spectrum of TiO_2[36,149] (ω_{L0}/ω_{T0} = 4.4) or a band with ω_{T0} = 569.5 cm^{-1} in the spectrum of Al_2O_3[36] (ω_{L0}/ω_{T0} = 1.6). Therefore, there are no grounds to consider the dispersion analysis version based on the factorized model (43) as providing more reliable results for the spectra of crystals obeying the $(\omega_{L0j}/\omega_{T0j}) - 1 \ll 1$ inequality than the classical model. As should be expected, perceptible differences in γ_{Tj} and γ_{Lj} magnitudes are only characteristic of the strongest oscillators. Indeed, for the above bands in the spectra of TiO_2 and Al_2O_3, the ratio γ_{Lj}/γ_{Tj} equals to 3.1 and 2.6, respectively, whereas for bands in the spectra of quartz and silicates, this ratio does not exceed 1.5 to 1.8. Generally, the expected relation $\gamma_{Lj} - \gamma_{Tj} \geq 0$ holds. In some cases, however, the opposite relation $\gamma_{Lj} - \gamma_{Tj} <, 0$, i.e., the unnatural one from the standpoint of the above, is obtained; this is probably explained by either large experimental $R(\omega)$ errors or by large computational errors for these two parameters due to the shortcomings of a particular algorithm involved.

The modification of Equation (43) presupposing that $\gamma_{Lj} \equiv \gamma_{Tj}$ is also used in the dispersion analysis. As was already mentioned, such an analytical form of Equation (43) is an equivalent of the classical model (17). However, according to Belousov and Pogarev,[151] the use of this form allows one to make the dispersion analysis procedure more rapidly, enabling better efficiency in fitting the $R(\omega)$ spectrum when varying ω_{L0j} as compared to that obtained when varying S_j in (17).

IV. DISPERSION ANALYSIS OF GLASS SPECTRA

A. SPECIFIC FEATURES OF DISPERSION ANALYSIS FOR GLASSES

As was mentioned in Section III.A, the dispersion analysis, when applied to multiband and poorly resolved spectra (such as those of most glasses), should acquire evident advantages over the Kramers-Kronig transform because it allows the reconstruction of optical constants and calculation of spectral component parameters with no additional computational procedure involved. Therefore, it is more reasonable (at least, in cases presupposing the calculation of band frequencies and intensities) to conduct the quantitative treatment of experimental glass spectra with the dispersion analysis rather than with Kramers-Kronig transforms. Data on band frequencies and band intensities are of more interest for glasses than for crystals, because there exists a possibility of studying band parameter vs. glass composition plots with continuously varying glass compositions. However, the opportunity to apply the dispersion analysis to glass spectra was not, strictly speaking, available until recently because all $\hat{\varepsilon}(\omega)$ models used in the standard versions of the dispersion analysis, as was argued above, are in apparent contradiction with the nature of glasses as the disordered solids. Nonetheless, as was mentioned in Chapter 2, Section III.A, there are a few studies devoted to the treatment of the fundamental glass spectra with these "crystal-oriented" versions of the dispersion analysis. As a rule, the best computer fits obtained in most studies involving the classical dispersion analysis are characterized by appreciably flatter band wings as compared to those in the experimental spectra. This results in substantial positive errors in reflectivity in regions between the reflection maxima (see, for example, Reference 101). Naturally, such fits cannot be treated as satisfactory. On the contrary, Ljundahl, Fellman, and Ribbing[102] obtained, with the classical dispersion analysis, rather good computer fits to the IR reflection spectra of five multicomponent silicate glasses in the 430 to 3330 cm^{-1} range. A good coincidence between the spectra of optical constants of these glasses obtained with the dispersion analysis and with the Kramers-Kronig transform from the same reflection spectra was demonstrated. In view of Chapter 2, Section III.A, the main reason for such good quality of the computer fits and of the spectra of optical constants obtained in Reference 102 should probably be sought in the specific features of the reflection spectra under study. As will be seen from Chapter 4, Section II.A.2, spectra studied in Reference 102 are much more smeared and have less pronounced structure than those of glasses having simpler chemical compositions; this is probably due to overlapping spectral features related to a broad variety of different vibrations. Naturally, such smooth spectra impose much less stringent requirements for the validity of the $\hat{\varepsilon}(\omega)$ model used for their description than is the case for the spectra of binary glasses and vitreous simple oxides having deep minima between separate bands. The numerical $n(\omega)$ and $\kappa(\omega)$ values obtained in Reference 102 are demonstrated in Chapter 4, Section II.A.2. Furthermore, the studies of Gervais et al.[103,104] conducted recently with the above version of dispersion analysis using the factorized $\hat{\varepsilon}(\omega)$ model (43) should be mentioned. In Reference 103, rather good computer fits to the experimental IR reflection spectra of various silicate and aluminosilicate glasses in the broad frequency range from 20 to 1500 cm^{-1} were obtained. Moreover, judging by graphs presented, satisfactory agreement between the $\varepsilon''(\omega)$ spectra found with this dispersion analysis version and with the Kramers-Kronig transform was observed. On this basis, the authors concluded that band shapes in the spectra of glasses are indeed Lorentzian and assumed the substantial degree of coherence of vibrations. Later, similar results and conclusions were presented by Parot-Rajaona, Vaills, Massiot, and Gervais[104] for some lithium aluminosilicate glasses. The idea of substantial coherence of vibrations is mentioned in Reference 104; also, band broadening is likened to that generated by thermal motion. In view of Chapter 2, Section II.B, the spectra of any disordered solids with coherence of vibrations kept intact would demonstrate the *asymmetric* (that is not observed experimentally) rather than the *symmetric* (that actually takes place) band broadening; moreover, there is no chance for thermal motion of structural groups, such as that observed for molecular liquids, in the three-dimensional glass network. Thus, both the main idea of Gervais' approach[103] and the addition to it in Reference 104 do not seem too convincing. Hence, another reason for explaining the good quality of computer fits and $\varepsilon''(\omega)$ spectra obtained in References 103 and 104 should be sought; evidently, this reason consists in the great flexibility of factorized model (43).

B. SPECIAL VERSION OF DISPERSION ANALYSIS FOR TREATING THE SPECTRA OF GLASSES

Based on the specific analytical model for the complex dielectric constant of glasses (50) described in Chapter 2, Section III.B, Efimov and Makarova[75,76,114] developed a new version of dispersion analysis specifically equipped to cope with multiband, poorly resolved glass spectra. As mentioned above, the procedure of the dispersion analysis comprises the computer simulation of an experimental reflection spectrum, $R_{exp}(\omega)$; a computer fit, i.e., the model spectrum $R_{mod}(\omega)$, is constructed with the help of a certain analytical model for the complex dielectric constant. The proof of fit is the magnitude of the error function, Q, this function being some norm of the $R_{mod}(\omega) - R_{exp}(\omega)$ difference. In the dispersion analysis version under discussion,[75,76] the error function was chosen in the following form:

$$Q = \sqrt{\frac{1}{b-a}\int_a^b \left[R_{mod}(\omega) - R_{exp}(\omega)\right]^2 d\omega}\ , \qquad \textbf{(60)}$$

a and b being, as above, the limits of frequency range studied. The algorithms developed by Makarova[114] for computing the convolution integral that enters into Model (50) and for minimizing the Q function (60) are described in References 76 and 80. In particular, the idea of Belousov and Pogarev[151] about a better efficiency of reflection spectrum fitting when varying ω_{LOj} instead of S_j (see Section III.A) was used, a transition from ω_{LOj} to S_j being conducted with Formula (45).

Implementing the dispersion analysis, as with other methods for resolving a complicated spectrum into components, involves a specific problem such as that of how to choose reliably the total number of oscillator types, J (i.e. of bands assumed for a spectrum under treatment). Because J is not a free parameter, it should be determined by an investigator prior to starting the dispersion analysis procedure. When analyzing the poorly resolved IR glass spectra, the choice of a reasonable J value is not a simple task. In a general case, J should not be equated to the number of distinct features (peaks or shoulders) in the experimental reflection spectrum, because many weak bands can be concealed by the wings of stronger neighboring ones. On the other hand, a voluntary increase in the number of spectral components not restricted by any consideration except minimizing the error function (practiced, for example, in some works on the Raman spectroscopy of glasses, such as Reference 72) is more unacceptable because it might invalidate the dispersion analysis procedure (see Section III.A). In particular, weak false "bands" caused by some local errors might be obtained. Therefore, the problem of criteria for choosing the total number of oscillator kinds is of principal importance for the dispersion analysis of glass spectra. According to References 76 and 79, an expediency to add an oscillator over the total number of distinct spectral features should be approved when simultaneously obeying the criteria as follows: (1) some independent information (such as that on the spectrum of corresponding crystal or on the spectra of other glasses from the same system) confirms the presence of an unappreciated band in a certain frequency range and (2) adding this oscillator substantially improves the computer fit in this range with no perceptible worsening of the fit in other ranges.

Dispersion analysis based on Model (50) was tested thoroughly with the IR reflection spectra of various borate glasses taken as examples.[75,76,80] Later, the IR reflection spectra of many other glasses such as vitreous silica,[79] binary (and, for some systems, more complex) silicate,[79] borate,[79] germanate,[79,152] and tellurite[79,153] glasses were studied quantitatively with this version. In these studies, the data on the numerical values of optical constants, band parameters, and (for some of the glasses) static dielectric constant ε_0 were obtained; the totality of these data is discussed in Chapter 4 and, moreover, presented in the Appendix. For glasses studied in References 75, 76, 79, 80, 152, and 153, the best fits to the experimental IR reflection spectra are characterized, as a rule, by small error function (60) values

A

B

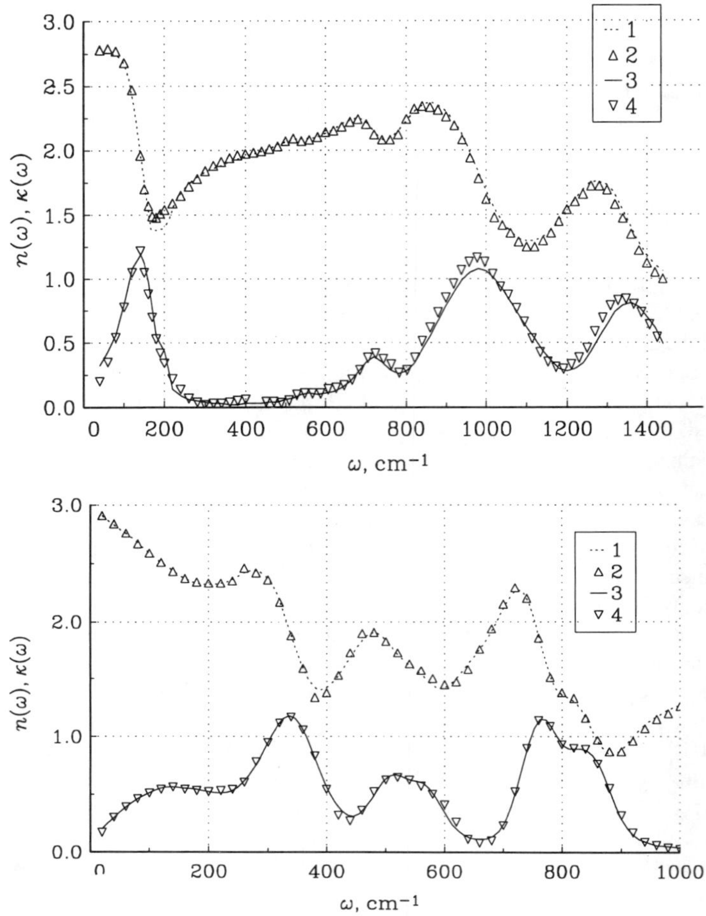

Figure 1 Computer fits to the experimental IR reflection spectra obtained with the dispersion analysis version based on Model (50) for various oxide glasses according to the data of Reference 79. (A) Vitreous silica; 1 is experimental spectrum and 2 is fit. (B) $23.6Na_2O \cdot 76.4SiO_2$ glass (mol% by analysis); 1 is experimental spectrum and 2 is fit. (C) $33.4Na_2O \cdot 66.6SiO_2$ glass (mol% by analysis); 1 is experimental spectrum and 2 is fit. (D) $36.0BaO \cdot 64.0B_2O_3$ glass (mol% by analysis); 1 is experimental spectrum and 2 is fit. (E) $32.5Na_2O \cdot 67.5GeO_2$ glass (mol% by synthesis); 1 is experimental spectrum and 2 is fit. (F) $35Na_2O \cdot 65TeO_2$ glass (mol% by synthesis); 1 is experimental spectrum and 2 is fit.

as (2.5 to 4.5) $\cdot$ 10^{-3}, thus demonstrating the potentialities of both Model (50) for the complex dielectric constant and the corresponding dispersion analysis version. Figure 1 shows examples of best fits to the IR reflection spectra of vitreous silica and binary silicate, borate, germanate, and tellurite glasses borrowed from Reference 79. Data on the optical constants and band parameters corresponding to these fits are presented in Chapter 4, the Appendix, and also in Figure 2.

C

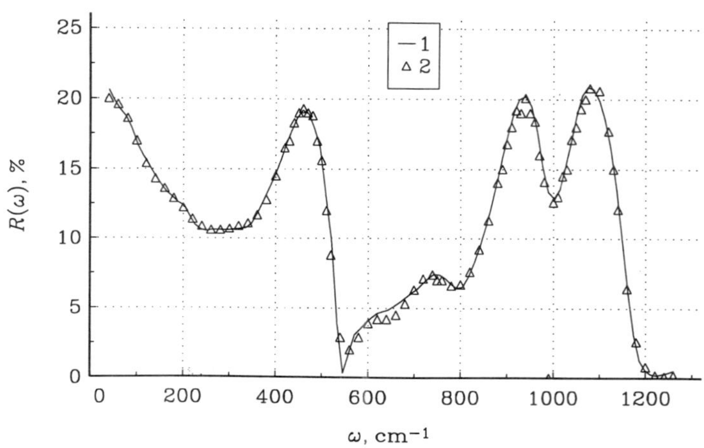

D

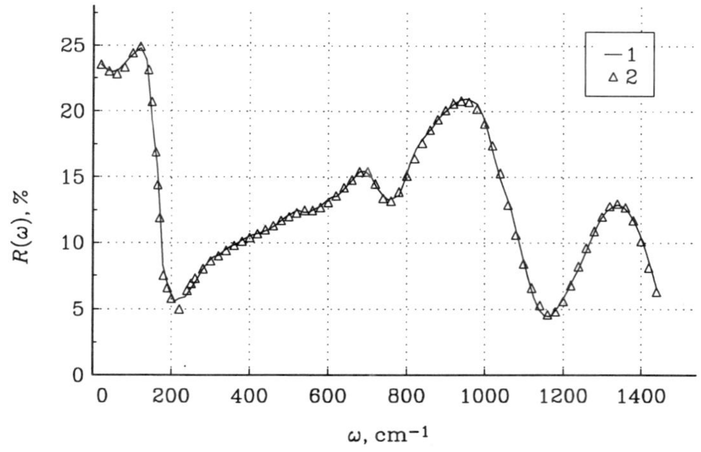

Figure 1 (continued)

Notably, Q values obtained in References 75, 76, 79, 80, 152, and 153 are even smaller than the absolute error, $\Delta R(\omega)$, of the reflectivity values used ($\pm 5 \cdot 10^{-3}$).[a] This means, evidently, that attempts to further improve the analytical model for the complex dielectric constant would be meaningless, because any further decrease in the $R_{mod}(\omega) - R_{exp}(\omega)$ difference would be indistinguishable against the $\Delta R(\omega)$ value mentioned.

Thus a good quality of best fits in a broad frequency range covering the frequencies of all the fundamental IR excitations speaks for the reliability of band parameter values and spectra of optical constants obtained with the dispersion

---

[a] The error mentioned takes into account all possible error sources, such as heating a sample by the IR beam, a spread in the incidence angle over the beam diameter, etc.[154] This somewhat unexpected inequality, $Q < \Delta R(\omega)$, is explained by the integrating nature of the Q norm (60) chosen.

E

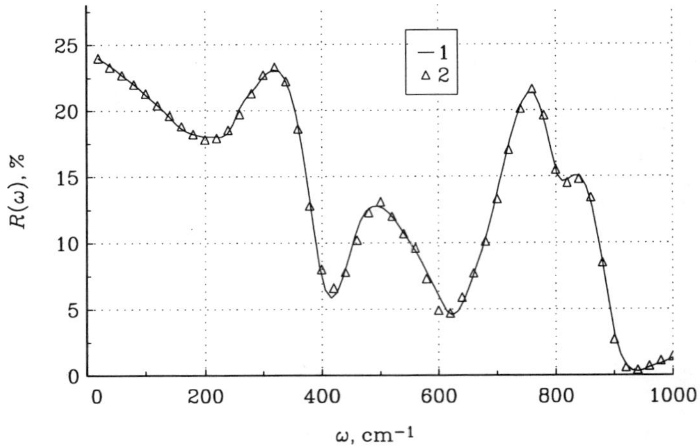

F

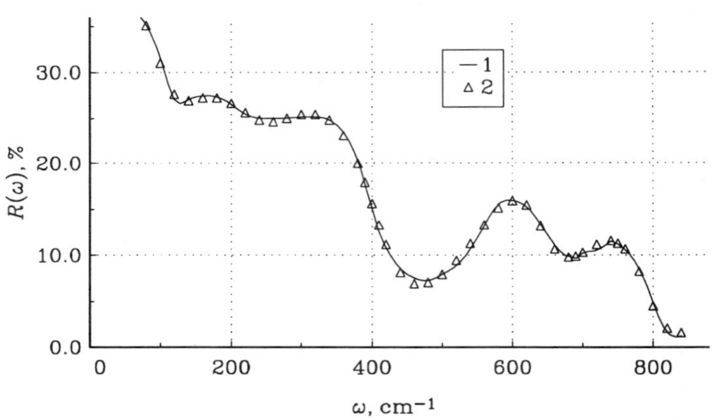

Figure 1 (continued)

analysis version for glasses described in References 75, 76, and 80. This is confirmed by error estimations.

Various methods, such as numerical experiments and a direct comparison with data obtained independently, were used to estimate the errors of dispersion parameters and optical constants calculated with the dispersion analysis version developed for glasses. Two kinds of numerical experiments were carried out by Makarova.[76,114] The first analyzed variations in the dispersion parameter magnitudes occurring in the course of the dispersion analysis when conducting this procedure for model spectra mimicking the experimental spectra of some borate glasses. It was shown that, with an increase in the number of cycles of fitting a spectrum, the absolute magnitudes of the parameter errors decrease successively, thus approaching some limiting values. The second studied the behavior of error function Q (60) in the vicinity of the minimum point for model spectra. It was shown that the finite exactness of $R(\omega)$ data (along with the inevitably approximate

A

B

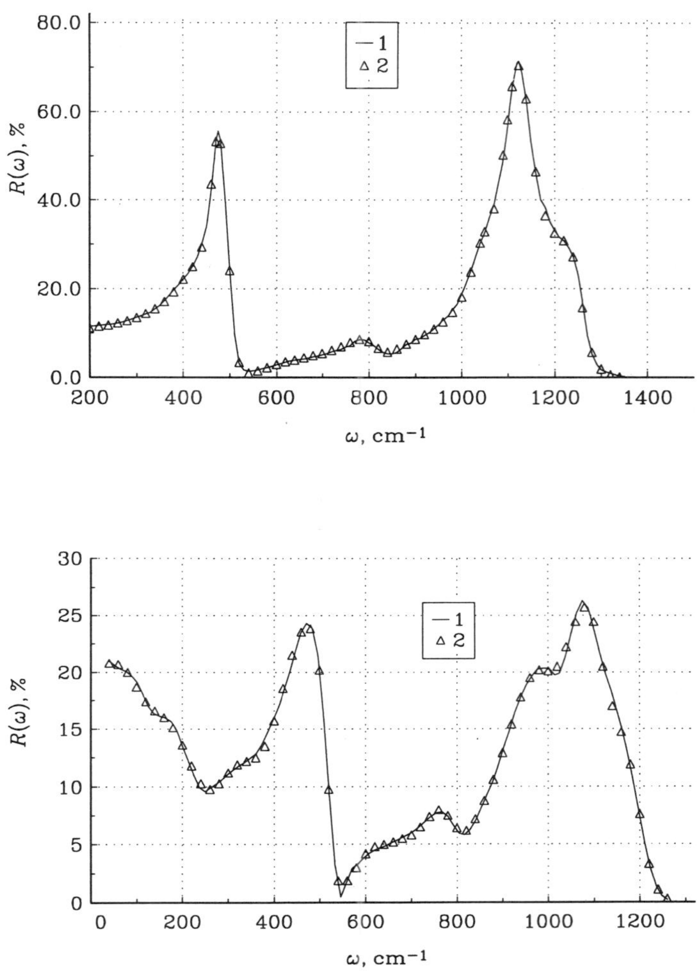

Figure 2 Comparison of the spectra of optical constants reconstructed (2,4) with the Kramers-Kronig transform and (1,3) and with the dispersion analysis version based on Model (50) for two binary glasses according to the data of Reference 79. (A) 36.0BaO · 64.0B_2O_3 glass (mol% by analysis); 1 and 2 are *n(ω)* ; 3 and 4 are *κ(ω)*. (B) 32.5Na_2O · 67.5GeO_2 glass (mol% by synthesis); 1 and 2 are *n(ω)* ; 3 and 4 are *κ(ω)*.

nature of any analytical model for the complex dielectric constant) results in the existence of some domain of oscillator parameter values similar in shape to the 4*J*-dimensional ellipsoid corresponding to *Q* values of about (3 to 4) · 10^{-3} and containing a point related to the actual *Q* minimum. The projections of the ellipsoid on the axes of particular dispersion parameters serve as the estimates of corresponding parameter errors. In general, the estimates of maximum parameter errors obtained with the two kinds of numerical experiment are as follows: $\delta\omega_j/\omega_j \leq \pm 0.8\%$; $\delta S_j/S_j \leq \pm 10\%$; $\delta\gamma_j/\gamma_j \leq \pm 20\%$; and $\delta\sigma_j/\sigma_j \leq \pm 20\%$. These maximum errors of dispersion parameters result in errors of the optical constants $n(\omega)$ and $\kappa(\omega)$ of about ±0.2 each. One should not see a contradiction in the fact that rather substantial magnitudes of errors for the dispersion parameters of a particular oscillator do not deprive one of obtaining the high quality of computer fits and optical constants spectra calculated with these parameters. The dispersion analysis procedure is accompanied by a certain correlation between errors of dispersion parameters for neighboring oscillators because the deviations of parameter values for a particular oscillator can be compensated to some extent by the deviations of parameter values for a neighboring oscillator having the opposite signs. This mutual compensation of errors for different oscillators allows the provision of quite a good quality of computer fits such as is shown in Figure 1.[b]

Naturally, a direct comparison between the spectra of optical constants reconstructed with the dispersion analysis version based on Model (50) and those found with the Kramers-Kronig transform, provides the most reliable estimates for the errors in these spectra.[76,79] The examples of such a comparison for the range of the fundamental IR excitations taken from Reference 79 are presented in Figure 2 (see also Chapter 5, Sections II.A.1 and IV.B.3). As is seen from Figure 2, the spectra obtained with these two methods are quite similar to one another in the entire frequency range studied. Indeed, the absolute differences between the $n(\omega)$ or $\kappa(\omega)$ curves compared that are determined by the sums of errors introduced by both the dispersion analysis and Kramers-Kronig transform are, for most frequencies, less than ±0.05, thus being in good agreement with the above theoretical estimates for the dispersion analysis errors.[c] In this connection, it is of importance if the two sets of data compared are independent of one another. Though reconstructed from the same experimental $R(\omega)$ spectra, the data obtained with the dispersion analysis and Kramers-Kronig transform should be considered to be rather independent of one another due to principal differences between the methods both in the kind of additional information used and in the relative sensitivity to different kinds of experimental errors (see Chapter 3, Sections II.A, II.B, and III.A). Therefore, the good coincidence between the $n(\omega)$ or $\kappa(\omega)$ spectra in a broad frequency range found with these two methods demonstrates the validity of the dispersion analysis version described in References 75 and 76 as a method for reconstructing the reliable data on the optical constants and, hence, on the IR band frequencies and intensities from *raw* IR reflection spectra of glasses.

Another independent experimental proof of reliability of the dispersion analysis version under discussion is the comparison between the "static" dielectric constant ε_0 calculated, using the dispersion parameters of a glass, with Equation (18) and the $\varepsilon'(\omega)$ value of this glass directly measured in the ultrahigh radio frequency (UHF) range. Frequencies corresponding to the UHF range are quite small compared to any of the vibrational frequencies; at the same time, these frequencies are much higher than those at which inelastic dielectric losses, such as the migration losses, begin to manifest themselves. Therefore $\varepsilon'(\omega)$ measured in this range can be taken, with good reason, to be equal to ε_0 determined by Equation (18). For providing a calculation of ε_0 of a particular glass with (18), the values of dispersion parameters

[b] The above holds, naturally, for any dispersion analysis version capable of providing a good quality of computer fits irrespective of particular nature of a material and $\hat{\varepsilon}(\omega)$ model used.

[c] Notably, the best coincidence between the spectra under discussion, such as that illustrated by Figure 2, was obtained when applying the particular algorithm of the Kramers-Kronig transform, namely, the one taking into account the actual structure of $R(\omega)$ spectrum in the VUV range of fundamental excitations.

Table 1 Comparison Between the ε_0 Values Calculated with Sum Rule (18) Using Band Parameters of Reference 79 and the Experimental $\varepsilon(\omega)$ Values in the UHF Range

Glass composition				ε_0	Experimental data for the UHF range		
by synthesis		by analysis		Calculated	Frequency,		
Na_2O	SiO_2	Na_2O	SiO_2	with (18)	Hz, $\cdot 10^{10}$	$\varepsilon'(\omega)$	Source
—	100	—	—	3.73	—	—	79
—	100	—	—	—	0.94–1.0	3.7–3.8	155
20	80	19.3	80.7	5.84	—	—	79
20	80	—	—	—	1.0	5.6	156
25	75	23.6	76.4	6.28	—	—	79
25	75	—	—	—	1.0	6.3	156
BaO	B_2O_3	**BaO**	B_2O_3				
35	65	36.0	64.0	7.53	—	—	79
35	65	—	—	—	3.56	7.8	157
Na_2O	TeO_2	Na_2O	TeO_2				
30	70	—	—	17.40	—	—	79
30	70	—	—	—	3.56	17.9	158

should, naturally, be known for all the oscillators making up the fundamental IR spectrum of the glass. In Table 1, ε_0 values of various glasses taken from Reference 79 are compared to the experimental $\varepsilon'(\omega)$ values in the UHF range according to References 155 through 157. As is seen from Table 1, relative differences between calculated and measured dielectric constants in Table 1 are no more than 3.5%. Bearing in mind that the errors of $\varepsilon'(\omega)$ measurements in the UHF range are usually about 1 to 3%, an agreement between the compared quantities should be taken as rather good, thus confirming the reliability of band parameters calculated with this version of the dispersion analysis.

Such a good coincidence between the data obtained with the dispersion analysis version based on Model (50) and those obtained with various independent methods speaks for the opinion that the analytical model for the complex dielectric constant (50) used as a basis for this dispersion analysis version can indeed be of physical significance. In this connection, it would be of interest to compare quantitatively the data of References 75, 76, and 79 with the above data of References 102 through 104 obtained with dispersion analysis versions based on the "crystal-oriented" $\hat{\varepsilon}(\omega)$ models. As was stated by the authors of Reference 102, differences between the calculated and experimental reflectivities did not exceed a relative experimental error estimated to be of 5% (which makes up, for peak reflectivity of 0.2, an absolute error of about $1 \cdot 10^{-2}$), and differences between the $n(\omega)$ or $\kappa(\omega)$ values obtained with the classical dispersion analysis and with the Kramers-Kronig transform were mostly no more than 0.1. Thus, the errors of the best computer fits and of $n(\omega)$ or $\kappa(\omega)$ spectra obtained in Reference 102 exceed those obtained in References 75, 76, and 79 only by 2. As was mentioned in Section IV.A, the classical dispersion analysis is undeniably inadequate for the spectra of glasses. Therefore, a good quality of data obtained with the classical dispersion analysis in Reference 102 is probably due to the smoothed band contours in the spectra of multicomponent glasses studied. Indeed, such smoothed contours make the task of accurate description of the spectra much easier. Unlike the data of Reference 102, those of References 103 and 104 concern binary and ternary glasses having highly structured IR spectra as those studied in References 75, 76, and 79. Unfortunately, the lack of numerical data in References 103 and 104 does not allow assessing if error function Q values for fits obtained by Gervais et al. are as small as those for fits obtained in References 75, 76, and 79. Judging by best fit plots presented in References 103 and 104, it seems possible that Q values for some of the latter might turn out, like those of best fits in References 75, 76, and 79, to obey the inequality $Q < \Delta R(\omega)$. If this is the case, any differences between the best fits constructed with the two dispersion analysis versions under discussion should be considered as indistinguishable against the experimental error. In such a case, there is no chance for discrimination between the two versions on the grounds of quality of the best fits obtained. Thus, a reason for such discrimination should be sought on another basis: first of all, on that of independent physical considerations concerning the validity of corresponding $\hat{\varepsilon}(\omega)$ models. As was stressed in Chapter 2, Sections III.A and III.B.3, and Section IV.A, such physical considerations, when applied to glass spectra, speak for the $\hat{\varepsilon}(\omega)$ Model (50) rather than for the factorized one (43). Thus, it is the dispersion analysis version of Efimov and Makarova[75,76] based on the $\hat{\varepsilon}(\omega)$ Model (50) that is the most appropriate for treating the fundamental spectra of glasses, especially in cases when the band parameter values are of interest. However, when only the spectra of optical constants of glasses are of interest, the above does not exclude the possibility of using the dispersion analysis version based on Model (43) or even (for spectra as smoothed as those analyzed in Reference 102) the classical dispersion analysis. As is seen from the above discussion, obtaining a good quality of computer fit to the experimental reflection spectrum in a broad frequency range covering the entire fundamental spectrum (or, at least, the predominant fraction of the latter) can result in rather reliable magnitudes of optical constants irrespective of the physical significance of the $\hat{\varepsilon}(\omega)$ model used as a basis. However, one should be cautious because, due to insufficient validity of the underlying physical models for the spectra of glasses, there is no guarantee that computer fits based on Equations (43) or (17) can be as good for the spectra of other glass types as those obtained in References 102 through 104 for the spectra of silicate glasses. For example, a comparison was conducted in References 75, 76, and 80 between the best fits to the IR reflection spectra of simple borate glasses obtained with the classical dispersion analysis and with the dispersion analysis based on Model (50). Appreciable deviations of best fits obtainable with the classical dispersion analysis were observed near $R(\omega)$ maxima and, especially, in the minima between the reflection peaks. The resulting error function values were several times greater than those obtained with the dispersion analysis version based on Model (50).

Along with treating the IR reflection spectra of various glasses, this dispersion analysis version was also used for reconstructing the spectra of optical constants from the raw VUV reflection spectrum of vitreous SiO_2.[159] A good best fit ($Q = 1.2 \cdot 10^{-3}$) was obtained; moreover, as is demonstrated in Chapter 4, Section III.A, good agreement between literature data on the spectra of optical constants and those reconstructed with this version from a similar array of $R(\omega)$ values was observed. Thus, this dispersion analysis version may indeed be considered as a reliable method for obtaining the spectra of optical constants of glasses in the VUV range. This seems to be of importance because applying the Kramers-Kronig transform for this range is not a simple task. For this, the asymptotic behavior of $R(\omega)$ function at highest frequencies should be approximated with the model function $C\omega^{-k}$ (see Section III.B), whereas reflectivity measurements at photon energies exceeding 20 to 25 eV that are necessary for fitting the k value can entail substantial methodological difficulties. One should bear in mind, of course, that the dispersion parameter values found for the VUV spectra of electronic transitions with the dispersion analysis can lack a distinct physical meaning because, in accordance with what was discussed in Chapter 2, Section I, separating the VUV spectrum into the contributions of individual oscillators can be dubious due to the principally continuous nature of the interband constituent of these spectra. Therefore, the dispersion parameters of the VUV spectrum of any solid should be considered, strictly speaking, to be useful effective quantities rather than actual parameters of individual spectrum features. Sometimes, the practice of applying the classical model (17) to the optical spectra of electronic excitations of typical semiconductors is substantiated qualitatively by the fact that exciton states (for which the idea of individual oscillators is expedient due to the localized nature of these excitations) participate, to some extent, in the formation of these spectra. For oxides and halides, including glasses, the role of excitonic transitions, as was mentioned in Chapter 2, Section III.C, is much greater than for typical semiconductors. For example, according to the approach of Laughlin[81] for SiO_2, almost all oscillator strengths of the interband transitions turn out to be transferred to the exciton-like transitions. Bearing this in mind, one may conclude that the use of the dispersion analysis version based on Model (50) for reconstructing the spectra of optical constants of oxide and halide glasses in the range of electronic excitations is, at any rate, more justified than is the case for the classical dispersion analysis of spectra of typical semiconductors.

Chapter 4

Spectra of Optical Constants in the Ranges of Fundamental Excitations for Particular Glass Types

I. INTRODUCTORY REMARKS

In this chapter, available data on the spectra of optical constants of various inorganic glasses in the ranges of the fundamental excitations are discussed. The range of the fundamental vibrational excitations, for which broader information has been obtained, is considered first; a few available data on the spectra of optical constants of glasses in the range of the fundamental electronic excitations are discussed separately in Section III. Special attention is paid to sources presenting data on the numerical values of optical constants, because it is these data that can be used by readers in their own quantitative studies. When quantitatively studying the IR spectra, the ultimate goal in most cases is not the spectra of optical constants themselves but the accurate values of band frequencies and the estimates of band intensities that can be obtained using these spectra. Therefore data on the numerical values of the IR band parameters (if any) are also discussed in this chapter. Data on the Raman spectra are considered to be beyond the scope of this book. For every particular glass type, the general pattern of the $n(\omega)$ and $\kappa(\omega)$ spectra (and also, in some cases, of the $\varepsilon'(\omega)$ and $\varepsilon''(\omega)$ spectra) is discussed first and the structural interpretation of the IR spectra is cited briefly. It is the structural interpretation of the spectra in terms of the internal valence coordinates that is cited, because this kind of interpretation is the most common now and, moreover, it provides the most appropriate basis for a comparison between the structures of different glasses. In the case where practically no information on the spectra of optical constants is available for a glass type of interest (as is the case for phosphate and most of heavy metal fluoride glasses), approximate estimates of band frequencies and band assignments are described briefly based on the qualitative experimental IR spectra. Available data on the numerical $n(\omega)$ and $\kappa(\omega)$ values vs. frequency are cited in detail;[a] if possible, a comparison is conducted between data obtained with different methods for samples of the same glass composition. Finally, available numerical data on the IR band parameters (if any) are cited. To avoid overloading the text of this chapter with large tables, the largest tabulations of the numerical $n(\omega)$ and $\kappa(\omega)$

[a] In all tables to follow, the $n(\omega)$ and $\kappa(\omega)$ numerical values are presented, according to a common practice, with two decimal figures, because the accuracy of all methods used for calculating the optical constants in the ranges of the fundamental excitations meets well-known severe limitations due to significant absorption (see Chapter 3). Theoretical estimates for the absolute errors of optical constants in these ranges characteristic of the dispersion analysis version for glasses (see Chapter 3, Section IV.B) are about 0.02 and the errors of the Kramers-Kronig transform are undoubtedly no less than these. Some literature sources, however, provide the $n(\omega)$ and $\kappa(\omega)$ values for the ranges of the fundamental excitations with three or even four decimal figures, thus ignoring the above limitations. When compiling tables for this book, such data were rounded off to two decimal figures (except for ranges in which $\kappa(\omega)$ values are less than 0.02, as is the case for Table 4).

values vs. frequency are placed in the Appendix. Included among data cited in the Appendix are the author's own data on the optical constants of a broad variety of simple glasses that have not been presented anywhere else until now except in Reference 79. All the data of Reference 79 cited in the Appendix were reconstructed with the above dispersion analysis version for glasses based on Model (50) (see Chapter 3, Section IV.B) from initial $R(\omega)$ spectra registered for sample surfaces polished with water-free organic liquids. The experimental $R(\omega)$ values from Reference 79 are also cited in the Appendix for some glasses, because it is useful to have such data at one's disposal for assessing if the resultant $n(\omega)$ and $\kappa(\omega)$ spectra are reliable. For the convenience of the reader, those data placed in the Appendix that should be at hand in the text for comparing the results of various authors with each other are repeated, in substantially abridged form, in this chapter, too.

Unlike Chapters 1 through 3 that were devoted to the most general problems, this chapter touches upon the problem of interrelation between the specific features of a particular glass spectrum and glass composition and/or glass structure. Some aspects of this problem should be specified briefly prior to discussing available experimental data on the optical constants of glasses.

As is known, the spectra of glasses of different types can differ substantially from each other in the location of frequency ranges occupied by the fundamental excitations of every kind. The origin of the differences can be understood based on Equation (11) for the classical damped oscillator. The frequency of the damped oscillator, ω_0, is determined by the well-known relation,

$$\omega_0 = \sqrt{k/m - \left(\frac{\gamma}{2}\right)^2}\,, \tag{61}$$

the meaning of k, m, and γ being the same as that specified in Chapter 1, Section II. Usually, k/m ratios are much greater than $(\gamma/2)^2$, so ω_0 values derived from (61) are rather close to those expected for the harmonic undamped oscillator.

In accordance with (61), the large differences between frequency ranges in which the fundamental spectra of the electronic excitations and those of the vibrational excitations are located are caused mainly by a difference between the effective electron masses and the reduced masses of structural groups involved in the vibrational excitations. The vibrational spectra, as is known, are always located in the IR range; the frequencies of both the stretching and bending vibrations vary substantially with the type of glass-forming compound, which also is due mostly to the role of reduced masses. The range of the highest frequency stretching vibrations can vary from 100 to 200 cm^{-1} for chalcogenide glasses containing heavy elements to 1200 to 1500 cm^{-1} for borate and phosphate glasses. In addition, band frequencies are also influenced by variations in force constants. As is known, the frequencies of the stretching vibrations are always much greater than the frequencies of the bending ones due to differences in force constants of about one order of magnitude.

According to the well-known selection rule by the dipole moment derivative, only vibrations involving variations in the dipole moments are active in the IR spectrum, whereas vibrations involving variations in the polarizabilities are active in the Raman spectrum. Naturally, the greater the permanent dipole moment the greater its variations caused by atomic displacements; hence the general interconnection between the total IR spectrum intensity and the ionicity degree of chemical bonds predetermining the effective atomic charges. Along with the permanent dipole moments, those resulting from atomic charges induced by an electric field can also contribute to the vibrational band intensities.

Frequency limits of the fundamental optical spectra related to the electronic excitations also can vary substantially depending on the glass type under study. As is known, the spectra of the electronic excitations are located in the VUV range for most of inorganic glasses such as oxide and halide; the most considerable long-wavelength shift of the absorption edge (from the UV to the visible or even to the IR range) is observed when passing from oxide or halide glasses to chalcogenide glasses. This shift is caused by substantial changes in the energy spectrum of a material that, strictly speaking, can be described correctly only in terms of rigorous electronic theories, such as that outlined in Reference 44, which takes into account both the energy spectrum of electrons in the material and the effects of mobilities of electrons and holes. As a crude approximation, this shift may be understood, in terms of the simplest band model, to result from a substantial decrease in optical band gap, i.e., in the energy gap between the top of the valence band and the bottom of the conducting band. In terms of Equation (61), this corresponds to a large decrease in the force constants (the changes in effective electron masses also taking place but playing a much less substantial role). Similar reasons are responsible for the absorption edge shift in the sequence of sulfide-selenide-telluride glasses. The nature of individual features in the optical spectra of electronic excitations that predetermine the kind of interrelation with the structure of a substance is outlined briefly in Chapter 2, Sections I and II.C.

II. SPECTRA OF OPTICAL CONSTANTS IN THE INFRA-RED RANGE AND INFRA-RED BAND PARAMETERS OF INORGANIC GLASSES

Seeking an interrelation between the specific features of a particular spectrum and the structure of a material was always the main topic of the vibrational spectroscopy of glasses (see, for example, the comprehensive monograph by Wong and Angell[160]). Starting with the review of optical constants and spectra of glasses in the IR range, it should be remembered that, as was specified in Chapter 2, Section II.B, the nature of the vibrational excitations in glasses is less understood than the nature of those in crystals due to the lack of translational symmetry in the glass network. For crystals, a rigorous dynamical theory (such as that in Reference 41) is available which can predict the type and number of modes active in the IR and/or Raman spectrum for a given symmetry of unit cell; this allows for reliable theoretical calculating of band frequencies and intensities (see, for example, References 20 and 161 through 162). For inorganic glasses, approaches capable of providing a satisfactory agreement between the calculated and experimental band parameters are not yet available. Due to this situation, the structural interpretation of multicomponent glass spectra is confined commonly to oversimplified considerations based on either the molecular-like and small cluster models or qualitative empirical approaches involving the well-known notion of characteristic frequencies of structural groups borrowed from the spectrochemistry of molecular substances. In most cases, experiment shows that the total number of individual fundamental bands and accurate values of their frequencies and intensities are characteristics that are too subtle to be described satisfactorily in terms of molecular-like or small cluster models. The empirical spectrochemical approaches assume that band frequencies related to the vibrations of particular structural groups remain approximately intact regardless of the nature of the surrounding matrix (thus ignoring coupling between different vibrations — see Chapter 1, Sections II.A and III). Moreover, attribution of the IR bands is made mostly as an empirical judgment by analogy with crystals similar in chemical composition (thus ignoring both the specific nature of vibrations in glass discussed in Chapter 2, Section II.B and the great differences in the experimental bandwidths between the spectra of a glass and a crystal). How crude these approaches are is evident, in general, from that said in Chapter 1, Sections II.A and III, and Chapter 2, Section II.B; this also depends on whether the idea of constancy of group frequencies is applied literally or not. When relating a particular vibration to a certain frequency range wide enough for covering band frequency variations due to mode coupling, it is usually possible to interpret strong bands in the IR glass spectra to be traceable to the vibrations of structural groups making a glass network. On the other hand, when treating the vibrational frequencies of structural groups to literally be constant, irrespective of whether a material is vitreous or crystalline and whether it is simple or complex in composition, principal errors in interpretation can arise. There is a large number of mistakes, both early and recent, originating from this in literature. Criticism to such oversimplified approaches was expressed, for example, in References 100 and 163.

There is also no unified view in literature as to what kind of structural fragment is responsible for particular vibrations in glass network. On the whole, the general approach most widely used now treats the higher frequency IR bands in glass spectra as traceable to the vibrations of the simplest structural groups described in terms of displacements with respect to the internal valence coordinates, such as bond lengths, bond angles, etc. Therefore, the totality of the vibrational frequencies of vitreous silica and germania is treated now by the majority of investigators as originating from the vibrations of the most simple structural fragments, such as the SiOSi oxygen bridge and also (for the bending vibrations) the OSiO atomic configuration (see Section II.A.1). For binary and more complex glasses, the vibrations of the terminal groups (such as the SiO^-, $Si(O^-)_2$, and $Si(O^-)_3$ in silicates) formed from breaking the SiOSi bridge by ionic oxide oxygen are taken into account along with the vibrations of this bridge and of OSiO group. This kind of structural interpretation of the IR spectra developed initially by Lazarev and his co-workers for complex oxide crystals[20,161,162] also was shown to be valid for the spectra of silicate, germanate, and similar glasses (see Sections II.A.2 and II.A.4). For the case of the IR spectra of glasses comprising the same element in different coordination states (such as three- and fourfold-coordinated boron, etc.), the coordination polyhedra (such as the BO_3 group or BO_4 tetrahedron) remain to be commonly taken for the simplest groups responsible for particular vibrations (see Section II.A.3).

Unlike the general approach mentioned above, various attempts to attribute bands in the IR spectra of glasses to the characteristic vibrations of greater structural motifs of particular chemical compositions approaching in size to the medium-range-order limits or even extending outside these limits lead mostly to dubious or wholly erroneous conclusions. For some specific types of vibrations, however, attribution based on the medium-range-order motifs can be valid, as is the case for the 808 cm^{-1} band in the Raman spectrum of vitreous boron oxide reliably identified as the "breathing" mode of the boroxol ring (see, for example, Reference 100).

A. OXIDE GLASSES

1. Vitreous Silica

In the field under discussion (as is also the case for practically all other fields of glass science), vitreous silica is the most studied glass. There are many literature sources on the optical constants of the latter in the range of fundamental vibrational excitations, such as References 43 and 164–176. Most of them, however, including References 43, 164–166, and 174–176, demonstrate only plots of the $n(\omega)$ and $\kappa(\omega)$ spectra, thus reducing information presented to a qualitative pattern. Among available studies, those related to the problem of surface layer influence on the optical constants of vitreous silica[174,175] should be mentioned. References 167 through 173 provide readers with the numerical $n(\omega)$ and $\kappa(\omega)$ values vs. frequency.

Due to quite a strong absorption in the IR range imposing severe limitations on the accuracy of direct measurements of $n(\omega)$ and $\kappa(\omega)$ values with thin film spectroscopy (see Chapter 3, Section I), such a technique (used, for example, in Reference 176) did not become widely used when obtaining the spectra of optical constants of vitreous SiO_2 and similar glasses. Almost all available data for vitreous silica were reconstructed from the reflection spectrum. Some early sources, such as References 164, 165, and 167, present data calculated with Fresnel equations from $R(\omega)$ values for two different incidence angles (usually, 20 and 70°), i.e., with the method that is not appropriate for the case of such a strong absorption (see Chapter 3, Section I). Most of the available $n(\omega)$ and $\kappa(\omega)$ spectra of vitreous silica were reconstructed with the Kramers-Kronig transform from either the near-normal incidence reflection spectroscopy data[43,166,169,170] or the internal reflection spectroscopy data.[171,172,174] At last, the recent information of Reference 79 on the $n(\omega)$ and $\kappa(\omega)$ numerical values was obtained with the dispersion analysis version for glasses described in Chapter 3, Section IV.B.

In view of the fact that no precise reference $n(\omega)$ and $\kappa(\omega)$ values can be available for the range of fundamental excitations, the relative reliability of data presented for vitreous silica in various sources should evidently be assessed with the help of thorough comparative analysis of these data. Large deviations of $n(\omega)$ and $\kappa(\omega)$ values given in a particular source from average estimates for the majority of studies should probably point out that the data of a source under discussion are erroneous. On the other hand, the significance of close similarity between the results of studies under comparison depends greatly on the methods used in these studies. The data of studies involving identical methods of spectrum registration and of data treatment should be characterized by systematic errors of similar origin and magnitudes; hence, satisfactory coincidence between such data would mean little concerning their reliability. On the other hand, a good agreement between two or, better yet, several sets of data obtained with substantially different methods speaks for the fact that the data under comparison are indeed reliable.

With some exceptions discussed below, the data of various authors show a rather similar general pattern of $n(\omega)$, $\kappa(\omega)$, $\varepsilon'(\omega)$, and $\varepsilon''(\omega)$ spectra of vitreous silica in the IR range. This pattern is illustrated by Figure 1A, B using, as an example, the data of Reference 79. Spectra of optical constants presented in Figure 1A, B are reconstructed from the $R(\omega)$ spectrum shown in Chapter 3, Figure 1A. As is seen from Figure 1B, the experimental IR spectrum of vitreous silica comprises three principal IR bands, the $\varepsilon''(\omega)$ maxima being located at about 1060 to 1080, 790 to 800, and 450 to 460 cm^{-1}; some weaker features are also seen in the same 400 to 1300 cm^{-1} range.

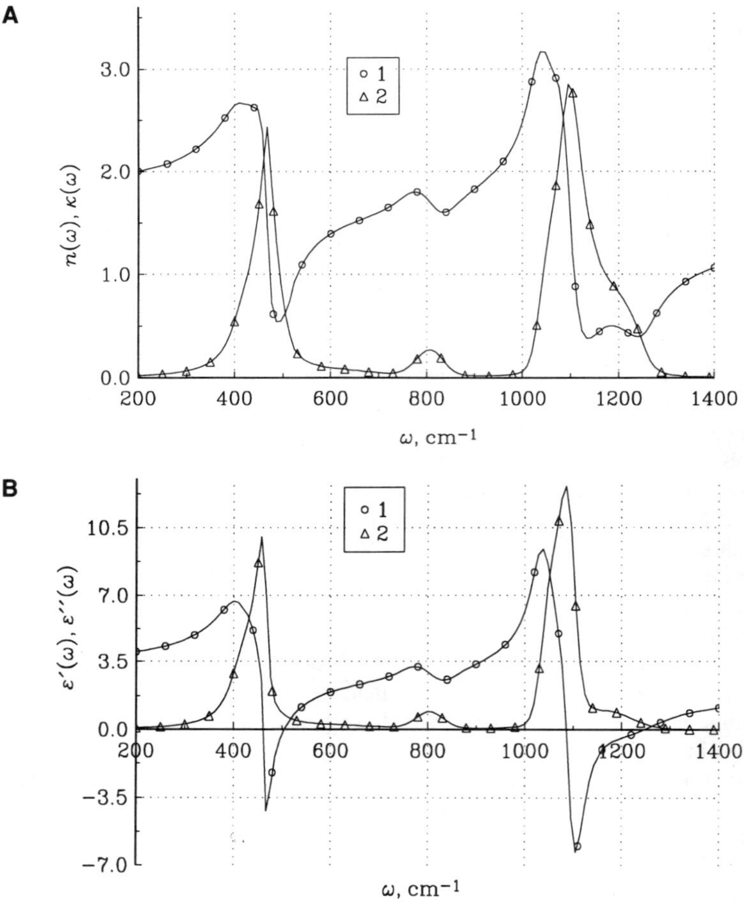

Figure 1 Spectra of the optical constants of vitreous silica according to the data of Reference 79. (A) 1 is *n(ω)* and 2 is *κ(ω)*. (B) 1 is *ε′(ω)* and 2 is *ε″(ω)*.

In accordance with the general approach mentioned above in Section I, these three principal bands are attributed commonly to various vibrations of the SiOSi and OSiO atomic configurations. First, the band at 1060 to 1080 cm^{-1} is acknowledged to originate from the asymmetric stretching vibration of the SiOSi bridge. There are, however, differences in the assignment of other spectrum features depending on the choice of a particular model of atomic displacements. One of the most widely used is the approach based on a simple model taking into account only a single SiOSi bridge between two adjacent SiO_4 tetrahedra and usually is referred to as Sen and Thorpe's central force model.[47,58,177] This model presupposes a single SiOSi asymmetric stretching vibration resulting in the above 1060 to 1080 cm^{-1} band. The band at 790 to 800 cm^{-1} is treated in terms of this model as the SiOSi symmetric stretching mode involving the displacement of bridging oxygen perpendicularly to the Si-Si direction in the SiOSi plane, whereas the band at 450 to 460 cm^{-1} is attributed to the SiOSi rocking mode related to the oxygen displacement out of the SiOSi plane. Thus, the estimations of band limits made with the central force model[47,58] are in accordance with the general pattern of the IR spectrum of vitreous silica. However, this model, due to its oversimplified character, presupposes the existence of only a single band for each type of vibrations mentioned above and therefore meets with difficulties when explaining more subtle features of a spectrum under discussion. Galeener, Leadbetter, and Stringfellow[58] attempted to explain the presence of additional weak spectrum features on the basis of intercomparison between the Raman, $\omega\varepsilon''(\omega)$, and the $\mathrm{Im}[-\hat{\varepsilon}^{-1}(\omega)]$ spectra for vitreous SiO_2, GeO_2, and BeF_2. They mentioned the similarity of the frequency locations of weak features in the Raman and $\omega\varepsilon''(\omega)$ spectra to the locations of strong peaks in the $\mathrm{Im}[-\hat{\varepsilon}^{-1}(\omega)]$ spectrum and discussed this similarity in terms of splitting between the transverse and longitudinal optical modes (so-called TO-LO splitting). In other words, the possibility of attributing tentatively weak features in the Raman and $\varepsilon''(\omega)$ spectra to the longitudinal optical modes associated with the neighboring principal transverse counterparts was considered. For the Raman features in the spectrum of vitreous GeO_2, this idea was then formulated definitely.[47] In general, the longitudinal optical modes can be indeed observed in the Raman spectra; in some favorable cases, the IR intensities of transverse modes in the spectra of crystals can even be calculated from their Raman spectra using the TO-LO splitting. However, it does not mean that every weak band in the Raman spectrum of a glass should be necessarily related to the longitudinal mode and not to one more transverse mode. Thus, a mere coincidence in the approximate locations of weak features in the $\varepsilon''(\omega)$ and $\mathrm{Im}[-\hat{\varepsilon}^{-1}(\omega)]$ spectra of glasses cannot be taken as proof of the violation of selection rules sufficient for providing the IR activity of longitudinal modes. As is known, the IR spectrum of E modes of α-quartz (in which, naturally, no longitudinal modes can be observed) also comprises at least one additional oscillator in the same 1170 to 1200 cm^{-1} range (see Chapter 2, Section III.B.3); this fact cannot be a mere coincidence. Therefore, the idea of Galeener et al.[58] is probably not applicable to the shoulder near 1170 to 1200 cm^{-1} located at the high-frequency wing of the principal $\varepsilon''(\omega)$ maximum of vitreous silica at 1060 to 1080 cm^{-1} (see Chapter 3, Figure 1A and Figure 1A, B). As follows from the above approach to the formation of the IR spectrum of glass based on the notion of inhomogeneous-like band broadening (see Chapter 2, Section II.B), a rather close similarity between the number, locations, and total intensities of the high-frequency IR bands in the spectra of a glass and a corresponding crystal is only natural. This is confirmed, as is shown below, by the comparison between the numerical values of band parameters obtained with the dispersion analysis of the IR spectra of vitreous silica and α-quartz. Hence, it is the information on the nature of the IR bands in the spectrum of α-quartz theoretically justified by the analysis of forms of vibrations[162,178] that can serve as the most solid basis for attributing the IR bands in the spectrum of vitreous silica; it is this kind of band attributing that will be followed below.

According to the data of References 162 and 178, the presence of six translationally inequivalent positions of the SiOSi bridges in the primitive cell of α-quartz results in the fact that there are three different SiOSi asymmetric stretching vibrations active in the IR spectrum, two of them being the E-type modes with transverse frequencies of about 1070 cm^{-1} (the stronger one) and 1160 cm^{-1} (the weaker one[b]). Hence it follows that the vibrational spectrum of silica glass also

[b] The nature of the excitation at about 1160 cm^{-1} will be discussed in more detail later when comparing band parameter values.

comprises at least two asymmetric stretching vibrations of the SiOSi bridge, the first of them resulting in the principal band at 1060 to 1080 cm^{-1} and the second resulting in the above shoulder near 1170 to 1200 cm^{-1}.

Similarly, the data of References 162 and 178 demonstrate the presence of three SiOSi symmetric stretching vibrations active in the IR spectrum, two of them being the E-type modes with transverse frequencies of about 700 cm^{-1} (quite weak) and 800 cm^{-1} (the stronger one). Therefore, the band at about 800 cm^{-1} in the IR spectrum of silica glass should be attributed to the symmetric stretching vibration of the SiOSi bridge; it is this assignment that is used in this book. Furthermore, the band at 450 to 460 cm^{-1} should be attributed, according to the data of Reference 162, to the OSiO bending vibration. As is seen, such an assignment of the 450 to 460 cm^{-1} band differs from that presupposed by the central force model (i.e., the assignment involving the SiOSi rocking mode). This difference results not from the actual difference in the kind of atomic displacements but merely from the fact that the central force model assumes all the force constants, except that of SiO stretch, to be zero and hence is principally inappropriate for treating the atomic displacements in terms of vibrations such as the OSiO bending ones.

Along with the bands mentioned, one more poorly distinguishable shoulder at about 1050 cm^{-1} having no counterparts in the spectrum of α-quartz is observed, as follows from the data of Reference 79, in the IR spectrum of vitreous silica (see Chapter 3, Figure 1A and Figure 1A, B). Experimental reflection spectra presented in some other sources (such as References 179 through 181) also show a slight deviation from the monotonical $R(\omega)$ change in the 1000 to 1050 cm^{-1} interval similar to the deviation seen in Chapter 3, Figure 1A; however, the occurrence of an additional shoulder at about 1050 cm^{-1} is not taken into account anywhere except in Reference 79. According to the data of References 162 and 178, this shoulder also falls into the range of asymmetric stretching vibrations of the SiOSi bridge (1000 to 1250 cm^{-1}). The specific origin of this feature will be discussed below when comparing the band parameters of silica glass to those of α-quartz.

As was mentioned above, the plots of $\kappa(\omega)$ and $\varepsilon''(\omega)$ spectra presented in some sources (mostly the early ones, such as References 43, 164–166, and 176) differ appreciably in the magnitudes of optical constants, the number and shape of spectrum features, or both from the data of the majority of studies. Evidently, it is the spectra with such differences that are determined with maximum errors. The maximum deviations in the absolute values of optical constants probably originate from large errors of initial experimental $R(\omega)$ spectra. As was pointed out by Saidov and Bernstein,[171] the initial $R(\omega)$ spectrum for the incidence angle as large as that of 70° used by Simon and McMahon[164] contains considerable errors in the 1150 to 1250 cm^{-1} range due to the occurrence of internal reflection; evidently this might also be the case for other studies using two R (ω) values for incidence angles of 20 and 70°, such as References 165 and 167. There are also deviations in the structure of spectra of optical constants observed in some studies from spectrum structure found by the majority of authors. For example, this is the case for the plots of the spectra in the 700 to 1300 cm^{-1} obtained in Reference 176 from the transmission spectra of thin blown film. The $n(\omega)$ spectrum of Reference 176, unlike the pattern of the spectra obtained in References 79 and 168 to 172 comprises two distinct maxima in the 1000 to 1100 cm^{-1} range located rather far from one another (instead of a single maximum with a high-frequency shoulder — see Figure 1A, B). Moreover, the location of the small highest-frequency $n(\omega)$ maximum in Reference 176 is shifted to 1250 cm^{-1} (instead of about 1200 cm^{-1}), the

Table 1 Locations of $n(\omega)$ and $\kappa(\omega)$ Maxima Related to the SiOSi Asymmetric Stretch and the Respective $n(\omega)$ and $\kappa(\omega)$ Maximum Values for Vitreous Silica According to Various Sources

	$n(\omega)$		$\kappa(\omega)$	
Source	ω_{max}, cm^{-1}	$n_{max}(\omega)$	ω_{max}, cm^{-1}	$\kappa_{max}(\omega)$
Girin et al.[167]	1050	2.40	1088	2.24
Zolotarev[168]	1050	2.88	1090	2.60
Popova et al.[169]	1040	3.01	1100	2.82
Steyer et al.[170]	1050	2.95	1091	2.56
Saidov et al.[171]	1045	3.11	1100	2.74
Mamedov et al.[172]	1050	3.25	1100	2.77
Efimov[79]	1045	3.18	1100	2.86
Zolotarev et al.[173]	1050	3.20	1095	2.85

$n(\omega)$ values in the range of this maximum being overestimated by a factor of about 2. Furthermore, some authors reported the presence of a weak peak in the $\kappa(\omega)$ and $\varepsilon''(\omega)$ spectra at about 900 to 920 cm^{-1} (Miler,[166] 902 cm^{-1} and Gaskell and Johnson,[43] 919 cm^{-1}). Miler[166] also found one more additional weak peak at 568 cm^{-1}. However, the occurrence of these bands was not confirmed by recent studies.[79,168-172] It should be noted that the initial experimental reflection spectra of References 43 and 166 do not reveal, like the spectrum shown in Chapter 3, Figure 1A, any signs of $R(\omega)$ features near these frequencies. Probably, both the 900 to 920 cm^{-1} peak in the $\kappa(\omega)$ or $\varepsilon''(\omega)$ spectra of References 43 and 166 and the 568 cm^{-1} peak in the spectrum of Reference 166 are indeed some false features due to the errors of the Kramers-Kronig transform procedure. The latter view is supported by the presence of other undoubtedly false peaks in the $\kappa(\omega)$ and $\varepsilon''(\omega)$ spectra of α-quartz obtained in Reference 43.

We turn now to discussing the numerical data on $n(\omega)$ and $\kappa(\omega)$ values vs. frequency, such as those presented in References 79 and 167 to 172. Data published in References 168 to 170 and 172 are cited also, mostly in the abridged form, in the review monograph by Zolotarev, Morozov, and Smirnova;[173] moreover, this monograph comprises the optical constants of a particular Russian commercial KI silica glass sample. Comparing the data of References 79 and 167 to 173 to each other reveals that variations in the $n(\omega)$ and $\kappa(\omega)$ numerical values presented by different authors, though not large, are rather perceptible. The spread of data is illustrated by Table 1 using the $n(\omega)$ and $\kappa(\omega)$ maxima locations and maximum magnitudes in the 1000 to 1100 cm^{-1} range as examples.

As is seen from Table 1, an early work of Girin et al.[167] shows the extreme deviations from average values of quantities under consideration. This is no wonder in view of the discussion in Chapter 3, Section I concerning the poor accuracy of $n(\omega)$ and $\kappa(\omega)$ calculations based on two $R(\omega)$ values for different incidence angles when applied to the ranges of strong absorption. With the data of Reference 167 thrown out, the estimates of maximum $n(\omega)$ and $\kappa(\omega)$ values vary in the 2.88 to 3.25 and 2.56 to 2.86 limits, and the estimates of frequency locations of the $n(\omega)$ and $\kappa(\omega)$ maxima vary in the 1040 to 1050 and 1090 to 1100 cm^{-1} limits, respectively. Studies using the samples of two or more silica glass types synthesized with substantially different methods[79,168,169,171] showed no perceptible differences in the reflectivity values of the samples. Hence it follows that variations in the actual reflectivity of silica glass types studied in these sources are too small to

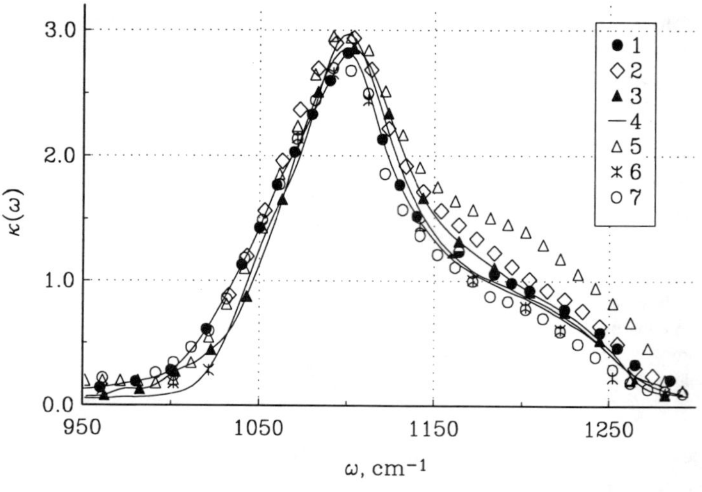

Figure 2 Comparison between the $\kappa(\omega)$ spectra of vitreous silica in the 900 to 1300 cm^{-1} range plotted with the data from various sources: (1) Zolotarev;[168] (2) Popova, Tolstykh, and Vorob'ev;[169] (3) Steyer, Day, and Huffman;[170] (4) Saidov and Bernstein;[171] (5) Mamedov, Mansurov, and Dubovikov;[172] (6) Efimov;[79] and (7) Zolotarev, Morozov, and Smirnova.[173]

be the cause of differences in the spectra of optical constants demonstrated by Table 1. On the contrary, the effect of surface layers formed when polishing particular silica glass samples[174,175] and the conditions of $R(\omega)$ spectrum registration[118,154] indeed can influence the reflectivity values registered in a study and, hence, the $n(\omega)$ and $\kappa(\omega)$ values. Moreover, as was specified in Chapter 3, Sections I, II.A, and IV, the thorough choice of computational procedure is also of significance for obtaining the reliable $n(\omega)$ and $\kappa(\omega)$ spectra. Hence it follows that the possible reasons responsible for variations in data cited in Table 1 comprise a spread in the deviations of "raw" $R(\omega)$ spectra from the spectrum characterizing the actual properties of bulk silica glass and also a spread in the errors of computational methods used for the reconstruction of optical constants.

For illustrating the general pattern of differences in the $\kappa(\omega)$ values of References 79 and 168 to 172 and also of Reference 173 for a particular KI silica glass sample, these values are plotted vs. frequency in the 950 to 1300 cm^{-1} range in Figure 2.

As is seen from Figure 2, the data of References 168 and 170 being rather close to one another show perceptibly lowered $\kappa(\omega)$ values as compared to the rest of data both in the vicinity of maximum at 1070 to 1100 cm^{-1} and in the range of high-frequency shoulder (1150 to 1270 cm^{-1}). Furthermore, it is these sources that present appreciably lowered peak $n(\omega)$ values (see also Table 1). The data of the rest of the five studies demonstrate good agreement in the $\kappa(\omega)$ peak values. In the range of a 1150 to 1270 cm^{-1} shoulder, however, the $\kappa(\omega)$ values of Reference 173 for a particular KI silica glass sample appreciably exceed those of other authors, the difference ranging from +0.2 to +0.5. Conversely, the $n(\omega)$ values of Reference 173 for this sample at frequencies exceeding 1240 cm^{-1} are appreciably lower than those of other authors, the difference ranging from –0.2 to –0.3. The data of Reference 173 for the KI silica glass sample are reconstructed from the internal reflection spectrum registered for the polished surface; probably, the differences mentioned are caused mostly by certain specific features of surface layer of this particular sample.

Thus, according to Figure 2, the list of data more or less compatible with each other in the 900 to 1300 cm^{-1} range comprises the studies of the following authors:

1. Popova, Tolstykh, and Vorob'ev[169] (Russian commercial KU and KI silica glass samples; polished sample surfaces; near-normal incidence reflection spectrum; Kramers-Kronig transform)
2. Saidov and Bernstein[171] (commercial KU, KV, and KI silica glass samples; polished sample surfaces; internal reflection spectrum; Kramers-Kronig transform)
3. Mamedov, Mansurov, and Dubovikov[172] (commercial KU-1 glass sample; broken sample surface; internal reflection spectrum; Kramers-Kronig transform)
4. Efimov[79] (samples of commercial KI silica glass and of high-purity KS4V silica glass synthesized in the Institute of Silicate Chemistry of Russian Academy of Sciences; sample surfaces polished thoroughly with water-free organic liquids; near-normal incidence reflection spectrum; dispersion analysis version described in Chapter 3, Section IV.B).

For further comparison in more detail, the plots of the spectra, such as those shown in Figure 2, are not suitable; therefore, the numerical data of these four studies on $n(\omega)$ and $\kappa(\omega)$ values are presented, in abridged form, in Table 2. As is seen from Table 2, the $n(\omega)$ spectrum presented in Reference 171, unlike the rest of spectra compared, lacks a weak but characteristic $n(\omega)$ maximum at about 1160 to 1220 cm^{-1}, showing only a smeared shoulder in this range. Moreover, the $\kappa(\omega)$ spectrum shows increased $\kappa(\omega)$ values in the 1040 to 1080 cm^{-1} range as compared to those of the rest of studies cited, maximum differences between the data of Reference 171 and those of References 79 and 168 being about +0.2 to +0.3. Conversely, the data of References 79, 169, and 172 agree rather satisfactorily with each other throughout the entire 880 to 1260 cm^{-1} range, differences in $n(\omega)$ and $\kappa(\omega)$ values being, in most cases, no more than 0.15 to 0.20. Such an agreement is meaningful, because studies described in References 79, 169, and 172 differ substantially from each other in most methodological aspects, such as preparation sample surface type, the kind of experimental spectrum used, and the computer treatment of data. As was mentioned above, the initial $R(\omega)$ spectrum used in Reference 172 was recorded from the broken surface and, hence, was not influenced by any surface layer.[c] On the other hand, the $R(\omega)$ spectrum used in Reference 169 was recorded at near-normal incidence; hence, the optical constants obtained are not influenced by any additional terms in the Kramers-Kronig relation for phase angle that might arise when treating internal reflection spectrum (see Chapter 3, Section II.B). Finally, Reference 79 (also using the initial $R(\omega)$ spectrum registered at the near-normal incidence) presents the $n(\omega)$ and $\kappa(\omega)$ spectra reconstructed not with the Kramers-Kronig transform, as is the case for the data of References 169 and 172, but with the dispersion analysis version for glasses described above in Chapter 3, Section IV.B. Therefore, a satisfactory agreement between the $n(\omega)$ and $\kappa(\omega)$ spectra obtained in References 79, 169, and 172 confirms their validity. The complete list of data on optical constants according to these three sources is presented in the Appendix (Table A1); the cited data of

[c] This is the reason why the data of Reference 172 are considered in Reference 174 to be closest to the actual optical constants of bulk vitreous silica.

Table 2 Comparison Between the Data on Optical Constants of Vitreous Silica in the 900 to 1260 cm^{-1} Range Presented in Various Sources

ω	$n(\omega)$ according to references				$\kappa(\omega)$ according to references			
(cm^{-1})	169	171	172	79	169	171	172	79
880	1.84	1.84	—	1.75	0.19	0.05	—	0.03
920	1.96	—	1.94	1.90	0.14	—	0.02	0.02
960	2.20	—	2.17	2.10	0.14	—	0.04	0.02
980	2.36	—	2.36	2.24	0.19	—	0.09	0.03
1000	2.62	—	2.56	2.48	0.28	—	0.23	0.06
1020	2.87	—	2.82	2.87	0.61	—	0.40	0.25
1040	3.01	3.11	3.19	3.18	1.13	1.14	0.82	0.87
1050	2.98	3.11	3.25	3.14	1.43	1.50	1.19	1.25
1060	2.83	3.03	3.24	3.02	1.77	1.89	1.59	1.56
1070	2.62	2.79	2.99	2.91	2.03	2.29	2.05	1.87
1080	2.40	2.42	2.64	2.73	2.33	2.61	2.43	2.30
1090	2.12	1.91	2.07	2.24	2.60	2.81	2.71	2.73
1100	1.50	1.40	1.50	1.52	2.82	2.85	2.77	2.86
1120	0.52	0.53	0.47	0.52	2.13	2.14	2.26	2.17
1140	0.41	0.44	0.31	0.38	1.52	1.65	1.60	1.49
1150	—	0.44	0.36	0.41	—	1.50	1.40	1.28
1160	0.47	0.41	0.39	0.45	1.18	1.39	1.26	1.12
1170	—	0.39	0.40	0.48	—	1.28	1.15	1.02
1180	0.51	0.36	0.39	0.50	1.00	1.17	1.05	0.95
1190	0.52	0.35	0.39	0.50	0.93	1.06	0.95	0.89
1200	0.50	0.35	0.38	0.49	0.87	0.96	0.86	0.82
1220	0.45	0.33	0.36	0.44	0.72	0.80	0.70	0.68
1240	0.41	0.27	0.32	0.40	0.54	0.59	0.48	0.48
1260	0.47	—	0.38	0.46	0.29	—	0.17	0.23

Note: Data of Reference 171 are the averages of magnitudes obtained for two directions of the IR beam polarization.

Reference 79 comprise, in addition, the initial $R(\omega)$ values and also the $\varepsilon'(\omega)$ and $\varepsilon''(\omega)$ magnitudes. Among these three sources, the data of References 79 and 172 demonstrate the best agreement with one another throughout the entire 900 to 1300 cm^{-1} range, differences in $n(\omega)$ and $\kappa(\omega)$ values being, in most cases, no more than 0.10. The data of Reference 169 differ from those of References 79 and 172 more perceptibly, showing slightly increased $\kappa(\omega)$ values for frequencies lower than 1020 cm^{-1} (see Table 2). Hence, the data of References 79 and 172 are probably the most reliable for the entire 900 to 1300 cm^{-1} range.

Unfortunately, the numerical data of Reference 172, like those of most studies cited above, do not extend to the lower frequency range. Among studies discussed, only those of Popova, Tolstykh, and Vorob'ev;[169] of Steyer, Day, and Huffman;[170] and of Efimov[79] provide information on $n(\omega)$ and $\kappa(\omega)$ values for frequencies lower than 870 to 900 cm^{-1}. The data in Reference 170 are limited by the frequency of 400 cm^{-1} and those of References 79 and 169 extend to 200 cm^{-1}. The full list of these data is placed in Table A2 in the Appendix. For comparison, these data are also presented, in abridged form, in Table 3.

As is seen from Table 3, the differences between the data of these three studies are not too substantial, except in the 400 to 500 cm^{-1} range. The location of the

Table 3 Comparison Between the Data on Optical Constants of Vitreous Silica in the 200 to 860 cm^{-1} Range Presented in Various Sources

Frequency	$n(\omega)$ according to references			$\kappa(\omega)$ according to references		
(cm^{-1})	170	169	79	170	169	79
200	—	2.06	2.00	—	0.03	0.02
300	—	2.19	2.16	—	0.11	0.06
350	—	2.33	2.35	—	0.16	0.16
400	2.53	—	2.64	0.20	—	0.54
430	2.86*	—	2.65	0.94*	—	1.06
440	—	2.77	2.62	—	1.18	1.31
450	2.74	2.68	2.56	1.83	1.52	1.69
460	—	2.45	2.07	—	1.92	2.29
465	1.71	—	1.56	2.31	—	2.38
470	—	1.84	1.09	—	2.30	2.23
475	—	1.37	0.78	—	2.39	1.93
480	0.72*	0.98	0.62	1.71*	2.16	1.62
490	—	0.52	0.53	—	1.39	1.12
500	0.67	0.64	0.58	0.75	0.87	0.75
540	—	1.12	1.09	—	0.32	0.19
550	1.23*	—	1.16	0.08*	—	0.16
580	—	1.35	1.33	—	0.19	0.11
590	1.41**	—	1.36	0.05**	—	0.10
740	1.70*	1.86	1.71	0.06*	0.12	0.05
800	1.77	1.82	1.74	0.34	0.37	0.27
860	1.68*	1.77	1.67	0.06*	0.20	0.05

Note: Asterisks denote values of optical constants related to frequencies by 1 to 3 cm^{-1} greater (*) or lower (**) than those listed in column 1 (for accurate frequencies corresponding to these values, see Appendix, Table A2).

$\kappa(\omega)$ peak in this range found in Reference 169 (475 cm^{-1}) is shifted to higher frequencies as compared to that according to both References 79 and 170 (465 cm^{-1}). Although this shift is as small as 10 cm^{-1}, it causes substantial differences in $\kappa(\omega)$ values at 450 to 460 and 475 to 490 cm^{-1} due to small peak width. In general, taking into account the above discussion of data for frequencies exceeding 900 cm^{-1}, it is the data of Reference 79 that probably best represent the optical constants of bulk vitreous silica in the entire 200 to 1300 cm^{-1} range.

The lack of strong bands is a characteristic of vitreous silica at frequencies below 200 cm^{-1} and, hence, no substantial variations in the $n(\omega)$ and $\kappa(\omega)$ spectra are observed here (see, for example, Reference 108). The attention of investigators is paid mostly to the 10 to 100 cm^{-1} range[181-184] due to a relatively weak broad shoulder observed in the absorption spectrum near 40 cm^{-1}. There is no commonly acknowledged view concerning the origin of this feature yet. In most cases, this feature is treated as the manifestation of the vibrational-density-of-states maximum attributed to the transverse acoustic excitation.[183,184] It is supposed that, for this case, the density-of-states value should be proportional to the $\alpha(\omega)/\omega^2$ ratio. Therefore, it is the plot of this ratio vs. frequency having the form of a smeared maximum at about 38 cm^{-1} that is discussed in References 181, 183, and 184. Data on the absorption coefficient $\alpha(\omega)$ in the 16 to 80 cm^{-1} range are presented, for

Table 4 *κ(ω)* of Vitreous Silica in the 16 to 80 cm^{-1} Range According to the Data of References 182 and 183

Frequency (cm^{-1})	κ(ω) values according to references 182	183
16.9	0.00104	0.00122
20	0.00175	0.00183
30	0.00477	0.00390
40	0.00659	0.00573
50	0.00804	0.00672
60	0.00926	0.00909
70	0.0100	0.00945
80	0.0108	0.00985

example, in References 182 and 183. These data reduced to the $\kappa(\omega)$ numerical values are cited in Table 4.

With minor features omitted, there are only three principal bands in the IR spectrum of vitreous silica located rather far from each other in the 300 to 500, 600 to 850, and 1000 to 1250 cm^{-1} ranges (see Figure 1A, B). Because of this, some information on the parameter values of these principal bands could be obtained from the data of the Kramers-Kronig transform (see Table 5).[43,73,166] Miler[166] estimated only band frequencies from the locations of maxima in the $\varepsilon''(\omega)$ spectra reconstructed with the Kramers-Kronig transform. Gaskell and Johnson[43] also used the $\varepsilon''(\omega)$ spectrum reconstructed with the Kramers-Kronig transform. They estimated the frequencies and also calculated the intensities of three distinct bands directly from the $\varepsilon''(\omega)$ spectrum by weighing the areas under the corresponding maxima; moreover, the estimates of corresponding bandwidths were given. Mitchell, Bishop, and Taylor[73] calculated the IR conductivity spectra, $\rho(\omega) = \omega\varepsilon''(\omega)$, of several chalcogenide glasses and also of vitreous silica. Using these spectra, they estimated the damping coefficients (from the shapes of band wings) and standard deviations for the Gaussian distribution of oscillators (from band shapes in the vicinity of band maxima). Band intensities were characterized in Reference 73 only with the peak conductivity values.

The total combination of band parameters for the IR spectrum of vitreous silica in the entire 200 to 1350 cm^{-1} range obtained with the dispersion analysis version for glasses that was described in Chapter 3, Section IV.B is given in Reference 79. It is these band parameter values that result in the numerical values of optical constants cited in Tables 2 and 3 and in Tables A1 and A2 of Appendix. Prior to considering data on band parameters, some specific features of the IR spectrum of silica at frequencies higher than 1100 cm^{-1} should be discussed.

As was specified in Chapter 1, Section III, an IR spectrum comprised of closely located oscillators can sometimes be described quantitatively, by means of the dispersion analysis, with two approaches differing in the sequence of ω_{T0j} and ω_{L0j} values chosen for oscillators, the validity of the approaches being indistinguishable experimentally. First, this is the case for the highest frequency interval of the IR spectrum of α-quartz. The earlier approach based on the classical version of the dispersion analysis was developed by Spitzer and Kleinman[13] and then also used

Table 5 Frequencies and Intensities of Principal Bands for Three Frequency Ranges Related to Different Types of Vibrations in Vitreous Silica Obtained with the Kramers-Kronig Transform[43,73,166] and Dispersion Analysis[79] Based on Model (50)

Type of vibration and frequency range	Kind of parameter	Kramer-Kronig transform data according to references			Dispersion analysis data[79] (later approach)	
		166	73	43	Band parameter values	Total intensity for the entire range
SiOSi asymmetric stretch, 1000 to 1250 cm^{-1}	ω_j, cm^{-1}	1061	1070	1076	1088	
	$\Delta\varepsilon_{0j}$	—	—	0.55	0.29	0.55
	γ_j, cm^{-1}	—	32	69*	15	
	σ_j, cm^{-1}	—	≤10	—	12	
SiOSi symmetric stretch, 600 to 850 cm^{-1}	ω_j, cm^{-1}	811	800	804	805	
	$\Delta\varepsilon_{0j}$	—	—	0.078	0.051	0.074
	γ_j, cm^{-1}	—	50	64*	9	
	σ_j, cm^{-1}	—	≤10	—	26	
OSiO bend, 350 to 500 cm^{-1}	ω_j, cm^{-1}	443	450	460	460	
	$\Delta\varepsilon_{0j}$	—	—	0.88	0.25	1.02
	γ_j, cm^{-1}	—	80	39*	11	
	σ_j, cm^{-1}	—	≤10	—	6	

Note: Values denoted by asterisk are the full widths taken directly from the $\varepsilon''(\omega)$ plots.

by Russell and Bell.[125] As is illustrated by Table 6, this approach implies the presence of two weak oscillators (denoted further as the j-th and (j + 1)-th) in the 1100 to 1250 cm^{-1} interval corresponding to the high-frequency wing of (j – 1)-th strong E mode with $\omega_{TO(j-1)}$ of about 1072 cm^{-1}. According to the approach used by Gervais and Piriou,[150] the 1100 to 1250 cm^{-1} interval comprises only a single weak oscillator with an ω_{TOj} value of about 1158 cm^{-1} that occupies an "enclosed" location with respect to the stronger E mode with $\omega_{TO(j-1)}$ estimated as 1065 cm^{-1} (see Table 6). In other words, the ω_{TOj} value of weak oscillator (1158 cm^{-1}) lies, in Gervais and Piriou's approach, within the range of LO-TO splitting for the stronger one. For providing the regular sequence of zeros and poles determined by Equation (43), the inequality $\omega_{LOj} < \omega_{TOj} < \omega_{LO(j-1)}$ should hold in such a case (see Chapter 1, Section III). Hence, the enclosed oscillator should give a negative contribution to the total IR intensity, thus mimicking a minimum in the spectrum between the actual and one more nonexistent (j + 1)-th band with an intensity determined by the combination of ω_{TOj} and $\omega_{LO(j-1)}$.[d] According to Lazarev, Mirgorodskii, and Ignat'ev,[162] a preference should be given to the approach of Gervais and Piriou.[150] Thorough theoretical calculations made in References 162 and 178 show that there is no fundamental mode with transverse frequency near 1200 cm^{-1} and that only one fundamental mode at about 1160 cm^{-1} exists in the 1100 to 1250 cm^{-1} range.

Studies of the silica glass spectrum based on the Kramers-Kronig transform data[43,73] did not touch at all upon the problem specified above, because the entire

[d] Concerning the ambiguity of separating the greatly overlapped bands when using the factorized analytical model for the complex dielectric constant (43), see Chapter 1, Section III.

Table 6 Frequencies and Intensities of the IR Oscillators Calculated for Silica Glass[79] and α-Quartz Crystal[13,150]

Type of vibration	Kind of parameter	Approaches to spectrum description: With an enclosed oscillator in the 1150 to 1230 cm^{-1} range		With two usual oscillators in the 1150 to 1230 cm^{-1} range	
		Glass[79]	Crystal[150]	Glass[79]	Crystal[13]
SiOSi asymmetric stretch	ω_j, cm^{-1}	—	—	1187	1227
	$\Delta\varepsilon_{0j}$	—	—	0.04	0.009
	ω_j, cm^{-1}	1178	1158	1166	1163
	$\Delta\varepsilon_{0j}$	0.05	0.01	0.002	0.010
	ω_j, cm^{-1}	1088*	1065*	1087*	1072*
	$\Delta\varepsilon_{0j}$	0.29	0.65	0.34	0.67
SiOSi symmetric stretch	ω_j, cm^{-1}	1057	—	1054	—
	$\Delta\varepsilon_{0j}$	0.21	—	0.17	—
	ω_j, cm^{-1}	805*	797*	804*	797*
	$\Delta\varepsilon_{0j}$	0.05	0.11	0.04	0.11
	ω_j, cm^{-1}	—	695	—	697
	$\Delta\varepsilon_{0j}$	—	0.02	—	0.02
OSiO and SiOSi bend, lower than 500 cm^{-1}	ω_j, cm^{-1}	635	—	633	—
	$\Delta\varepsilon_{0j}$	0.02	—	0.02	—
	ω_j, cm^{-1}	460*	450*	460*	450*
	$\Delta\varepsilon_{0j}$	0.25	0.83	0.20	0.85
	ω_j, cm^{-1}	445	—	447	—
	$\Delta\varepsilon_{0j}$	0.11	—	0.16	—
	ω_j, cm^{-1}	424	—	424	—
	$\Delta\varepsilon_{0j}$	0.66	—	0.67	—
	ω_j, cm^{-1}	—	394	—	394
	$\Delta\varepsilon_{0j}$	—	0.33	—	0.36
	ω_j, cm^{-1}	—	—	—	263
	$\Delta\varepsilon_{0j}$	—	—	—	0.05

Note: Asterisks point out the principal oscillator for a range under consideration.

envelope of the spectrum in the 1000 to 1250 cm^{-1} range was treated inevitably as a single band. In Reference 79, on the other hand, both approaches specified above for α-quartz i.e., the one involving the enclosed location of a single higher-frequency oscillator at about 1160 to 1180 cm^{-1} and the one using two usual weak oscillators in the 1100 to 1250 cm^{-1} range, were tested with the dispersion analysis version developed for glasses. The ensemble of oscillators distinguished in the spectrum of vitreous silica with this version comprises as much as eight or nine oscillators depending on the choice between the two approaches. It is shown in Reference 79 that these approaches provide quite similar values of band frequencies and intensities (see Table 6) and practically identical $n(\omega)$ and $\kappa(\omega)$ spectra, such as those presented in Figure 1A, Table 2, and Table 3. Thus, it is indeed reasonable to do without an additional oscillator in the 1180 to 1230 cm^{-1} interval for vitreous silica, too.

It is of interest to compare the dispersion analysis data on band parameter values for vitreous silica with those found in sources mentioned[43,73,79,166] (see Table 5).[e] In all tables to follow and in the text, band intensities are expressed in terms of contributions, $\Delta\varepsilon_{0j}$, to the "static" dielectric constant ε_0 in accordance with Equation (18).

As is seen from Table 5, the data on the frequencies of three principal bands obtained in References 43 and 79 agree rather well with one another. This holds for both the above approaches to the choice of oscillators; Table 5 presents the numerical data corresponding to the latest approach, i.e., to that using an enclosed high-frequency oscillator. On the contrary, band frequencies found in Reference 73 and, especially, in Reference 166 show substantial deviations from those of References 43 and 79. A special comment should be made concerning a comparison between the band intensities given in References 43 and 79. In Reference 43 the intensities of bands under discussion were estimated using the areas under corresponding $\varepsilon''(\omega)$ maxima; hence, the contributions of unresolved weak bands turned out to be included in the intensity values obtained for three principal bands. As a result, Reference 43 presents the total intensities of all bands within each of three principal frequency ranges rather than the individual band intensities. When calculating total intensities for all the bands found in Reference 79 in the ranges of asymmetric and symmetric SiOSi stretching vibrations, respectively, the results turn out to coincide almost exactly with the data of Reference 43 (see the last column of Table 5). On the other hand, the total intensity of bands in the range of bending vibrations, as is seen, is appreciably greater than that attributed in Reference 43 to the strong band at 450 cm^{-1}. This discrepancy is connected, probably, to the fact that Gaskell and Johnson had at their disposal only the approximate estimate of contribution of the 0 to 300 cm^{-1} frequency range taken from the literature sources and not the experimental magnitude of this contribution obtained from the same $R(\omega)$ spectrum as the rest of the data presented in Reference 43.

Table 5 shows also that the damping coefficients found in Reference 73 greatly exceed those calculated in Reference 79 by a factor varying from 2 to 7. Similarly, full bandwidths estimated in Reference 43 also are much greater than those expected from damping coefficients and standard deviations calculated in Reference 79. Such a tendency to overestimate bandwidths is an inevitable consequence of neglecting weak bands in Reference 43, as shown above. Also, this might be caused by appreciable experimental errors in the minima of the initial $R(\omega)$ spectrum due to the influence of either the surface layer of polished samples (see, for

[e] In Table 5 and others to follow, the band parameter values of Reference 79 are rounded off to the number of decimal figures correlated to the magnitude of methodological errors specified in Chapter 3, Section IV.B. For accurate computing of the reliable $n(\omega)$ and $\kappa(\omega)$ values with Equations (50) through (52), band parameter values should be used without rounding off (in particular, band frequencies should be taken with no less than two decimal figures).

example, References 97 and 174) or the thermal radiation of samples heated by the IR beam of a spectrophotometer.[154] The crudely approximate character of the procedure used by Mitchell, Bishop, and Taylor[73] for obtaining the estimates of damping coefficients also might contribute to this effect.

Table 6 illustrates a comparison between the dispersion analysis data on band frequencies and intensities for vitreous silica[79] and those for α-quartz[f] taken from literature[13,150] for both cases of choosing the set of oscillators in the 1100 to 1250 cm^{-1} range specified above. As is seen from Table 6, the transverse frequencies of oscillators related to three principal bands in the IR spectrum of silica glass (1060 to 1090, 790 to 810, and 450 to 460 cm^{-1}) are rather close (within a 10 to 15 cm^{-1} deviation) to the frequencies of E modes of α-quartz.[g] This confirms, for the case of silica, a certain similarity in the origin of major vibrational excitations in the vitreous and crystalline states specified above in Chapter 2, Sections II.B and III.B.3. Also, the dispersion analysis reveals additional oscillators in the range of the asymmetric stretching vibrations of the SiOSi bridge (1000 to 1250 cm^{-1}) responsible for two shoulders on the wings of the principal $R(\omega)$, $\kappa(\omega)$, and $\varepsilon''(\omega)$ maxima shown in Chapter 3, Figure 1A and Figure 1A, B. The frequency of the higher frequency oscillator at about 1160 to 1180 cm^{-1} (irrespective of whether the latter is treated as enclosed or not) is almost the same as that in the crystal spectrum, thus speaking definitely against the above idea of Reference 58 to interpret this feature as the longitudinal mode associated with the principal transverse mode at 1060 to 1090 cm^{-1}.

On the contrary, a rather intense oscillator at 1050 to 1060 cm^{-1} corresponding to the above poorly resolved shoulder at the lower frequency wing of principal band at 1080 to 1090 cm^{-1} has no counterparts among the normal modes of α-quartz. The frequency of this oscillator falls into the range of vibrations of SiO^- terminal group (see Section II.A.2) that might arise as a defect of SiO_2 network. However, the intensity of the oscillator seems to be too large to relate it to the local defect mode. Moreover, the total magnitudes of oscillator intensities in the entire 1000 to 1250 cm^{-1} range for the spectra of α-quartz and of silica glass (0.66 and 0.55, respectively) differ only insignificantly; this difference is caused mostly by differences in density, because the ratio between these total magnitudes is strictly the same as that between the densities of the crystal and glass (1.20). Therefore the oscillator with a frequency of about 1050 to 1060 cm^{-1} probably has nothing in common with the local defect modes.

In the range of the symmetric stretching vibrations (600 to 850 cm^{-1}), the frequency of the principal oscillator in the spectra of α-quartz and silica glass also coincides closely (see Table 6). At the same time, the weak E mode of α-quartz at 697 cm^{-1} has no counterpart in the vitreous silica spectrum. On the other hand, the

[f] Usually, it is the highest temperature phase, i.e., β-cristobalite, that is assumed to be the closest, as compared to the rest of crystalline SiO_2 modifications, to the structure of vitreous silica. Unfortunately, the IR spectroscopic data for the single crystal of neither cristobalite nor any other silica modification except α- and β-quartz were not available until now. On the other hand, both the experimental data on the IR reflection spectra of the polycrystalline samples of various silica modifications (see, for example, References 165 and 179) and the data of theoretical calculations[185] speak for rather insignificant differences in the IR spectra of β-cristobalite and α-quartz. Therefore, a comparison between the IR band parameter values of vitreous silica and α-quartz seems to be of interest.

[g] Concerning the preferability of the E-mode spectrum (rather than the totality of all modes) for comparison between α-quartz and silica glass, see Chapter 2, Section III.B.3.

latter comprises, unlike the α-quartz spectrum, a quite weak oscillator at about 633 cm^{-1} that corresponds, quite probably, to the local defect mode (see Chapter 2, Section II.B).

The most substantial differences in the ensemble of oscillators between the spectra of α-quartz and silica glass, as is seen from Table 6, are characteristic of the range lower than 500 cm^{-1}. As is the case for the ranges of the stretching vibrations discussed above, the frequencies of the principal oscillator in these spectra are also in good agreement (450 and 460 cm^{-1}, respectively). At the same time, neither the strong E mode with frequency of 394 cm^{-1} nor the two lower frequency modes of α-quartz specified in Reference 125 have counterparts in the vitreous silica spectrum. The latter, at frequencies lower than 450 cm^{-1}, consists only of a smooth curve (see Chapter 3, Figure 1A and Chapter 4, Figure 1A, B). This curve, however, cannot be interpreted simply as the wing of the principal maximum at 460 cm^{-1}, because two oscillators with frequencies of about 445 and 425 cm^{-1} having nothing in common with the modes of α-quartz are necessary to simulate the spectrum. The sum of intensities of these two oscillators is much higher than that of three lower frequency α-quartz modes. However, the total magnitudes of oscillator intensities in the entire 200 to 500 cm^{-1} range for the spectra of α-quartz and of silica glass are rather similar (1.16 and 1.02, respectively), the ratio between them being close (as is also the case for the 1000 to 1250 cm^{-1} range) that between the densities of the crystal and glass.

Thus, the results of comparison between the parameters of principal oscillators in the IR spectra of vitreous silica and α-quartz fit quite well within the concept of glass spectrum formation specified above in Chapter 2, Section II.B. However, the presence of some additional oscillators in the vitreous silica spectrum such as those with frequencies of about 1050 to 1060, 445, and 425 cm^{-1}, shown unequivocally with the dispersion analysis, extends outside this pattern. As was mentioned above, these oscillators have no counterparts in the α-quartz spectrum (nor in the spectra of other crystalline forms of silica); at the same time, they are too intense to relate them to local defect modes. Therefore, two other possible reasons for the appearance of such oscillators should be taken into account (see Chapter 2, Section II.B). First, the idealized (i.e., freed of the random distribution of parameters) structure of the localization regions of vibrational excitations in vitreous silica can correspond to a greater and less symmetric elementary unit than the unit cell of α- or β-quartz, hence increasing the number of normal modes. Second, these oscillators can correspond to no particular kind of near-zero-wavevector phonon, i.e., they can merely be the effective oscillators which serve to approximate the effect of the asymmetric broadening of most strong principal bands originating from the density-of-states spectrum activation. In other words, the necessity of introducing these oscillators into the dispersion analysis can be caused by the need to take into account the contributions of vibrational states with nonzero wavevectors. To discriminate between these two possibilities, further studies into the problem of glass spectrum formation are necessary.

2. Silicate Glasses

There is a great number of studies of the IR spectra of silicate glasses in the literature; those published before 1973 were reviewed in detail in Reference 160. However, most of these studies are of qualitative character and do not extend to the calculation of optical constants. Literature sources comprising information on the optical constants of these glasses present this information mostly in the form of

A

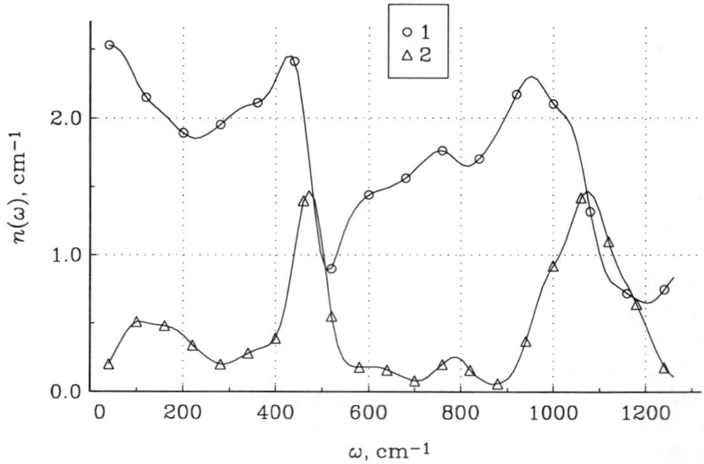

B

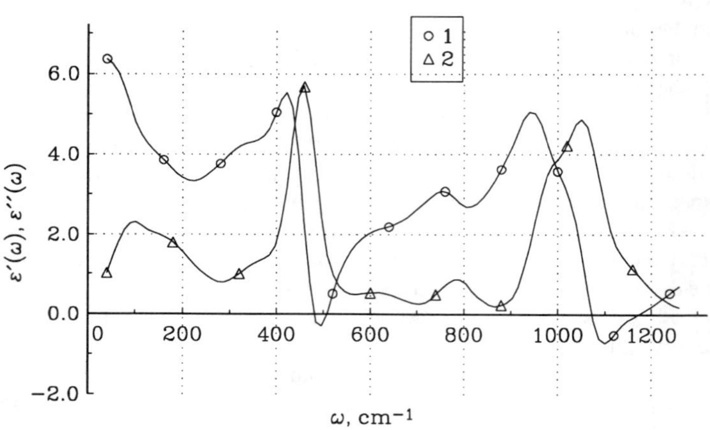

Figure 3 Spectra of optical constants of some sodium silicate glasses in the 40 to 1300 cm^{-1} range according to the data of Reference 79. Glass compositions in mol% by analysis. (A) $19.3Na_2O \cdot 80.7SiO_2$ glass; 1 is *n(ω)* and 2 is *κ(ω)* spectra. (B) $19.3Na_2O \cdot 80.7SiO_2$ glass; 1 is *ε′(ω)* and 2 is *ε″(ω)* spectra. (C) $23.6Na_2O \cdot 76.4SiO_2$ glass; 1 is *n(ω)* and 2 is *κ(ω)* spectra. (D) $23.6Na_2O \cdot 76.4SiO_2$ glass; 1 is *ε′(ω)* and 2 is *ε″(ω)* spectra. (E) $33.4Na_2O \cdot 66.6SiO_2$ glass; 1 is *n(ω)* and 2 is *κ(ω)* spectra. (F) $33.4Na_2O \cdot 66.6SiO_2$ glass; 1 is *ε′(ω)* and 2 is *ε″(ω)* spectra.

plots of the *n(ω)* and *κ(ω)* spectra (see, for example, References 103, 104, 176, and 186 to 191); the amount of numerical data is rather limited.[79,102,167,173,192]

The typical *n(ω)*, *κ(ω)*, *ε′(ω)*, and *ε″(ω)* spectra of alkali silicate glasses are shown, using the data of Reference 79 as examples, in Figures 3A through F. As is known, the introduction of ionic oxides, such as alkali and alkaline earth metal denoted further by Me_2O and MeO, respectively, results in a rather similar general character of changes in the high-frequency region of the IR spectra of silicate glasses. These changes consist in gradually splitting the principal band in the range of the asymmetric stretching vibrations into a distinct doublet (see, for example, Chapter 3, Figure 1B, C) and also the spectra of References 180, 193, and 194). The

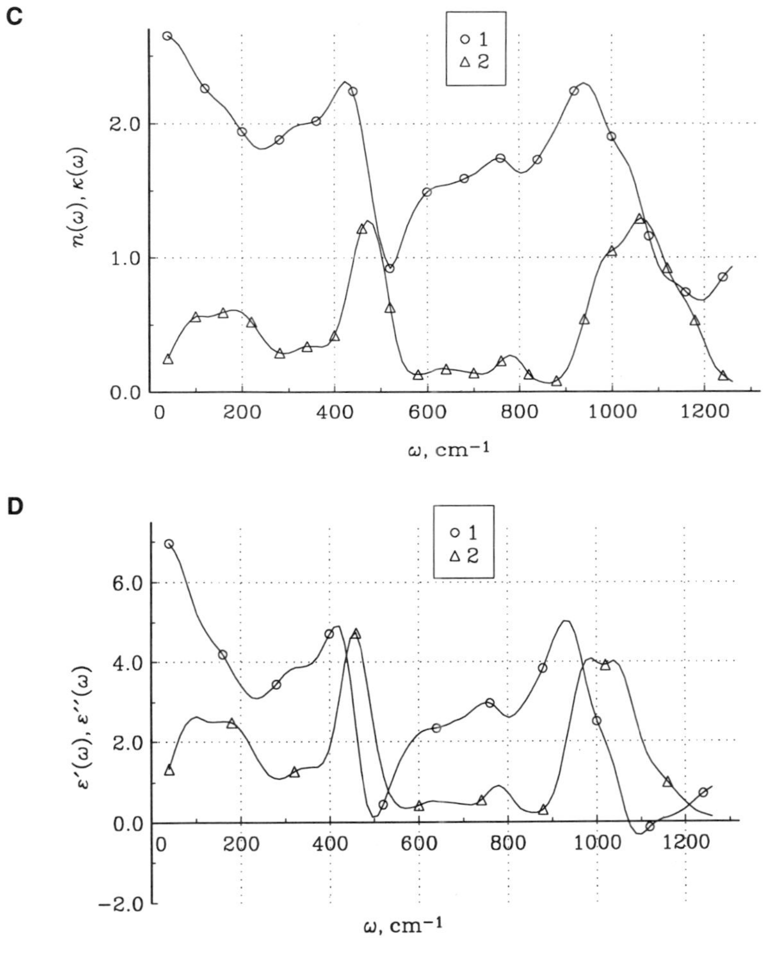

Figure 3 (continued)

corresponding changes in the spectra of optical constants are seen in Figure 3A through F; for convenience, the general pattern of successive variations in the $\kappa(\omega)$ spectrum with an increase in alkali content is also illustrated by Figure 4A (spectra 1 through 4). As is seen, distinct separation of the lower frequency component from the higher frequency one occurs somewhere in the Na_2O content range from 24 to 28 mol% by analysis. The frequency location of the higher frequency component of the doublet (1000 to 1150 cm^{-1}) remains practically unchanged for a particular glass-forming system regardless of the Me_2O or MeO content of glasses. On the contrary, the lower frequency component gradually shifts from 1000 cm^{-1} to much lower frequencies; resultant differences between the $\kappa(\omega)$ spectra of glasses differing in Na_2O content are illustrated by Figure 4B. Along with the 1000 to 1150 cm^{-1} doublet, two other principal features inherent in the ranges of the SiOSi symmetric stretching vibrations and of the OSiO bending vibrations at about 760 to 800 and 450 to 480 cm^{-1}, respectively, are also seen in Figures 3 and 4; these features are rather similar in location to those observed in the spectrum of vitreous silica.

E

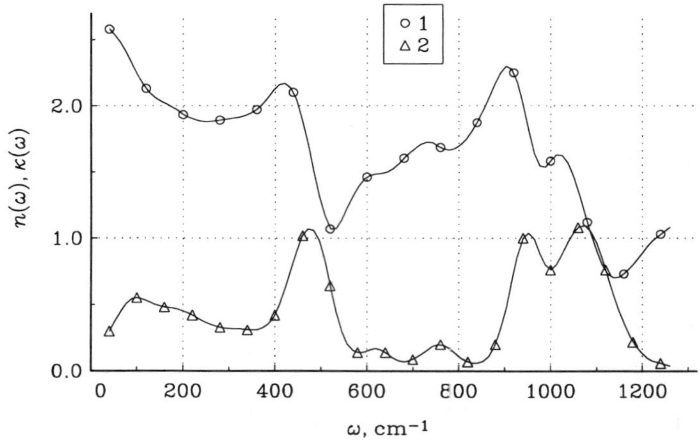

F

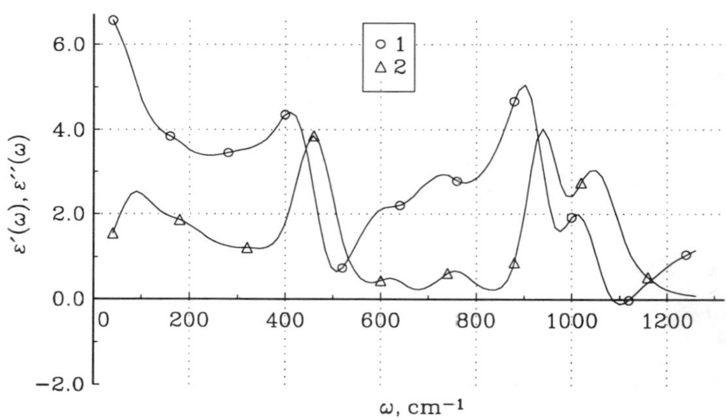

Figure 3 (continued)

The plots of the *n(ω)* and *κ(ω)* spectra presented in various sources in most cases agree rather satisfactorily with each other; there are, however, a few cases with rather great differences. For example, the pattern of *n(ω)* spectra presented in Reference 176 showing two distinct *n(ω)* high-frequency maxima in the spectra of vitreous silica and glasses with an Na_2O content of about 20 to 25 mol% is incompatible with both the *n(ω)* spectra known from other sources and the pattern of the corresponding *κ(ω)* spectra presented in the same reference. Moreover, the data in Reference 176 show splitting the principal high-frequency *κ(ω)* maximum at an appreciably lower Na_2O content than is the case for the data of recent studies. Probably, the initial experimental spectra of Reference 176 were influenced by the hydrated layer on the surfaces of sodium silicate glass films due to the chemical reaction with atmospheric moisture.

Studies conducted from 1950s to the 1970s, including those in References 176 and 180, interpreted the variations in IR spectra of silicate glasses in the 400 to 1200 cm^{-1} range with glass composition mostly in terms of the molecular-like approach based on the pattern of normal modes of an isolated SiO_4 tetrahedron. A

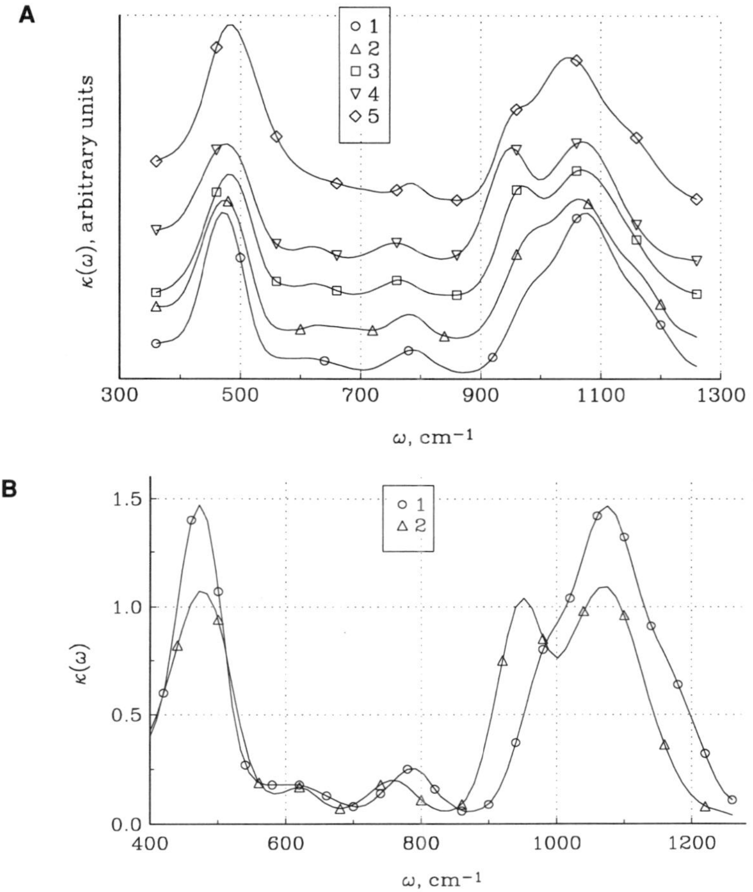

Figure 4 Variations in the *κ(ω)* spectra with alkali silicate glass composition in the 400 to 1300 cm^{-1} range according to the data of Reference 79. Glass compositions in mol% by analysis. (A) General pattern of the spectra depending on Me_2O content; 1 is $19.3Na_2O \cdot 80.7SiO_2$, 2 is $23.6Na_2O \cdot 76.4SiO_2$, 3 is $28.4Na_2O \cdot 71.6SiO_2$, 4 is $33.4Na_2O \cdot 66.6SiO_2$, and 5 is lithium disilicate (by synthesis). (B) Comparison of the spectra of glasses with maximum difference in Na_2O content. 1 is $19.3Na_2O \cdot 80.7SiO_2$ and 2 is $33.4Na_2O \cdot 66.6SiO_2$.

specific case was the practice of interpreting these spectra based on the crystallite hypothesis of glass structure (see, for example, Reference 193) that was widely used from 1950s to the 1960s, especially in Russian literature, but is now completely obsolete. Later, the above approach of interpreting the IR spectra of silicate polyanions in terms of vibrations of the SiOSi bridge and the SiO^-, $Si(O^-)_2$, and $Si(O^-)_3$ terminal groups (see Section II) was put into wide use in glass spectroscopy. Among various terminal groups, it is the SiO^- group (this group being typical of the disilicate structure) that plays the predominant role in the network of most silicate glasses and, hence, in the formation of their spectra. Many investigators of silicate glasses tend to attribute the higher frequency component of the 1000 to 1150 cm^{-1} doublet to the asymmetric stretching vibration of the SiOSi bridge and the lower frequency component to the stretching vibration of the SiO^- terminal group (see, for example, References 186, 191, and 194 to 196) or, correspondingly, to the SiO_3O^- tetrahedron.[103] Similar relative locations of these two types of vibrations were assumed initially for crystalline silicates, too (see, for example, Reference 161). However, the thorough theoretical calculations of the IR spectra of crystalline silicates excellently coinciding with experiment[20,162] unambiguously showed that the stretching vibrations of the SiO^- terminal group can be located at frequencies as high as those of SiOSi bridge or even higher due to differences in force constants. For alkali disilicates, the SiO^- stretching vibrations always correspond to frequencies higher than 1000 cm^{-1}, whereas the lower frequency limit of asymmetric stretching vibrations of the SiOSi bridge can be shifted to frequencies as small as those of about 900 cm^{-1}, the shift probably being the result of mode coupling. For example, the asymmetric stretching vibrations of the SiOSi bridge and the stretching vibrations of the SiO^- terminal group in the lithium disilicate spectrum were shown to correspond to 910 to 1200 and 1000 to 1140 cm^{-1} ranges, respectively.[20] Evidently, similar relative locations of these types of vibrations should hold for Me_2O-SiO_2 glasses, too. Hence, it is the lower frequency component of the doublet in the 900 to 1000 cm^{-1} range that is related solely to the lower frequency vibrations of the SiOSi bridge. On the contrary, the higher frequency component of the doublet is probably of a complex nature, i.e., it is formed by overlapping features related to both the stretching vibrations of the SiO^- terminal group and to some highest frequency vibrations of the SiOSi bridge. Available data confirm the lack of direct connection between the degree of splitting the doublet in the 900 to 1100 cm^{-1} range into two components and the concentration of the SiO^- terminal groups. As is seen from Figure 4A, the spectrum of vitreous lithium disilicate in the 900 to 1100 cm^{-1} range (spectrum 5) is only poorly resolved into two components and looks more like the spectrum of glass with a rather low Na_2O content (spectrum 2) than like that of vitreous sodium disilicate (spectrum 4). Thus, the degree of splitting this doublet into two components depends not only on the ionic oxide content but also on the kind of corresponding cation, more so for MeO-containing glasses.

The triatomic $Si(O^-)_2$ terminal group exhibits, unlike SiO^-, both types of stretching vibrations, i.e., the asymmetric and symmetric vibrations. The frequencies of the asymmetric vibrations of the $Si(O^-)_2$ group in the IR spectra of alkali metasilicates, according to Reference 162, are of the same order as the frequencies of the stretching vibrations of the SiO^- group in disilicates, whereas the frequencies of the symmetric vibrations of the $Si(O^-)_2$ group are considerably lower. Moreover, the frequencies of the SiOSi asymmetric stretching vibrations in metasilicates (for lithium metasilicate, 850 to 970 cm^{-1}) are much lower than those of disilicates.[162] Thus, bands related to the $Si(O^-)_2$ symmetric stretching vibrations are located in between bands related to the asymmetric stretching vibrations of the same group and of the SiOSi bridge. This explains the appearance of three distinct $R(\omega)$ maxima (such as those discussed in Reference 180) in the IR spectra of glasses close to the metasilicate composition.

Some authors attribute the band at about 750 to 800 cm^{-1} in the IR spectra of silicate glasses to the OSiO bending vibration and that at about 450 to 480 cm^{-1} to the SiOSi bending vibration (see below). However, such an assignment contradicts the results of the above thorough theoretical calculations of the IR spectra of crystalline disilicates and metasilicates.[20,162] These calculations confirm that the frequencies of the IR active symmetric SiOSi stretching vibrations usually correspond to the 600 to 850 cm^{-1} range.

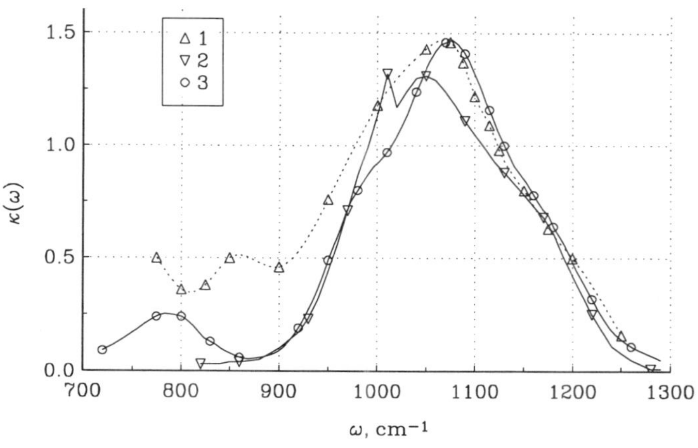

Figure 5 Comparison of the *κ(ω)* spectrum of glasses of $20Na_2O \cdot 80SiO_2$ as-batched composition in the 700 to 1300 cm^{-1} range plotted with data of various sources: (1) Girin et al.;[167] (2) Zolotarev et al.,[173] and (3) Efimov.[79]

Along with the principal bands inherent in the ranges of the SiO^- and SiOSi stretching vibrations and also of OSiO bending vibrations discussed above, the *κ(ω)* and *ε″(ω)* spectra shown in Figures 3A through F reveal at least five weaker bands. The possible origin of the latter is touched upon below when considering band parameter values.

As already mentioned, there is not much data in literature on the numerical values of optical constants of binary and more complex glasses allowing for the direct comparison between the sources. For alkali silicate glasses, the data of References 79, 167, and 173 can be compared.

In Reference 167, the *n(ω)* and *κ(ω)* spectra of sodium silicate glasses with Na_2O content ranging from 5 to 50 mol% by synthesis[h] were calculated from the *R(ω)* spectra recorded at two different incidence angles (20 and 70°). The minimum relative error of *n(ω)* and *κ(ω)* values was assessed to be as large as 10% (with a substantial increase in the error for the case of high *n(ω)* and low *κ(ω)* values). In Reference 173, *n(ω)* and *κ(ω)* spectra of $20Na_2O \cdot 80SiO_2$ glass (mol% by synthesis) were reconstructed (probably, with the Kramers-Kronig transform) from the experimental internal reflection data obtained at an incidence angle of 72°. In Reference 79, numerical values of optical constants for four sodium silicate glasses and also for some other (Me_2O, MeO) SiO_2 glasses were obtained in the same way as the above data on the optical constants of vitreous silica, i.e., with the dispersion analysis version described in Chapter 3, Section IV.B; the near-normal incidence reflection spectra registered from surfaces polished thoroughly with water-free organic liquids were used as starting data.

The *κ(ω)* spectra obtained in References 79, 167, and 173 are compared with each other in Figures 5 and 6. As is seen from Figure 5, the spectra of glasses with 20 mol% Na_2O by synthesis taken from References 79 and 173 coincide rather well with one another only in the limited 860 to 980 and 1150 to 1290 cm^{-1} ranges,

[h] It should be remembered that glass compositions with Na_2O content up to 18 mol% fall into the range of metastable phase separation, thus data for such glasses lacking accompanying information concerning the degree of phase separation development seem to be of no use.

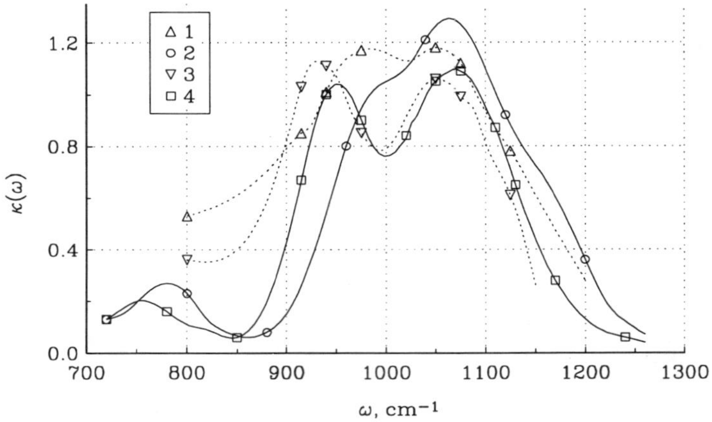

Figure 6 Comparison of the *κ(ω)* spectra of glasses of (1,2) $25Na_2O \cdot 75SiO_2$ and (3,4) $35Na_2O \cdot 65SiO_2$ as-batched compositions plotted for the 700 to 1300 cm^{-1} range according to two available sources: (1,3) Girin et al.[167] and (2,4) Efimov.[79]

whereas the differences in the 1000 to 1120 cm^{-1} range covering the *κ(ω)* maximum itself are quite substantial. First, the *κ(ω)* spectrum of Reference 173 comprises a quite narrow lower frequency component at 1010 cm^{-1}, which seems to be improbable because neither the complementary *n(ω)* spectrum nor *R(ω)* spectra of similar glasses given in other sources[179,193] show any corresponding feature. Probably, such a narrow *κ(ω)* feature is an artifact caused by some error of the Kramers-Kronig transform. Also, the location of the main broad *κ(ω)* maximum found in Reference 173 is shifted by 25 cm^{-1} to lower frequencies and *κ(ω)* values in the vicinity of this maximum are appreciably lowered as compared to those observed in the other two studies;[79,167] probably, these differences are caused by the influence of the above false peak at about 1010 cm^{-1}. On the contrary, the *κ(ω)* spectra presented in References 79 and 167 agree rather well with one another both in the vicinity of the maximum at 1075 cm^{-1} and at higher frequencies up to 1250 cm^{-1}. However, the lower frequency region of the *κ(ω)* spectrum from Reference 167 differs crucially from those published in References 79 and 173 in the presence of a broad maximum at about 850 cm^{-1} that has no counterpart in the experimental *R(ω)* spectrum. In addition, the spectrum of Reference 167 differs from that of Reference 79 in such a way as if the intensity of the lower frequency component of the above doublet is appreciably higher and the frequency of the latter is lower than is the case for the spectrum of Reference 79. This specific feature of data presented in Reference 167 is seen even more clearly in the spectra of other glasses with higher Na_2O content (see Figure 6). In other words, the *κ(ω)* spectra of Reference 167 look like the spectra related to glasses with higher concentrations of the SiO^- terminal groups than those following from their Na_2O content. Such a pattern (similar to the above case of plots presented in Reference 176) can probably result from the contribution of a greatly hydrated surface layer due to the rather low chemical durability of sodium silicate glasses. The flatness of lower frequency wings of the 930 to 980 cm^{-1} band in the spectra of Reference 167 resulting in substantially increased absorption index values in the vicinity of minimum in the 800 to 900 cm^{-1} range also points out the influence of the greatly affected surface. On the contrary, the data of Reference 79 are based, as was mentioned above, on

Table 7 IR Band Frequencies and Intensities for Some Sodium Silicate Glasses[79]

Type of vibrations	Kind of parameter	Band parameter values for glasses (compositions by analysis, mol%) $19.3Na_2O \cdot 80.7SiO_2$	$23.6Na_2O \cdot 76.4SiO_2$	$28.4Na_2O \cdot 71.6SiO_2$	$33.4Na_2O \cdot 66.6SiO_2$
SiO^- and highest SiOSi stretching modes	ω_j, cm^{-1}	1137	1139	1142	1145
	$\Delta\varepsilon_{0j}$	0.09	0.07	0.04	0.02
	ω_j, cm^{-1}	1055	1050	1050	1048
	$\Delta\varepsilon_{0j}$	0.26	0.22	0.25	0.23
SiOSi asymmetric stretching modes	ω_j, cm^{-1}	981	973	957	938
	$\Delta\varepsilon_{0j}$	0.16	0.20	0.18	0.20
SiOSi symmetric stretching modes	ω_j, cm^{-1}	784	780	763	757
	$\Delta\varepsilon_{0j}$	0.07	0.06	0.04	0.05
	ω_j, cm^{-1}	624	627	634	623
	$\Delta\varepsilon_{0j}$	0.04	0.03	0.03	0.02
OSiO and SiOSi bend and other internal modes of	ω_j, cm^{-1}	458	459	465	463
	$\Delta\varepsilon_{0j}$	0.67	0.66	0.63	0.59
polyanion	ω_j, cm^{-1}	365	349	(this spectrum was not studied at frequencies lower than 400 cm^{-1})	343
	$\Delta\varepsilon_{0j}$	0.19	0.17		0.19
Na^+O^- external vibrations and other lowest	ω_j, cm^{-1}	178	191		193
	$\Delta\varepsilon_{0j}$	1.49	1.52		1.48
frequency ones	ω_j, cm^{-1}	93	93		91
	$\Delta\varepsilon_{0j}$	0.68	1.10		1.03
High-frequency dielectric constant, ε_∞		2.196	2.220	2.206	2.243
"Static" dielectric constant, ε_0, calculated with (18)		5.84	6.28	—	6.06

the reflection spectra registered from sample surfaces polished thoroughly with water-free organic liquids, so errors caused by the effect of surface layers are minimized. Thus, an above comparison speaks in favor of data on the optical constants of sodium silicate glasses given in Reference 79. A list of the $n(\omega)$, $\kappa(\omega)$, $\varepsilon'(\omega)$, and $\varepsilon''(\omega)$ numerical values vs. frequency for these glasses in the entire 40 to 1300 cm^{-1} range borrowed from Reference 79 is presented in Table A3 of the Appendix. These values correspond to the spectra shown in Figures 3A through F.

Numerical data on band frequencies and intensities for some sodium silicate glasses according to Reference 79 are cited in Table 7; these parameter values correspond to the spectra of optical constants shown in Figures 3A through F and Figure 4. As is seen from Table 7, an increase in the Na_2O content of glasses from 19 to 33 mol% practically does not influence the frequency of the oscillator at about

1048 to 1055 cm^{-1} related mostly to the stretching vibrations of the SiO^- terminal group, the variations in this frequency remaining practically within the estimation error. Also, an additional weak oscillator at about 1137 to 1145 cm^{-1} is revealed by means of the dispersion analysis. In accordance with the general pattern of the spectra shown in Figures 3 and 4, the variation in glass composition under consideration results in substantial and gradual shifting of the frequency of the oscillator related to the main fraction of the SiOSi asymmetric stretching vibrations from 981 to 938 cm^{-1}. This shift should be attributed, in accordance with the above, to the effect of mode coupling. A shift similar in direction but weaker in magnitude (from 784 to 757 cm^{-1}) is also characteristic of the oscillator related to the SiOSi symmetric stretching vibration. A weak increase in the frequency of the oscillator related to the OSiO bending vibration from 458 to 463 to 465 cm^{-1} lies almost within the estimation error ±0.8%.[76]

It is of interest to compare the data on band frequencies presented in Table 7 to the plots of principal band locations vs. glass composition estimated by Ferraro and Manghnani from the transmission spectra obtained with the KBr and polyethylene pellet technique.[194] As follows from such a comparison, trends characterizing variations in band frequencies with glass composition according to Table 7 and to Reference 194 mostly coincide; the only difference is in the fact that Table 7 demonstrates a less pronounced decrease in the 750 to 800 cm^{-1} band frequency as compared to that of the 900 to 1000 cm^{-1} band, whereas the plots from Reference 194 related to these bands seem to be of almost equal slope. However, as was already mentioned, the band assignment proposed in Reference 194 is among those relating the 1050 to 1100 cm^{-1} band to the SiOSi asymmetric stretching vibration and the 900 to 1000 cm^{-1} band to the SiO^- stretching vibration, thus turning out to be the subject matter for the criticism expressed above on the basis of the data of References 20 and 162.

An additional conclusion can be made from data on band intensities presented in Table 7. Table 7 shows that the variations in the intensity of the lower frequency component of the doublet in the 900 to 1000 cm^{-1} range do not exceed the error of calculation mentioned in Chapter 3, Section IV.B. This supports the abovementioned view that the lower frequency component of the doublet cannot be related to the stretching vibration of the SiO^- terminal group, because the concentration of these groups varies substantially in the studied range of glass compositions.

Among the oscillators with frequencies lower than 700 cm^{-1} mentioned in Table 7, those at about 458 to 465 cm^{-1} are evidently related (like a similar band in the spectrum of vitreous silica) to the OSiO bending vibration. The data of References 20 and 162 also show that a weak oscillator observed in the spectra of silicate glasses near 623 to 634 cm^{-1} should be related to one more SiOSi symmetric stretching vibration and that at about 343 to 365 cm^{-1} to the SiOSi bending vibration.

All above oscillators are related to various internal vibrations of silicate polyanions forming the network of glasses under study. On the contrary, two oscillators at frequencies lower than 300 cm^{-1} (see Table 7) have nothing in common with the internal vibrations of silicate polyanions. The first of them with a frequency of about 178 to 193 cm^{-1} is evidently related to the vibration of the Na^+ ion with respect to its oxygen cage; this vibration is external against the silicate polyanion. Many studies of the low-frequency IR spectra of various glasses using the standard qualitative IR absorption techniques (see, for example, References 197 through

199) showed the occurrence of similar bands related to the cage-like vibrations of the modifying cations in their network sites (denoted further for simplicity as the Me^+O^- or $Me^{2+}O^-$ bands), the band frequencies depending in a regular way on the Me^+ cation mass in accordance with Equation (61). The Na^+O^- band was observed in the 180 to 230 cm^{-1} range, the frequency of the band increasing slightly with an increase in Na_2O content of glasses.[198]

The origin of the last oscillator near 91 to 93 cm^{-1} forming a not too distinct maximum in the $\kappa(\omega)$ and $\varepsilon''(\omega)$ spectra (see Figures 3A through F) cannot be specified now because of the lack of additional information. A few other studies comprising data on the low-frequency IR spectra of silicate glasses are not too informative in this respect. Gervais et al.,[103] whose $\varepsilon''(\omega)$ spectrum of $33Me_2O \cdot 67SiO_2$ glass is rather similar to that shown in Figure 3F in both the high- and low-frequency ranges, tend to relate the entire envelope in the 0 to 300 cm^{-1} to the external Me^+O^- vibrations (and also, for ternary glasses, to $Me^{2+}O^-$). Hanna[200] did not find any distinct bands in the absorption spectra of sodium silicate glasses in the 50 to 330 cm^{-1} range. Probably, the bands at about 170 to 200 and 90 to 95 cm^{-1} shown in Figures 3A through F are too broad to be resolved with standard qualitative IR absorption techniques. The low-frequency IR absorption spectrum of $10Na_2O \cdot 10CaO \cdot 80SiO_2$ glass, obtained in Reference 181 with Fourier-transform spectroscopy, covers the 10 to 90 cm^{-1} range; only a broad asymmetric $\alpha(\omega)/\omega^2$ maximum at about 40 to 45 cm^{-1} (similar to that in the IR spectrum of silica glass discussed in Section II.A.1) is seen in this spectrum.

Together with the data on the optical constants of binary alkali silicate glasses as discussed above, there is also information in the literature on the spectra of optical constants of various ternary Me_2O-MeO-SiO_2, Me_2O-Al_2O_3-SiO_2, MeO-Al_2O_3-SiO_2,[103,104] and some multicomponent silicate glasses of basically soda-lime-silica or soda-lime-alumina-silica compositions.[102,173,191,192] As was mentioned above, Gervais et al.[103,104] reported only plots of the spectra of optical constants obtained with the dispersion analysis based on Model (43). On the contrary, Ljundahl, Fellman, and Ribbing[102] presented both the numerical values of optical constants and the band parameter values found with the classical dispersion analysis. Also, the optical constants of window-type glass were given in References 173 and 192.

As was already mentioned in Chapter 3, Section IV.A, Ljundahl, Fellman, and Ribbing[102] obtained, with both the classical dispersion analysis and the Kramers-Kronig transform, the spectra of optical constants of five silicate glasses in the 3 to 23 μ (about 430 to 3330 cm^{-1}) range. Four of the glasses were synthesized from minerals such as diabase and dolomite and the fifth one was basically of soda-lime-silica composition with rather small additions of B_2O_3, Al_2O_3, MgO, and K_2O. A rather good coincidence between the spectra of optical constants obtained with both methods from the same $R(\omega)$ spectra was demonstrated. When considering these results, one should bear in mind that the compositions of glasses studied in Reference 102 comprise as many as 7 to 8 components; among the latter there are oxides such as Al_2O_3 (in all five glasses), FeO and TiO_2 (in four diabase-based ones), and B_2O_3 (in soda-lime-silica-based glass). As a result, both the experimental $R(\omega)$ spectra and the $\kappa(\omega)$ spectra of these glasses are somewhat broadened and lack deep minima between separate bands (such as those characteristic of the spectra of binary and ternary glasses) due to overlapping the large number of spectrum features related to different structural groups. This is illustrated by Figure 7, which compares the $\kappa(\omega)$ spectrum of soda-lime-silica-based glass from

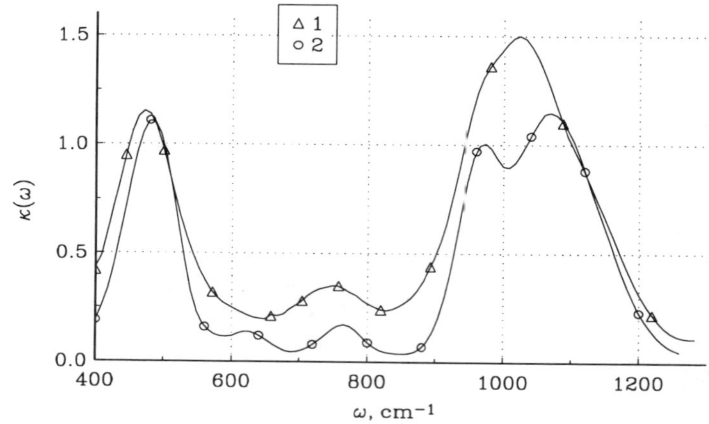

Figure 7 Comparison of the $\kappa(\omega)$ spectra of multicomponent silicate glass and binary sodium silicate: (1) multicomponent (Me_2O, MeO)-SiO_2-based glass no. 5 according to the data of Reference 102 and (2) $28.4Na_2O \cdot 71.6SiO_2$ glass according to the data of Reference 79.

Reference 102 to that of $28.4Na_2O \cdot 71.6SiO_2$ glass from Reference 79 characterized by a coinciding value of the SiOSi asymmetric stretch frequency of 957 cm^{-1} — see Table 7). In spite of a rather low SiO_2 content of glasses studied in Reference 102 (ranging from 44.7 to 64.2 wt%), a single smooth $\kappa(\omega)$ maximum is observed in the entire 900 to 1200 cm^{-1} range, so that the occurrence of two or even three oscillators in the range is revealed only with the dispersion analysis. Also, the $R(\omega)$ and $\kappa(\omega)$ values of these glasses in the 500 to 700 cm^{-1} range do not decrease lower than 13 to 17% and 0.2 to 0.4, respectively, thus providing an opportunity to describe this range with a single oscillator instead of two in the spectra of binary glasses (see Table 7). As already mentioned in Chapter 3, Section IV.A, the smoothed character of the reflection spectra of these glasses is probably the reason why the authors of Reference 102 obtained such good computer fits and reliable spectra of optical constants with such an inappropriate version of the dispersion analysis, as the classical one.

There is also some information in the literature on the spectra of optical constants of binary lead silicate glasses[173,188-190] and various lead-containing multicomponent commercial glasses.[173,189,201] These sources, as a rule, present only plots of $n(\omega)$ and $\kappa(\omega)$ spectra; only Reference 173 has numerical data for a single lead-containing commercial glass with unspecified chemical composition. The general pattern of variations in the $\kappa(\omega)$ spectra of lead silicate glasses in the 400 to 1300 cm^{-1} range with an increase in lead oxide content reveals a low-frequency shift for both the lower and higher frequency components of the doublet in the 800 to 1200 cm^{-1} range.

3. Borate Glasses

As is known, the structures of crystalline borates and borate glasses are greatly complicated when compared to those of crystalline silicates and silicate glasses. This complexity is caused by a change in the coordination state of the boron atom at the expense of ionic oxide oxygen which results in the formation of the $[BO_4]^-$ tetrahedra from the initial trigonal BO_3 groups; these two kinds of polyhedra combine with one another, thus forming various ring structural motifs (see, for example, References 202 and 203).

There are many IR spectroscopic studies of boron oxide and borate glasses designed to obtain information on their structure; most of these studies are confined to the raw experimental reflection spectra[189,204,206] or to qualitative transmission spectra obtained with thin films, KBr pellets, and similar techniques.[205-209] A comprehensive review of such data published before 1973 is presented in Reference 160 and a review of more recent data of this kind is given in Reference 210. The integrated absorptivities for the main IR bands of vapor-deposited thin film of vitreous boron oxide were calculated by Tenney and Wong.[211] The $n(\omega)$ and $\kappa(\omega)$ spectra of alkaline earth borate, zinc borate, and various more complex borate glasses in the 500 to 1600 cm^{-1} range reconstructed with the Kramers-Kronig transform were published in Reference 154; for the 20 to 500 cm^{-1} range, the spectra of optical constants of some of these glasses were given in Reference 212. Recently, the optical constants of the same alkaline earth borate glasses were redetermined (see also below).[79]

According to References 204 to 206, 210, and 211, the high-frequency region of the IR spectrum of vitreous boron oxide comprises a strong band at about 1250 to 1260 cm^{-1} and also a more broad band that forms a shoulder in the 1350 to 1400 cm^{-1} range. These bands are considered to usually originate from two components of the asymmetric stretching vibration of the BO_3 trigonal group (in an isolated BO_3 group, this vibration would be doubly degenerated[205,206,210]). As is known, the predominant structural motif of vitreous boron oxide is the B_3O_6 boroxol ring;[213] therefore, the band near 1250 cm^{-1} is often considered as characteristic of those BO_3 groups that participate in the formation of this ring. Unlike the asymmetric stretching vibrations, the symmetric “breathing” mode of the boroxol ring is inactive in the IR spectra of both vitreous boron oxide and borate glasses and results only in the narrow 808 cm^{-1} Raman band (see, for example, Reference 100). In the 600 to 800 cm^{-1} range, the IR spectrum of vitreous boron oxide is composed of two weaker bands at 718 and 665 cm^{-1} that usually are considered to be traceable to the components of OBO bending vibration of the BO_3 trigonal group.

The presence of the above mixed ring motifs lacking the elements of symmetry in the structures of various borates results in great complexity of their IR spectra (see, for example, Reference 214). The spectra of borate glasses, however, consist of a rather limited number of broad bands. In general, these broad bands are also traceable to the vibrations of the above simple groups, including bands attributed to the BO_3 trigonal group with frequency locations inherited from the spectrum of vitreous B_2O_3 and bands related to the components of the asymmetric stretching vibration of the $[BO_4]^-$ tetrahedron. An example of $n(\omega)$ and $\kappa(\omega)$ spectra of borate glass in the entire 20 to 1600 cm^{-1} range was presented above in Chapter 3 in Figure 2A. The variations in the general pattern of the $\kappa(\omega)$ spectra in the 600 to 1600 cm^{-1} range with glass composition are illustrated by Figure 8A using the data of Reference 79 on barium borate and mixed borate glasses as an example. As is seen, all bands inherited from the spectrum of vitreous boron oxide (except a weak band at about 665 cm^{-1}) are present in the spectra of glasses containing less than 30 mol% of ionic oxide. The band at 1230 to 1250 cm^{-1} related to the lower frequency component of the asymmetric stretching vibration of the trigonal BO_3 group gradually decreases in intensity and vanishes somewhere near the glass composition with MeO content of about 30 mol% by analysis, whereas that at 1340 to 1370 cm^{-1} related to the higher frequency component of the same vibration

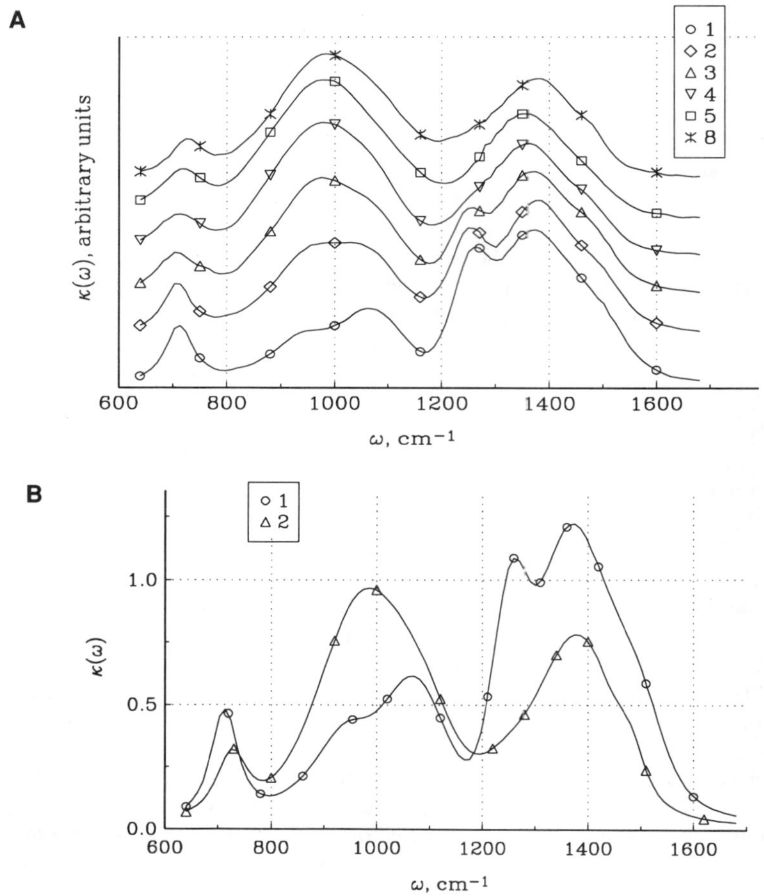

Figure 8 The $\kappa(\omega)$ spectra of barium borate and mixed borate glasses according to the data of Reference 79. Glass compositions in mol% by analysis. (A) Dependence of the general pattern of the spectra depending on the total (Me_2O + MeO) content. 1 is $2.9K_2O \cdot 9.0CaO \cdot 3.1BaO \cdot 85.0B_2O_3$, 2 is $21.1BaO \cdot 78.9B_2O_3$, 3 is $25.9BaO \cdot 74.1B_2O_3$, 4 is $31.1BaO \cdot 68.9B_2O_3$, 5 is $36.0BaO \cdot 64.0B_2O_3$, and 6 is $6.2K_2O \cdot 23.5CaO \cdot 7.6BaO \cdot 62.7B_2O_3$. (B) Comparison of the spectra of glasses with maximum difference in (Me_2O + MeO) content. 1 is $2.9K_2O \cdot 9.0CaO \cdot 3.1BaO \cdot 85.0B_2O_3$ and 2 is $6.2K_2O \cdot 23.5CaO \cdot 7.6BaO \cdot 62.7B_2O_3$.

remains practically unchanged. Moreover, new bands at about 930 to 950 and 1050 to 1070 cm^{-1} related to the components of the asymmetric stretching vibration of the $[BO_4]^-$ tetrahedron appear in glass spectra, the intensity of the former increasing substantially with MeO content up to 30 mol%. On the whole, Figure 8A demonstrates that the main common feature of the IR spectra of borate glasses distinguishing them from the spectra of silicate glasses (and also, as is shown below, from the spectra of germanate glasses) is the practical constancy of all principal band frequencies in the entire 650 to 1500 cm^{-1} range irrespective of the Me_2O or MeO content of glasses. This is also illustrated by Figure 8B, which clearly shows that the frequency locations of all spectrum features remain practically

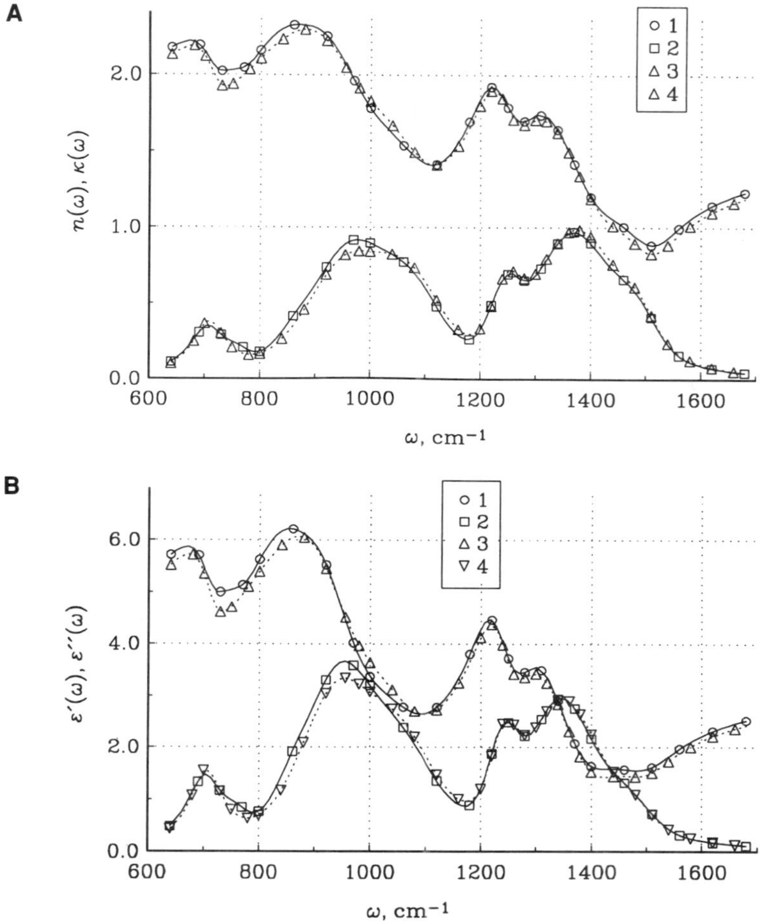

Figure 9 Comparison of the spectra of optical constants of two different borate glasses with similar B_2O_3 content according to the data of Reference 79. Glass compositions in mol% by analysis. (A) (1, 2) $25.9BaO \cdot 74.1B_2O_3$ and (3, 4) $4.7K_2O \cdot 15.4CaO \cdot 5.0BaO \cdot 74.9B_2O_3$ glasses; (1, 3) $n(\omega)$ and (2,4) $\kappa(\omega)$ spectra. (B) (1, 2) $25.9BaO \cdot 74.1B_2O_3$ and (3, 4) $4.7K_2O \cdot 15.4CaO \cdot 5.0BaO \cdot 74.9B_2O_3$ glasses; (1, 3) $\varepsilon'(\omega)$ and (2, 4) $\varepsilon''(\omega)$ spectra.

intact with the Me_2O + MeO content of glasses increasing from 15 to 37 mol%. Moreover, as follows from Figure 8A and, especially, Figure 9, the nature of ionic oxide also does not perceptibly influence both band frequencies and band intensities. As is demonstrated by Figure 9, the $n(\omega)$, $\kappa(\omega)$, $\varepsilon'(\omega)$, and $\varepsilon''(\omega)$ spectra of two glasses with similar B_2O_3 content belonging to different systems coincide almost completely. Such a constancy of band locations is combined with the above pronounced concentration dependence of 930 to 950 and 1230 to 1250 cm^{-1} band intensities. Thus, the case of the 1230 to 1250 and 930 to 950 cm^{-1} bands is a good example of so-called two-mode band behavior (i.e., that characterized by the independent vibrations originated from different structural groups), which is rather rare among the IR spectra of simple oxide glasses because of the absence of

Table 8 Comparison Between Reflectivities and Optical Constants of 21.1BaO · 78.9B_2O_3 Glass in the 640 to 1600 cm^{-1} Range Obtained in Reference 79 and Reference 154

	Ref. 79			Ref. 154			Deviation of data of Ref. 154		
ω, cm^{-1}	$R(\omega)$, %	$n(\omega)$	$\kappa(\omega)$	$R(\omega)$, %	$n(\omega)$	$\kappa(\omega)$	$\delta R(\omega)$, %	$\delta n(\omega)$	$\delta\kappa(\omega)$
640	13.6	2.17	0.11	15.6	2.29	0.19	+2.0	+0.12	+0.08
680	14.8	2.21	0.25	17.1	2.34	0.36	+2.3	+0.13	+0.11
720	12.6	2.04	0.32	14.3	2.13	0.38	+1.7	+0.09	+0.06
760	12.3	2.03	0.23	14.4	2.16	0.34	+2.1	+0.13	+0.11
800	13.7	2.16	0.18	15.9	2.27	0.34	+2.2	+0.11	+0.16
840	16.2	2.30	0.31	18.7	2.40	0.51	+2.5	+0.10	+0.20
880	18.1	2.32	0.51	20.6	2.38	0.75	+2.5	+0.06	+0.24
920	19.0	2.25	0.74	21.6	2.23	0.97	+2.6	–0.02	+0.23
960	18.6	2.03	0.90	21.2	1.99	1.08	+2.6	–0.04	+0.18
1000	16.7	1.78	0.90	19.0	1.74	1.04	+2.3	–0.04	+0.14
1040	13.9	1.62	0.82	16.7	1.57	0.92	+2.8	–0.05	+0.10
1080	10.7	1.47	0.70	12.4	1.46	0.78	+1.7	–0.01	+0.08
1120	7.0	1.41	0.48	8.2	1.40	0.58	+1.2	–0.01	+0.10
1160	5.5	1.56	0.29	6.4	1.51	0.39	+0.9	–0.05	+0.10
1200	10.0	1.84	0.32	11.1	1.80	0.51	+1.1	–0.04	+0.19
1240	13.5	1.86	0.65	14.7	1.78	0.79	+1.2	–0.08	+0.14
1280	12.5	1.70	0.65	13.7	1.61	0.81	+1.2	–0.09	+0.16
1320	14.0	1.73	0.79	15.0	1.55	0.89	+1.0	–0.18	+0.10
1360	16.4	1.50	0.96	17.4	1.37	1.01	+1.0	–0.13	+0.05
1400	14.8	1.20	0.90	16.6	1.10	0.93	+1.8	–0.10	+0.03
1500	6.0	0.89	0.47	6.9	0.83	0.48	+0.9	–0.06	+0.01
1600	0.3	1.10	0.10	0.4	1.09	0.10	+0.1	–0.01	0

substantial differences in the reduced masses and force constants of different structural groups.

The numerical $n(\omega)$ and $\kappa(\omega)$ values for binary barium borate and K_2O-CaO-BaO-B_2O_3 glasses obtained in Reference 79 with the dispersion analysis version based on Model (50) are presented in Tables A4 through A6 of the Appendix; these data correspond to the spectra shown in Chapter 3 in Figure 2A, and Chapter 4, Figures 8 and 9. As mentioned above, the $R(\omega)$ spectra of the same samples of BaO-B_2O_3 and K_2O-CaO-BaO-B_2O_3 glasses as those studied in Reference 154 were redetermined recently in Reference 79 with the Perkin-Elmer 580 spectrophotometer[i] and the optical constants were reconstructed from the redetermined spectra with the dispersion analysis version based on Model (50). A comparison between the data of References 79 and 154 on the $R(\omega)$, $n(\omega)$, and $\kappa(\omega)$ values for one of the barium borate glasses is illustrated by Table 8. As is seen from Table 8, the $R(\omega)$ values given in Reference 154 slightly exceed those given in Reference 79 throughout the entire spectrum. This results in some positive deviations of the $\kappa(\omega)$ values given in Reference 154 from those given in Reference 79, the maximum deviation of about 0.2 being observed in the 800 to 1000 cm^{-1} range. The deviations of $n(\omega)$

[i] Some overestimation of the $R(\omega)$ values used for calculations in Reference 154 was found to be caused by the perceptible contribution of thermal radiation of glass samples due to appreciable heating of the samples by a beam from the IR source of the Perkin-Elmer 621 spectrophotometer when registering the $R(\omega)$ spectra.

Table 9 IR Band Frequencies and Intensities for Some Barium Borate Glasses in the 650 to 1500 cm^{-1} Range According to Reference 79

Type of vibration	Kind of parameter	Band parameter values for glasses (compositions by analysis, mol%)			
		21.1BaO · 78.9B_2O_3	25.9BaO · 74.1B_2O_3	31.1BaO · 68.9B_2O_3	36.0BaO · 64.0B_2O_3
Asymmetric stretching vibrations of BO_3	ω_j, cm^{-1}	1466	1467	1466	1460
	$\Delta\varepsilon_{0j}$	0.05	0.03	0.03	0.04
	ω_j, cm^{-1}	1357	1348	1341	1333
	$\Delta\varepsilon_{0j}$	0.23	0.28	0.24	0.19
	ω_j, cm^{-1}	1243	1241	1248	—
	$\Delta\varepsilon_{0j}$	0.12	0.05	0.03	—
Asymmetric stretching vibrations of $[BO_4]^-$ tetrahedron	ω_j, cm^{-1}	1047	1054	1047	1061
	$\Delta\varepsilon_{0j}$	0.24	0.19	0.19	0.20
	ω_j, cm^{-1}	938	945	943	946
	$\Delta\varepsilon_{0j}$	0.24	0.35	0.41	0.43
	ω_j, cm^{-1}	853	860	862	861
	$\Delta\varepsilon_{0j}$	0.02	0.04	0.06	0.05
OBO bending vibrations	ω_j, cm^{-1}	706	704	710	716
	$\Delta\varepsilon_{0j}$	0.13	0.15	0.21	0.22

values presented in Reference 154 from those given in Reference 79 can be of both signs and are substantially less in absolute magnitude than is the case for $\kappa(\omega)$ values. Differences between the data of References 79 and 154 on $n(\omega)$ and $\kappa(\omega)$ values shown in Table 8 are due solely to differences in the initial $R(\omega)$ spectra and not by using the different methods of data treatment. When applied to the same array of the $R(\omega)$ values, both methods demonstrate an excellent coincidence (such as that illustrated by Figure 2 in Chapter 3 regardless of whether this array is chosen from Reference 154 or 79. In view of the above, the $R(\omega)$ values and optical constants of BaO-B_2O_3 and K_2O-CaO-BaO-B_2O_3 glasses from Reference 79 cited in Tables A4 to A6 of the Appendix should be considered to be more reliable than those from Reference 154. The data on the optical constants of binary zinc borate and ternary lanthanum-oxide-containing glasses also presented in Reference 154 were not redetermined; one should bear in mind that these data can comprise some small errors similar to those illustrated by Table 8.

The numerical values of band frequencies and band intensities in the IR spectra of BaO-B_2O_3 and K_2O-CaO-BaO-B_2O_3 glasses borrowed from Reference 79 are presented in Tables 9 and 10, respectively; it is these parameter values that correspond to the $n(\omega)$ and $\kappa(\omega)$ spectra shown in Figure 2A in Chapter 3, and Figures 8 and 9 in Chapter 4. The data of Tables 9 and 10 quantitatively confirm the above pattern of $\kappa(\omega)$ spectrum variations with borate glass compositions. Table 9 shows, for example, that the variations in oscillator frequencies in the IR spectra of barium borate glasses either lie within the limits of calculation error or exceed these limits only negligibly. In addition to oscillators related to the distinct

Table 10 IR Band Frequencies and Intensities for Mixed K_2O-CaO-BaO-B_2O_3 Glasses in the 650 to 1500 cm^{-1} Range According to Reference 79 (Glass Compositions in Mol% by Analysis[a])

Type of vibration	Kind of parameter	$15.0(Me_2O+MeO) \cdot 85.0B_2O_3$	$25.1(Me_2O+MeO) \cdot 74.9B_2O_3$	$37.3(Me_2O+MeO) \cdot 62.7B_2O_3$
Asymmetric stretch of BO_3 trigonal group	ω_j, cm^{-1}	1465	1472	1474
	$\Delta\varepsilon_{0j}$	0.05	0.03	0.02
	ω_j, cm^{-1}	1345	1352	1360
	$\Delta\varepsilon_{0j}$	0.32	0.27	0.19
	ω_j, cm^{-1}	1245	1243	1248
	$\Delta\varepsilon_{0j}$	0.12	0.06	0.03
Asymmetric stretch of $[BO_4]^-$ tetrahedron	ω_j, cm^{-1}	1062	1049	1053
	$\Delta\varepsilon_{0j}$	0.22	0.25	0.21
	ω_j, cm^{-1}	935	937	946
	$\Delta\varepsilon_{0j}$	0.16	0.30	0.42
	ω_j, cm^{-1}	851	848	848
	$\Delta\varepsilon_{0j}$	0.01	0.01	0.01
OBO bend	ω_j, cm^{-1}	709	706	725
	$\Delta\varepsilon_{0j}$	0.15	0.15	0.11

[a] Ionic oxide content of glasses is $2.9K_2O \cdot 9.0CaO \cdot 3.1BaO$, $4.7K_2O \cdot 15.4CaO \cdot 5.0BaO$, and $6.2K_2O \cdot 23.5CaO \cdot 7.6BaO$, respectively.

bands in the spectra shown in Chapter 3, Figures 2A and Chapter 4, Figures 8 and 9, the dispersion analysis reveals the presence of two weaker oscillators at about 1460 to 1480 cm^{-1} and 850 to 860 cm^{-1}. The first of these is probably related to some summation excitation, whereas the second has a counterpart in the spectra of many crystalline borates.[214] In accordance with the pattern of spectrum variations demonstrated by Figure 8, the numerical value of intensity of the 1240 to 1250 cm^{-1} oscillator falls down to zero with an increase in BaO content up to about 30 mol%. On the contrary, the intensity of the 930 to 950 cm^{-1} oscillator increases substantially with an increase in BaO content up to 30 mol%. As follows from Table 10, the same regularities are characteristic of band parameter values for mixed K_2O-CaO-BaO-B_2O_3 glasses; the same generally holds for sodium borate glasses.[79]

Such a similarity of band parameters and resultant spectra of optical constants irrespective of the particular kind of ionic oxide allows one to construct the plots of band intensities vs. ionic oxide content in the broad range of glass compositions using data on various glass-forming systems. An example of such a plot for the 930 to 950 cm^{-1} band is shown in Figure 10. As is seen from Figure 10, this plot, for the Me_2O + MeO content of glasses below 30 mol%, mimics approximately the pattern of an increase in the fraction of fourfold-coordinated boron presented, for example, by Bray[215] for binary sodium silicate glasses. The tendency to saturation revealed by the plot shown in Figure 10 at Me_2O + MeO content exceeding 30 mol% is also in accordance with the NMR data of Bray et al.[202,203] on the fraction of fourfold-coordinated boron. Such a correlation between the 930 to 950 cm^{-1}

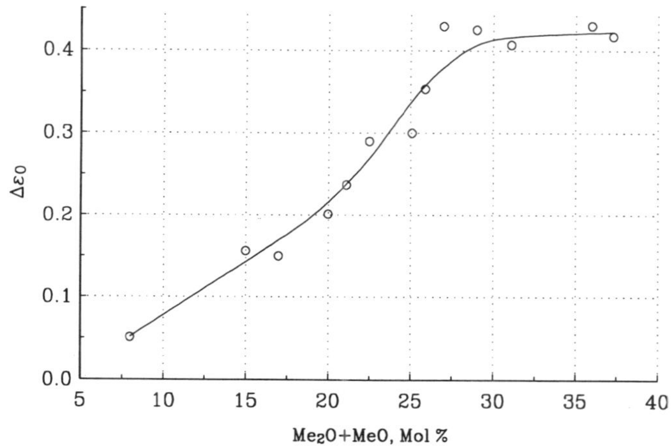

Figure 10 The 930 to 950 cm^{-1} band intensity (expressed in terms of contributions to the "static" dielectric constant) vs. total ionic oxide content of Na_2O-B_2O_3, BaO-B_2O_3, and K_2O-CaO-BaO-B_2O_3 glasses according to the data of Reference 79.

band intensity and the fraction of fourfold-coordinated boron confirms that this band is indeed related to the vibration of the $[BO_4]^-$ tetrahedron. It also follows from the monotonical character of variation in the intensity of the 930 to 950 cm^{-1} band in the entire composition range covered by Figure 10 that this band cannot be connected to a vibration inherent only in some structural fragment greater in size than the $[BO_4]^-$ tetrahedron itself. The concentrations of all possible fragments of this kind (such as the tetraborate or diborate group) change from almost zero to maximum in much narrower composition ranges.

Table 10 also shows the reappearance of the weak 1240 to 1250 cm^{-1} band (Me_2O + MeO) content of glass as high as 37 mol%. This probably results from an increase in the total concentration of BO_3 groups (at the expense of the formation of the ionized BO_3^- groups instead of some amount of $[BO_4]^-$ tetrahedra) at high ionic oxide content of glasses.[202,203]

The IR spectra of borate glasses at frequencies lower than 700 cm^{-1} also comprise several oscillators. The frequencies and intensities of these oscillators for the $36.0BaO \cdot 64.0B_2O_3$ glass according to Reference 79 are given in Table 11; it is these parameter values that correspond to the lower frequency range of the $n(\omega)$ and $\kappa(\omega)$ spectra shown in Chapter 3, Figure 2A. A weak oscillator located in this spectrum at about 550 cm^{-1} is probably related to some low-frequency internal vibration of borate polyanion. According to Reference 197, the frequency of external $Ba^{2+}O^-$ vibration lies near 130 cm^{-1}; hence, the strong 132 cm^{-1} oscillator contributing substantially to the static dielectric constant of $36.0BaO \cdot 64.0B_2O_3$ glass is related to the $Ba^{2+}O^-$ vibration. However, as is seen from Table 11, the spectrum of this glass also comprises a rather intense oscillator at 182 cm^{-1}; it seems probable that the latter corresponds to the $Ba^{2+}O^-$ vibration involving Ba^{2+} cation in a different coordination state. At last, the physical origin of the oscillator at 66 cm^{-1} is as unclear now as is the case for the oscillator at about 90 cm^{-1} in the spectra of silicate glasses (see Section II.A.2). The IR spectra of sodium borate glasses at frequencies lying outside the range of the internal vibrations of borate polyanion comprise only a single oscillator related to the Na^+O^- external vibration

Table 11 IR Band Frequencies and Intensities for $36.0BaO \cdot 64.0B_2O_3$ Glass (Mol% by Analysis) in the 20 to 650 cm^{-1} Range According to Reference 79

	Kind of parameter	Band parameter values
Bending and other multiatomic internal vibrations of borate polyanion, 600 to 800 cm^{-1}	ω_j, cm^{-1}	609
	$\Delta\varepsilon_{0j}$	0.01
	ω_j, cm^{-1}	551
	$\Delta\varepsilon_{0j}$	0.03
$Ba^{2+}O^-$ external vibrations and other lowest frequency ones	ω_j, cm^{-1}	197
	$\Delta\varepsilon_{0j}$	0.03
	ω_j, cm^{-1}	182
	$\Delta\varepsilon_{0j}$	0.22
	ω_j, cm^{-1}	132
	$\Delta\varepsilon_{0j}$	1.74
	ω_j, cm^{-1}	66
	$\Delta\varepsilon_{0j}$	1.70
High-frequency dielectric constant, ε_∞		2.65
"Static" dielectric constant, ε_0 calculated with (18)		7.53

(similar in location to the Na^+O^- band discussed above for sodium silicate glasses) and also the above oscillator near 60 to 70 cm^{-1}.

4. Germania and Germanate Glasses

Most studies devoted to the IR spectrum of vitreous germania deal only with the experimental IR transmission[216,218] or reflection[179,217,218] spectrum and do not extend to the calculation of optical constants. The experimental data of various authors differ somewhat in the form of the principal band at about 900 cm^{-1} and/or in the peak $R(\omega)$ value ranging from 48 to 60%. Several sources, such as References 47, 58, and 168 demonstrate the plots of optical constants of vitreous germania vs. frequency. The $n(\omega)$ and $\kappa(\omega)$ numerical values are presented only in References 79 and 219 for the 600 to 1100 and 400 to 1100 cm^{-1} ranges, respectively.

As was mentioned by many authors, the IR reflection spectrum of vitreous germania is rather like that of vitreous silica shifted to lower frequencies by approximately 200 cm^{-1}. Most sources (see, for example, References 216 and 220) relate the 800 to 1000 cm^{-1} range of the spectrum to the asymmetric stretching vibrations of the GeOGe bridge, the 400 to 600 cm^{-1} range of the spectrum to the symmetric stretching vibrations of the GeOGe bridge, and the range of the spectrum lying lower than 400 cm^{-1} to the OGeO bending vibrations; this is in accordance with the data of Reference 162 for the hexagonal quartz-like modification of GeO_2. The pattern of $n(\omega)$ and $\kappa(\omega)$ spectra of vitreous germania in the range of the principal GeOGe asymmetric stretching vibration is shown, using the data of Reference 79, in Figure 11A. As is seen from Figure 11A, the range of the GeOGe asymmetric stretching vibrations comprises, in addition to the principal $\kappa(\omega)$ peak near 900 cm^{-1}, a weak shoulder at the high-frequency wing of the peak and another shoulder at the low-frequency wing of the latter. Thus, the pattern of the $\kappa(\omega)$ spectrum of vitreous germania is indeed similar to that observed for vitreous silica (see Section II.A.1). The origin of the two shoulders will be discussed below when considering band parameter values.

As is seen from Figures 11B and C, the spectra of optical constants of vitreous germania presented in References 79 and 219 differ rather perceptibly. The $n(\omega)$ and $\kappa(\omega)$ spectra of Reference 219 exhibit lowered peak values and increased minimum values of optical constants as compared to the spectra of Reference 79. Moreover, the $\kappa(\omega)$ spectrum of Reference 219 differs from that of Reference 79 in a shift in the location of the principal peak by 10 to 15 cm^{-1} to lower frequencies and in the presence of a weak shoulder near 800 cm^{-1}. These differences may be tentatively explained by the poorer quality of sample surface used in Reference 219. The reaction of the surface layer with atmospheric moisture might cause breaking of the GeOGe bridges, thus resulting in the IR spectrum resembling the spectrum of low-alkali binary glasses. As is seen from data presented below, the peak $n(\omega)$ and $\kappa(\omega)$ values and also the location of the $\kappa(\omega)$ peak reported in Reference 219 are close to those obtained for $2.0Na_2O \cdot 98.0GeO_2$ glass. Therefore, the data of Reference 79 probably better represent the actual spectra of optical constants of vitreous germania. The $R(\omega)$, $n(\omega)$, and $\kappa(\omega)$ numerical values according to Reference 79 are presented in Table A7 of the Appendix.

There is not much about the IR spectroscopic studies of germanate glasses in the literature. Among such studies, those described in References 79, 152, 158, 189, and 216 are devoted to Me_2O-GeO_2 glasses. Furthermore, there are data on the IR spectra of some other binary glasses, such as PbO-GeO_2,[179,188,189] SiO_2-GeO_2,[220] Ag_2O-GeO_2, and Tl_2O-GeO_2,[221] and also of some ternary glasses (for example, borogermanate[209,222] and aluminogermanate[221]). Those studies that involve the calculation of spectra of optical constants of binary germanate glasses are confined to References 79, 152, 188, and 189. The numerical $n(\omega)$ and $\kappa(\omega)$ values for some sodium germanate glasses borrowed from Reference 79 are cited in Table A8 of the Appendix. The examples of $n(\omega)$ and $\kappa(\omega)$ spectra of two sodium germanate glasses according to Reference 79 are presented in Chapter 3, Figure 1E, and Chapter 4, Figure 12.

As follows from available data, a general trend characterizing variations in the IR spectra of germanate glasses with a successive increase in Me_2O content up to 18 to 20 mol% consists mostly of a substantial decrease in the frequency of the principal band in the 700 to 900 cm^{-1} range. This shift is attributed by Murthy and Kirby[216] to the specific features of germanate glass structures. As is known, the lattices of some germanates and, admittedly, the networks of germanate glasses comprise the $[GeO_6]^{2-}$ octahedra that, like the $[BO_4]^-$ tetrahedra in borates, are formed at the expense of oxygen introduced by ionic oxides.[j] However, no specific bands that might be traceable to the vibrations of $[GeO_6]^{2-}$ octahedra were found in the spectra of germanate glasses, which are usually connected to rather low concentrations of $[GeO_6]^{2-}$ octahedra in the networks of the glasses and to large

[j] For crystalline germanates, the maximum possible fraction of octahedra is reached for compounds such as $2Na_2O \cdot 9GeO_2$ (see, for example, Reference 223) and $BaO \cdot 4GeO_2$, thus consisting about 1/5 or 1/4 of the total amount of oxygen-germanium polyhedra. This limit, probably, also holds for sodium germanate glasses; as a result, the $GeO_2(O^-)_2$ metagermanate-type tetrahedra should mostly present in the networks of Na_2O-rich glasses instead of the $[GeO_6]^{2-}$ octahedra.[25]

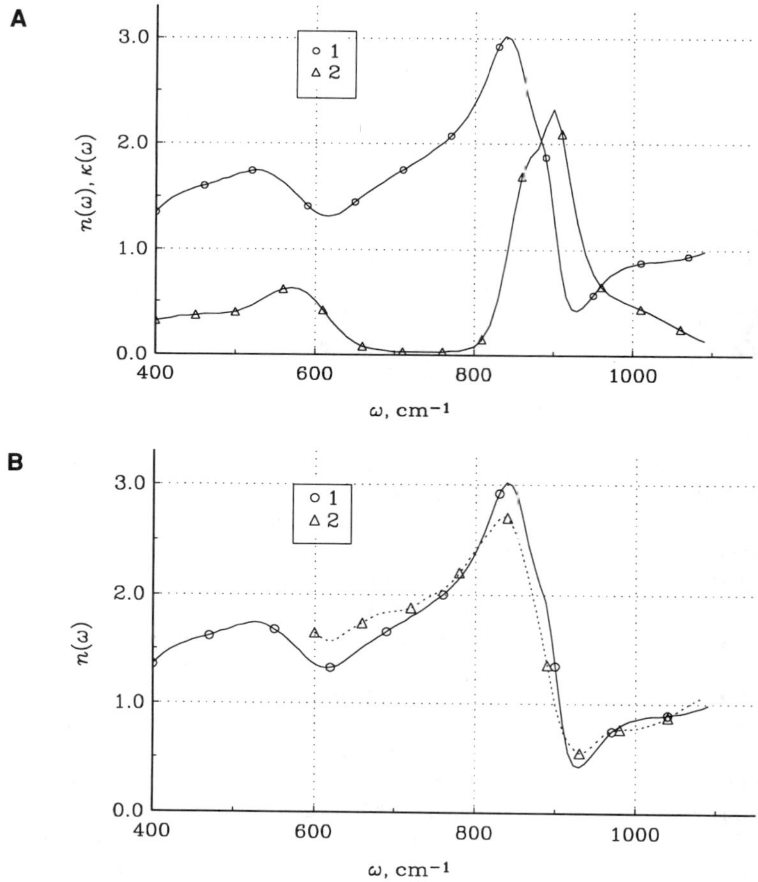

Figure 11 Spectra of optical constants of vitreous germania. (A) 1 is $n(\omega)$ and 2 is $\kappa(\omega)$ spectra according to the data of Reference 79. (B) Comparison of the $n(\omega)$ spectra of vitreous germania according to the data of (1) Reference 79 and (2) Reference 219. (C) Comparison of the $\kappa(\omega)$ spectra of vitreous germania according to the data of (1) Reference 79 and (2) Reference 219.

bandwidths characteristic of these vibrations. According to Roessler and Albers,[224] the IR bands of the rutile-like GeO_2 phase with the octahedral Ge coordination are not only much more broad but are also shifted considerably to lower frequencies as compared to those of hexagonal GeO_2. In view of this, Murthy and Kirby[216] interpreted the above successive low-frequency shift of the principal 700 to 900 cm^{-1} band with an increase in Me_2O content of alkali germanate glasses in terms of influence of the $[GeO_6]^{2-}$ octahedra. Similarly, Chryssikos et al.[209] assume that stretching vibrations involving the Ge-O pairs of both the GeO_4 tetrahedron and the $[GeO_6]^{2-}$ octahedron can result in the variety of overlapping bands located in the same 710 to 810 cm^{-1} range.

In the spectra of germanate glasses with Me_2O content no less than 25 mol%, the principal 700 to 900 cm^{-1} band discussed above splits into two components, thus making these spectra similar in character to those of silicate glasses. This

C

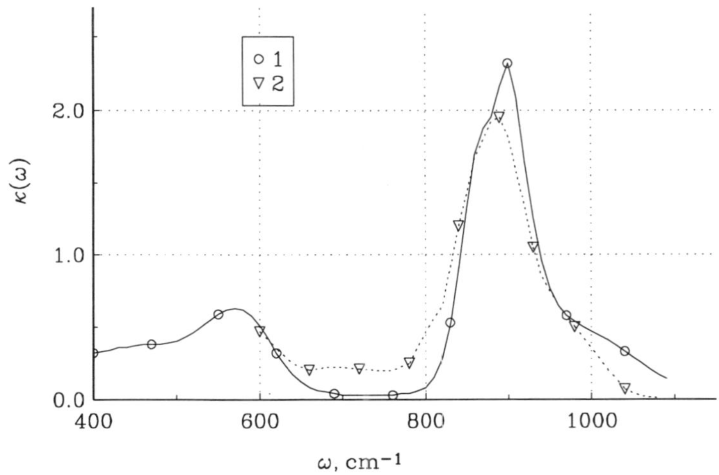

Figure 11 (continued)

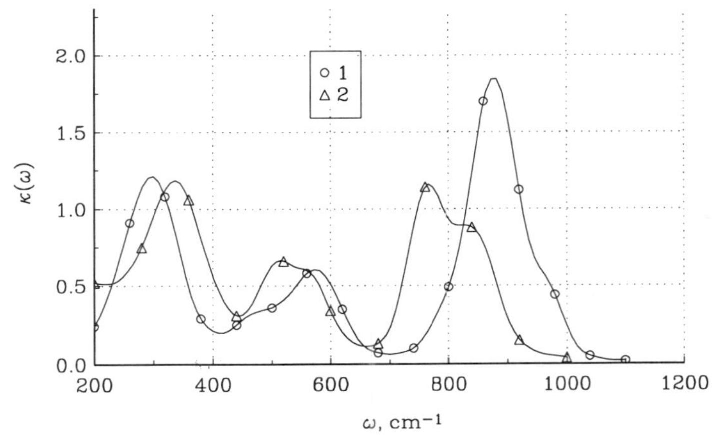

Figure 12 Comparison of the spectra of two sodium germanate glasses with the maximum difference in Na_2O content according to the data of Reference 79. (1) $2.0Na_2O \cdot 98.0SiO_2$ glass (mol% by analysis) and (2) $32.5Na_2O \cdot 67.5SiO_2$ glass (mol% by synthesis).

tendency is illustrated clearly by comparison between the $\kappa(\omega)$ spectra of two sodium germanate glasses with Na_2O content equal to 2 and 32 mol% shown in Figure 12. Figure 12 also shows that the spectra of alkali germanate glasses comprise two not so intense bands in the 400 to 600 cm^{-1} range and one much stronger band in the 200 to 400 cm^{-1} range. Both bands in the 400 to 600 cm^{-1} range should be attributed, according to the data of Reference 162 on the spectra of various germanates, to the symmetric stretching vibrations of the GeOGe bridge and that in the 200 to 400 cm^{-1} range to the OGeO and O^-GeO^- bending vibrations. On the contrary, Reference 209 relates the band at about 550 to 570 cm^{-1} to the OGeO bending vibration in the GeO_4 tetrahedron and that at about 410 to 460 cm^{-1} to the same vibration in the $[GeO_6]^{2-}$ octahedron. This is hardly the case because,

Table 12 IR Band Frequencies and Intensities for Some Sodium Germanate Glasses According to Reference 79

Na_2O content, mol% by analysis	Band parameters for the ranges of vibrations: $Ge(O^-)_2$ and highest GeOGe stretches, 800 to 1000 cm^{-1}		Main GeOGe asymmetric stretch, 700 to 900 cm^{-1}		GeOGe symmetric stretch, 400 to 600 cm^{-1}		OGeO bend, Na^+O^- external mode, and other ones below 400 cm^{-1}	
	ω_j, cm^{-1}	$\Delta\varepsilon_{0j}$	ω_j, cm^{-1}	$\Delta\varepsilon_{0j}$	ω_j, cm^{-1}	$\Delta\varepsilon_{0j}$	ω_j, cm^{-1}	$\Delta\varepsilon_{0j}$
0	989	0.09	894	1.37	569	0.20	**	
(GeO_2)			860	0.33	464	0.31	—	
2	959	0.03	859	0.44	570	0.25	285	1.60
			797	0.07	475	0.11	*	
5	939	0.05	842	0.46	559	0.33	284	1.76
			784	0.04	463	0.06	*	
15	889	0.10	797	0.45	565	0.29	292	2.42
					493	0.18	*	
18	883	0.07	785	0.43	565	0.29	**	
					482	0.25	—	
25	824	0.18	760	0.26	578	0.16	**	
					496	0.31	—	
30	829	0.16	755	0.26	581	0.16	**	
					494	0.32	—	
32.5	827	0.18	752	0.23	561	0.17	326	1.19
					501	0.14	213	1.60
							121	1.76

Note: Asterisks point out that lower frequency bands could not be revealed because the experimental spectrum studied did not extend to frequencies lower than (*) 160 or (**) 400 cm^{-1}.

as is seen from Figure 12, the 410 to 460 cm^{-1} band reaches maximum intensity in the spectrum of glass with the highest Na_2O content which should contain mostly the $GeO_2(O^-)_2$ metagermanate-type tetrahedra and only a negligible amount of the $[GeO_6]^{2-}$ octahedra (see Reference 25).

The numerical values of band frequencies and band intensities in the IR spectra of vitreous germania and sodium germanate glasses according to Reference 79 obtained with the dispersion analysis version of References 75 and 76 are presented in Table 12; it is these values that correspond to the spectra of optical constants shown in Chapter 3, Figure 2B and Figures 11A and 12 in this chapter.

For vitreous germania, Table 12 shows that the range of the GeOGe asymmetric stretching vibrations comprises, quite similar to the case of vitreous silica (see Section II.A.1), not only the principal 890-cm^{-1} oscillator but also the higher (at about 990 cm^{-1}) and lower frequency (at about 860 cm^{-1}) oscillators, these oscillators forming shoulders at the wings of the principal $\kappa(\omega)$ maximum (see Figure 11A). The frequencies and intensities of the principal and higher frequency oscillators agree reasonably with frequencies and intensities of two GeOGe asymmetric stretching vibrations in the spectrum of crystalline quartz-like GeO_2 found by Scott[225] from the Raman spectrum[k] and calculated theoretically in Reference 162, so the assignment of these oscillators to two components of GeOGe asymmetric stretching vibration is beyond question. In this connection, an attempt made in

[k] Band intensities were calculated in Reference 225 based on the LO-TO splitting.

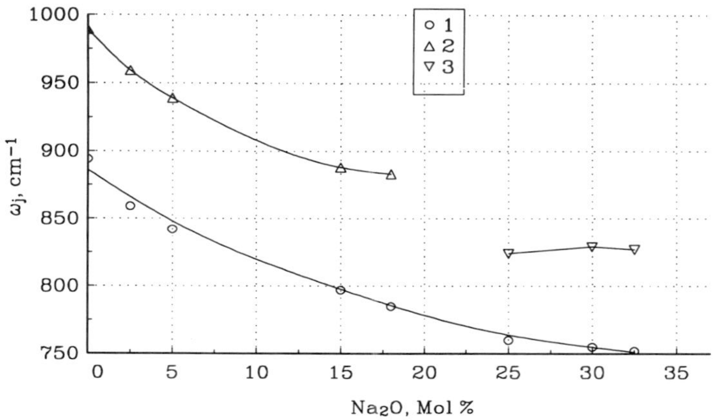

Figure 13 The frequencies of three features in the 750 to 1000 cm^{-1} range vs. Na_2O content of sodium germanate glasses according to the data of Reference 79. (1) the main component of the GeOGe asymmetric stretch, (2) the weak highest-frequency component of the GeOGe asymmetric stretch, (3) the $Ge(O^-)_2$ stretch.

References 47 and 58 to attribute the highest frequency feature in the Raman and, implicitly, in the $\varepsilon''(\omega)$ spectra of vitreous germania at about 990 cm^{-1} to the longitudinal optical excitation associated with the principal transverse 890 cm^{-1} excitation (see Section II.A.1) does not seem too convincing.

On the contrary, the oscillator at about 860 cm^{-1} forming the lower frequency shoulder has no counterpart in the spectrum of hexagonal GeO_2 crystal. The origin of this oscillator probably should be considered in terms of the same possibilities as those mentioned above for the case of the similar lower frequency satellite of the principal IR oscillator of vitreous silica in the range of the SiOSi asymmetric stretching vibrations (see Section II.A.1).

The frequencies of two oscillators in the range of the GeOGe symmetric stretching vibrations at about 570 and 460 cm^{-1} also resemble those found in References 162 and 225 for the spectrum of hexagonal GeO_2 crystal. Unfortunately, the initial $R(\omega)$ spectrum of vitreous germania used in Reference 79 did not extend to frequencies lower than 400 cm^{-1}. Therefore, the parameter values for the lower frequency oscillators could not be obtained.

For binary glasses, the data of Table 12 confirm that, in accordance with the data of previous studies, there are no additional bands in their spectra that might be traceable to the vibrations of the $[GeO_6]^{2-}$ octahedron. Probably, a substantial coupling between the GeOGe vibrations involving germanium atoms in different coordination states indeed takes place in the spectra of alkali germanate glasses. As is seen from Table 12, the frequencies of all oscillators in the 700 to 1000 cm^{-1} range decrease substantially with an increase in Na_2O content, thus complicating their assignment. The situation becomes much clearer when considering Figure 13, which illustrates the successive variations in the frequencies of oscillators related to this range. The frequencies of a strong oscillator at about 750 to 760 cm^{-1} in the spectra of glasses containing 25 to 32.5 mol% Na_2O fall in a curve characteristic of the principal GeOGe asymmetric stretch (Figure 13, curve 1), whereas the frequencies of the oscillator at about 820 to 830 cm^{-1} in the spectra of the same glasses clearly do not belong to this curve. Thus, a situation similar to that described above for silicate glasses is observed: it is the lower frequency band in the spectra of Na_2O-rich glasses that is related to the main component of the GeOGe asymmetric stretch. Hence, the higher frequency band should be attributed mostly to the stretching vibrations of the $Ge(O^-)_2$ terminal group (probably, with the highest frequency weak component of the GeOGe asymmetric stretch participating).

The data of Table 12 also show that the intensity of the oscillator related to the main component of the GeOGe stretching vibration decreases substantially when introducing a small Na_2O addition and then remains practically constant for all glasses with Na_2O content up to 18 mol%, whereas the oscillator at about 860 cm^{-1}, having no counterpart in the spectrum of hexagonal GeO_2 crystal, practically vanishes in the spectra of glasses with Na_2O content greater than 5 mol%.

According to Table 12, there are three oscillators in the IR spectra of sodium germanate glasses at frequencies lower than 400 cm^{-1}. The band related to the OGeO bending vibration is quite intense and its frequency lies in the 280 to 330 cm^{-1} range. Moreover, as follows from the data of Table 12 for the $32.5Na_2O \cdot 67.5GeO_2$ glass, there are also two more oscillators with frequencies of about 210 and 120 cm^{-1}. The oscillator at about 210 cm^{-1}, similarly to the oscillator near 180 to 200 cm^{-1} in the spectra of silicate glasses discussed above, is evidently related to the external Na^+O^- vibration.[197-199] The location of the lowest frequency oscillator at about 120 cm^{-1} is similar to that of the lowest frequency IR active mode of hexagonal GeO_2.[162,225] However, the intensity of this oscillator is by several orders of magnitude higher than that of the mode of the hexagonal GeO_2 crystal, which points out the principally different nature of the oscillator. Probably, this oscillator can comprise a contribution similar in nature to the oscillator of unclear origin observed in the spectra of alkali silicate glasses near 90 cm^{-1} (see Section II.A.1).

5. Tellurite Glasses

The IR spectra of tellurite glasses are less studied than those of glasses discussed above; most available data, such as those of References 189 and 226 through 228, are of a qualitative character. Moreover, there is also the lack of data on the crystal structures and the IR spectra of alkali tetratellurites and ditellurites, i.e., of the compounds corresponding to the composition ranges of glassforming regions in tellurite systems. Therefore, available data on band assignments for the IR spectra of tellurite glasses are mostly of a hypothetical nature. For example, band assignments proposed in Reference 226 for TeO_2 are in clear contradiction with the data of References 229 and 230 on the IR spectrum of α-TeO_2 (paratellurite) single crystal. The IR reflection spectra of alkali tellurite glasses were described initially in Reference 189 and then, in more detail, in Reference 227. These spectra, as is seen from an example presented in Chapter 3, Figure 1F, consist of two groups of reflection peaks in the 500 to 800 and 20 to 450 cm^{-1} ranges well separated from one another; the pattern of the spectra is only weakly dependent on the kind of Me^+ cation and on Me_2O content of glasses. Ovcharenko et al.[227] attributed the observed $R(\omega)$ maxima in the 600 to 800 cm^{-1} range to the stretching vibrations of the TeOTe bridge between adjacent trigonal dipyramidal TeO_4 groups and of the TeO^- terminal group; also, they connected an increase in reflectivity near a small high-frequency shoulder at about 750 to 780 cm^{-1} to an increase in the contribution of TeO^- groups when substituting Na_2O for TeO_2. Conversely, Heo et al.[228] interpreted their data on the variations in the IR transmission and Raman spectra of these

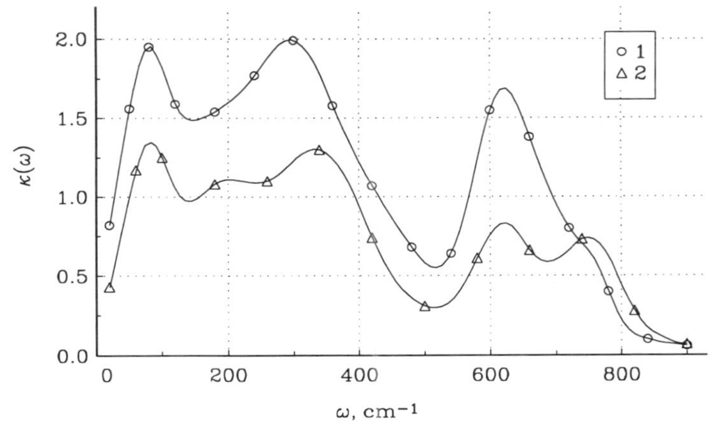

Figure 14 Comparison of the κ(ω)spectra of two sodium tellurite glasses with maximum difference in Na_2O content according to the data of Reference 79. (1) $12Na_2O \cdot 88TeO_2$ and (2) $35Na_2O \cdot 65TeO_2$ (mol% by synthesis).

glasses with an increase in alkali content in terms of substituting the trigonal TeO_3 pyramidal groups for the TeO_4 dipyramides.

Recently,[79,153] the data of Reference 227 on the $R(\omega)$ values were treated with the dispersion analysis version based on Model (50); the spectra of optical constants and band parameter values were obtained. The numerical values of optical constants for some sodium tellurite glasses according to Reference 79 are cited in Table A9 of the Appendix. The examples of the $\kappa(\omega)$ spectra of two sodium tellurite glasses according to Reference 79 are presented in Figure 14. These spectra demonstrate that an increase in the intensity of the shoulder at about 750 cm^{-1} is accompanied by a considerable decrease in the peak absorption index at about 600 to 620 cm^{-1} and by an appreciable decrease in the overall absorption throughout the 20 to 450 cm^{-1} interval of the spectrum.

Band frequencies and intensities in the spectra of sodium tellurite glasses obtained in Reference 79 with the dispersion analysis based on Model (50) are cited in Table 13. As is seen from Table 13, these spectra comprise seven oscillators with frequencies remaining practically intact in the entire composition range studied. There exist three bands in the 500 to 800 cm^{-1} range with inherent frequencies at about 600 to 610, 670 to 700, and 740 to 760 cm^{-1} (instead of the two assumed in Reference 227 for the experimental $R(\omega)$ spectra). An increase in Na_2O content leads to an increase in the intensity of both the 740 to 760 cm^{-1} and the 670 to 700 cm^{-1} oscillator, whereas the intensity of the 600 to 610 cm^{-1} oscillator decreases. The intensities of the oscillators in the 20 to 450 cm^{-1} range also tend, in general, to decrease with an increase in Na_2O content.

A comparison between the oscillator frequencies borrowed from Reference 79 for sodium tellurite glasses and those found in References 229 and 230 for α-TeO_2 (paratellurite) is illustrated by Table 14. As is seen from Table 14, almost every oscillator in the spectra of these glasses may be associated to a corresponding IR active mode or to the pair of closely located modes in the spectrum of α-TeO_2. First, an oscillator at about 600 to 610 cm^{-1} (the strongest one in the 500 to 800 cm^{-1} range) is undoubtedly an analog of the strong A_2 mode in the spectrum of α-TeO_2; the frequency of this mode (605 cm^{-1}) is practically the same as that of the oscillator

Table 13 IR Band Frequencies and Intensities for Some Sodium Tellurite Glasses According to References 153 and 79

Na_2O content, mol % by synthesis	Band parameters for the frequency ranges			
	500–800 cm^{-1}		20–500 cm^{-1}	
	ω_j, cm^{-1}	$\Delta\varepsilon_{0j}$	ω_j, cm^{-1}	$\Delta\varepsilon_{0j}$
12	746	0.06	405	0.86
	677	0.10	285	1.78
	602	0.72	191	8.30
			74	10.9
20	747	0.08	397	0.98
	680	0.08	283	1.15
	604	0.65	192	8.23
			78	6.33
24	762	0.07	396	0.32
	680	0.15	287	1.34
	603	0.45	185	8.40
			73	7.48
30	751	0.10	400	0.23
	694	0.19	311	1.21
	602	0.34	192	5.82
			72	6.12
35	747	0.12	374	0.20
	696	0.17	329	1.56
	607	0.26	201	3.77
			77	6.23

at about 600 to 610 cm^{-1} in glass spectra[229] (see Table 14). This mode, according to Reference 230, is inactive in the Raman spectrum of α-TeO_2 single crystal; similarly, the 600 to 610 cm^{-1} band was not registered in the Raman spectra of alkali tellurite glasses studied in Reference 231. The pattern of the experimental Raman spectra of alkali tellurite glasses presented by Sekiya, Mochida, Ohtsuka, and Tonokawa[232] also does not show the explicit signs of a band near 600 cm^{-1}, though they do assume the presence of the 615 cm^{-1} band (the weakest among those found in the 500 to 800 cm^{-1} range) based on the deconvolution of the spectra into a set of individual bands. For this deconvolution, purely Gaussian band contours were assumed, thus probably overestimating the narrowness of observed bands at the expense of neglecting the Lorentzian contribution to the total bandwidth (see Chapter 2, Section III.B.1). As a result, the number of components necessary for fitting spectra might exceed the number of actual bands. On the whole, available data speak against the presence of a strong Raman band near 600 to 620 cm^{-1} being a counterpart to the 600 to 610 cm^{-1} band in the IR spectra of alkali tellurite glasses. Rather, these data show that the selection rules predetermining the inactivity (or, at least, quite low intensity) of certain Raman bands in the IR spectrum and vice versa remain to be valid for glasses, too.

In addition to the 600 to 610 cm^{-1} oscillator, the 600 to 800 cm^{-1} range of the IR spectra of sodium tellurite glasses comprises, as was already mentioned, two

Table 14 Comparison Between Band Frequencies in the IR Spectra of Sodium Tellurite Glasses and α-TeO_2

Band parameters for sodium tellurite glasses according to References 153 and 79		Data for α-TeO_2 crystal			
		Band parameters			
		ω_j, of Ref. 229 (cm^{-1})	Data of Ref. 227		Mode type
ω_j, cm^{-1}	$\Delta\varepsilon_{0j}$		ω_j, cm^{-1}	$\Delta\varepsilon_{0j}$	
754 ± 8	0.07–0.1	770	769	0.20	E
688 ± 9	0.1–0.2	655	643	1.28	E
602 ± 5	0.3–0.8	605	575	3.86	A_2
394 ± 11	0.2–0.7	—	—	—	—
—		—	379	0.01	E
—		356	330	0.31	E
—		340	315	4.70	A_2
302 ± 16	1.2–1.7	310	297	4.09	E
—		258	259	0.77	A_2
—		220	210	2.16	E
192 ± 7	3.8–7.7	180	174	8.99	E
—		120	121	1.11	E
?		91	82	12.95	A_2
73 ± 3	6.1–10.9	—	—	—	?

more oscillators with frequencies of about 740 to 760 and 680 to 700 cm^{-1} that have counterparts in the α-TeO_2 spectrum at about 770 and 650 cm^{-1}, respectively. This makes the above assignment of the 740 to 760 cm^{-1} band to the vibrations of either TeO^- groups[227] or TeO_3 pyramids[228] more disputable; the contribution of mere redistribution of intensities between two components of the stretching vibrations analogous to those existing in the α-TeO_2 spectrum should probably also be taken into account.

According to Table 14, the only evident exception from such a correspondence between oscillators in the spectra of glasses and of α-TeO_2 crystal is the 390 to 400 cm^{-1} oscillator which has no strong counterparts in the spectrum of the latter. Also, the lowest frequency oscillator at about 70 cm^{-1} seems to be located too far from the nearest mode of the α-TeO_2 spectrum at about 90 cm^{-1}. On the other hand, its intensity is too great to relate this oscillator to an excitation similar in nature to the lowest frequency excitations observed in the IR spectra of silicate and germanate glasses (see Sections II.A.2 and II.A.4). Both of the mentioned factors can probably contribute to this spectrum feature.

The pattern of oscillator frequencies and intensities in the IR spectra of other alkali tellurite glasses also considered in References 79 and 153 is, in general, similar to that described above for sodium tellurite glasses.

6. On the Infra-Red Spectra of Other Oxide and Mixed Oxyhalide Glasses

There is a great number of qualitative IR spectroscopic studies of oxide glasses of other types, such as vitreous P_2O_5, phosphate, aluminate, arsenate, vanadate, tungstate, etc. glasses. Much of the early data, including those of References 233 through 236 for vitreous P_2O_5 and phosphate glasses, were cited by Wong and Angell;[160] some examples of more recent data are presented, mostly for phosphate glasses, in References 100, 197, 198, and 237 through 243; however, none of these studies extends to the calculation of optical constants. First, this is predetermined by the qualitative nature of chosen experimental spectroscopic methods. Studies such as References 197, 198, 233, and 237 through 243 only discuss experimental data on the IR transmission spectra (obtained with the KBr or polyethylene pellets, Nujol mulls, and, at best, thin blown or vapor-deposited films) that are considered to be sufficient for the purpose of qualitative structural interpretation. However, as was mentioned often above, spectra obtained with various pellets and mulls are wholly qualitative in nature and, hence, are inappropriate for further quantitative analysis. The use of thin film spectra for reconstructing the optical constants entails various difficulties and can result in great errors. A thorough comparison of the experimental IR spectra of various metaphosphate glasses obtained with different techniques, including the reflection spectroscopy, was conducted by Shih and Su;[234] however, the spectra of optical constants were also not reconstructed in this study.

Another reason for neglecting the studies of the spectra of optical constants is probably a current preference by many investigators for Raman spectra due to a substantial progress in the laser Raman spectroscopic technique. When one is interested only in obtaining some band locations for qualitative structural considerations, Raman spectra of glasses may indeed be more convenient, because neither special precautions against changes in the composition of the surface layer when sample polishing nor complicated mathematical data treatment are required. However, as was mentioned above (see Section II.A.5), certain strong IR bands indicative of the substantial features of glass structure can be inactive in the Raman spectra of glasses. Moreover, Raman spectroscopy of glasses, unlike the case for crystals, provides no possibility for calculating band intensities for vibrations active both in the IR and Raman spectra: the resolution of the Raman spectra of glasses is insufficient for estimating the TO-LO splitting because bandwidths are quite great, thus resulting in strong band overlapping.

Thus, studies of the Raman spectra of glasses, no matter how thorough, cannot serve as substitutes for the studies of their IR spectra extending to the reconstruction of the spectra of optical constants. This is especially true for phosphate glasses because of the extreme complexity of their vibrational spectra.

As is known from both the experimental data[161,244,245] and theoretical calculations,[162] the vibrational spectra of crystalline phosphates, due to the complexity of their lattices, usually comprise a large number of bands. Typically, each structural group in phosphates is characterized by no less than two bands related to the asymmetric and two bands related to the symmetric stretching vibrations, both of the types being active in the IR and Raman spectra. This probably is also the case for phosphate glasses, as follows from the pronounced complexity of their IR spectra (see, for example, References 234, 235, and 238 to 241). As a result, strong band overlapping takes place, thus hindering the estimation of not only accurate band frequency values but even the total number of bands in a spectrum under study.

There are also some other specific features inherent in the IR spectra of phosphates as a whole that are worth mentioning here. First, the PO_4 tetrahedron in pure P_2O_5, unlike the SiO_4 in SiO_2, is asymmetric because one of the oxygen atoms is connected to phosphorus by the "double" P=O bond (the latter being, in fact, not the double bond in the rigorous sense but the donor-acceptor one in nature). As a result, a particular highest frequency band related to the stretching vibration of the P=O terminal group, in addition to those traceable to the vibrations of the P-O-P bridge, is observed in the IR and Raman spectra of both crystalline and vitreous P_2O_5.[162,233,236,237,244,245] For the crystalline P_2O_5 modification constructed from the P_4O_{10} molecules, this band is located at about 1370 to 1390 cm^{-1}.[162,244,245] According to the comprehensive review of Corbridge,[245] the frequency of the P=O stretching vibration is greatly influenced by the chemical nature of substituent groups on the P atom and also by the formation of hydrogen bonding. Therefore, this frequency can vary in the IR spectra of phosphorus compounds from 1080 to 1450 cm^{-1}, most typical values falling into the 1200 to 1350 cm^{-1} range.

This band, however, is not observed in the spectra of metaphosphates,[161,162,244,245] because two nonbridging oxygen atoms of the same PO_4 tetrahedron differing in origin (such as the nonbridging oxygen related to the double P=O bond and the nonbridging oxygen appearing from P-O-P bridge breaking by Me_2O or MeO) are spectroscopically indistinguishable from each other. In other words, the formation of the unified $(PO_2)^-$ terminal group takes place; it is this group that is responsible for the high-frequency bands in the IR spectra of metaphosphates. Being triatomic, the $(PO_2)^-$ terminal group exhibits the asymmetric and symmetric stretching vibrations. According to References 162, 244, and 245, the asymmetric and symmetric stretching vibrations of the $(PO_2)^-$ terminal group fall into the 1140 to 1320 and 1000 to 1170 cm^{-1} ranges, respectively; usually, no less than two bands are observed in every range. The $(PO_3)^{2-}$ terminal group resulting from further depolymerization of the phosphate network is characterized by appreciably lower vibrational frequencies. According to References 162 and 244 the asymmetric and symmetric stretching vibrations of this group in the spectra of various phosphates correspond to the 1030 to 1170 and 940 to 1025 cm^{-1} ranges; hence, the respective IR bands only partially overlap with those of the $(PO_2)^-$. Similar to the case of the $(PO_2)^-$ group, no less than two bands are usually observed for the $(PO_3)^{2-}$ group in every range mentioned, thus contributing to the complexity of the observed pattern of the IR spectra of phosphates.

One more specific feature of the vibrational spectra of phosphates is the fact that the frequencies of the asymmetric and symmetric stretching vibrations of the P-O-P bridge, unlike the above case for silicate and germanate glasses, are appreciably lower than those of the $(PO_2)^-$ and $(PO_3)^{2-}$ terminal groups. According to Corbridge,[245] the overall values of these frequencies fall within the 850 to 1060 and 650 to 800 cm^{-1} ranges, respectively.

On the whole, available data on the IR spectra of crystalline phosphates clearly show that the band related to the P=O stretching vibration can probably be observed in the spectra of only those of phosphate glasses that are close to pure P_2O_5 in composition. Thus, it is the $(PO_2)^-$ terminal group, rather than the separate P=O and $P\text{-}O^-$ terminal groups, that participates in the formation of vibrational spectra of phosphate glasses with ionic oxide content close to metaphosphate composition; when approaching the pyrophosphate composition, the vibrations of the $(PO_3)^{2-}$ group should dominate in glass spectra.

In general, most studies devoted to the vibrational spectra of phosphate glasses, such as those described in References 238 through 241, assume band assignment in accordance with the data on crystalline phosphates cited above. The 1300 to 1460 cm^{-1} range comprising (if any) a separate band is commonly related to the stretching vibration of the P=O terminal group and the 600 to 1300 cm^{-1} range is related to the totality of the asymmetric and symmetric stretching vibrations of the rest of structural groups forming the phosphate polyanions, such as the P-O-P bridge, $(PO_2)^-$ and/or $(PO_3)^{2-}$ terminal groups (and also, for some Me_2O- or MeO-rich glasses, of the isolated $(PO_4)^{3-}$ anions). The estimates of frequency location of the band related to the stretching vibration of the P=O group in the IR spectrum of vitreous P_2O_5 and binary P_2O_5-SiO_2 films[233,236,237,246] vary substantially from study to study ranging from 1325 cm^{-1} (according to Reference 246) to 1230 cm^{-1} (according to Reference 237). In view of the above concerning the sensitivity of the P=O group frequency to the formation of hydrogen bonding, lowering this frequency for vitreous P_2O_5 as compared to that mentioned above for the P_2O_5 crystal and also variations in this frequency from study to study probably can be caused by the influence of water contaminations rather than by the difference between the three-dimensional polymeric structure of P_2O_5 glass and the molecular P_2O_5 crystal described in References 162, 244, and 245. Gray and Klein[242] point out an increase in water content of glasses much above 2 mol% when approaching to P_2O_5; Kolesova[233] reported the presence of several strong bands in the 1300 to 1600 cm^{-1} range related to the OH groups in the spectra of glasses, including that of P_2O_5.

There are also some approaches to the interpretation of phosphate glass spectra in literature differing from that specified above. Some early studies such as those of References 233 and 235 assumed that the P=O band is also observed in the spectra of most binary glasses, vanishing gradually when approaching pyrophosphate compositions. As follows from the above, this idea openly contradicts available data on the spectra of phosphates and phosphate glasses. First, the frequency range of the P=O band can overlap with that characteristic of the asymmetric stretching vibrations of the $(PO_2)^-$ group. Moreover, even the qualitative IR transmission spectra of metaphosphate glasses whose structure comprises only POP bridges and $(PO_2)^-$ terminal groups reveal as many as seven[235] or eight[234] bands in the 600 to 1300 cm^{-1} range, thus demonstrating that each of the two structural groups provides (as is also the case for crystalline phosphates) no less than one pair each of the asymmetric and symmetric stretching vibrations.[1] Probably, it was the higher frequency component of the asymmetric stretching vibrations of the $(PO_2)^-$ group that was taken in References 233 and 235 for the P=O stretching mode in the spectra of metaphosphate glasses.

Some other particular differences in the interpretation of the IR bands in the spectra of phosphate glasses can also be found in literature. This is natural now because reliable attribution of such a great number of overlapping IR bands based solely on the qualitative experimental spectra is practically impossible.

The 200 to 600 cm^{-1} range of the IR spectra of phosphate glasses usually comprising at least two strong bands is commonly related to the OPO and POP bending vibrations; moreover, as is the case for other glass types discussed above, the external Me^+O^- or $Me^{2+}O^-$ vibration is known to result in the low-frequency band,[197,198,243] its frequency being dependent on the cation mass.

The qualitative character of the available IR spectroscopic data for phosphate glasses holds true also for the vibrational spectroscopy of the rest of the oxide glasses mentioned in the beginning of this section and of mixed oxyhalide glasses, such as fluorophosphate. In spite of the substantial applied significance of fluorophosphate glasses for developing optical glasses with the abnormal partial

---

[1] For example, the total number of normal modes calculated for the metaphosphate chain in barium metaphosphate crystal[162] is as great as 20, the number of stretching modes being equal to 8. The total number of normal modes found by Shih and Su[234] in the molecular approximation for the zigzag model of metaphosphate chain with alternating positions of nonbridging oxygen pairs also equals 20.

dispersions (see Chapter 5, Section IV.B.3), available studies of their IR spectra do not involve the calculation of optical constants. Some information on the qualitative IR spectra of these glasses is given below, the list of sources by no means complete. The IR transmission spectra obtained with the KBr pellet technique were investigated by Murthy[247] for $NaPO_3$-NaF glasses; by Tarlakov, Khalilev, and Pogosyan[248] for $NaPO_3$-based glasses containing LiF, MgF_2, CaF_2, and AlF_3; and also by Pronkin, Tarlakov et al.[249] for $Ba(PO_3)_2$-AlF_3-RF_2 (R being Mg or Ca) glasses. Smirnova and Urusovskaya[250,251] presented the IR reflection spectra of $(Al(PO_3)_3$-MeF (Me being Li, Na, K) and $Al(PO_3)_3$-BaF_2 glasses. The qualitative IR transmission spectra of some $Al(PO_3)_3$-(MeF, MeF_2) glasses (Me being Li, Sr, Ba) were presented also by Gan Fuxi.[252] According to References 247 through 251 the introduction of halides can lead both to the depolymerization of metaphosphate chains and to the formation of the mixed $(PO_3F)^{2-}$ tetrahedra and also, for $Al(PO_3)_3$-based glasses, of $(AlF_6)^{3-}$ anions. Murthy[247] mentioned that the $(PO_3F)^{2-}$ tetrahedron manifests itself mostly in the additional 1010 cm^{-1} band because the rest of the vibrational frequencies of the tetrahedron fall into the ranges coinciding with those of various phosphate groups. This assignment of the additional 1010 cm^{-1} band is not, however, an unambiguous one. As follows from the above data of References 162 and 245, the frequency of 1010 cm^{-1} does not come out of the range characteristic of the asymmetric stretching vibrations of the POP bridge and, moreover, falls into the range of the asymmetric stretching vibrations of the $(PO_4)^{3-}$ anion.

Thus, in general, the IR spectra of fluorophosphate glasses are rather like the spectra of purely phosphate glasses. This, according to Almeida and Mackenzie,[253] also holds for chlorophosphate glasses. No bands other than those traceable to the vibrations of known phosphate groups were observed in the qualitative IR transmission spectra of $NaPO_3$-$ZnCl_2$ glasses, the influence of $ZnCl_2$ being confined to some shifts in band frequencies. According to a rather specific scheme of the structural changes in $NaPO_3$-$ZnCl_2$ glasses with glass composition assumed by Almeida and Mackenzie,[253] the network of glasses with a relatively low $ZnCl_2$ content admittedly comprises the $ZnCl_4$ tetrahedra built in the sodium metaphosphate chains, whereas an increase in $ZnCl_2$ content results in the formation of $(PO_2Cl)^{2-}$ terminal groups accompanied by the appearance of the Zn^{2+} modifying cations. Reference 253, like some of the early studies,[233,235] attributes the 1250 to 1280 cm^{-1} band to the P=O stretching mode (instead of relating it to the higher frequency component of the asymmetric stretching vibrations of the $(PO_2)^-$ group), thus being the subject matter for the criticism expressed above.

In view of all of the above, the vibrational spectroscopy of both purely phosphate and halogen-containing phosphate glasses is rather far from completion. Naturally, quantitative studies of the IR spectra of these glasses involving the calculation of optical constants and, especially, the estimation of the IR band parameter variations with glass composition might be quite helpful for clarifying band assignments. One should probably expect such studies for phosphate (and some other not too well-studied glasses) to be developed in the future.

B. NON-OXIDE GLASSES

1. Halide Glasses

In spite of a broad interest in the studies of various halide glasses (especially heavy metal fluoride) aroused by the prospects of using them as materials for the low-loss waveguides and optical fiber amplifiers, what was said in Section II.A.6 concerning the qualitative nature of the IR spectroscopic studies of phosphate and some other oxide glasses also holds to a substantial extent for halide glasses. There are several comprehensive reviews of the properties of halide glasses, including the IR spectra.[254-256] On the whole, however, the fundamental vibrational spectra of halide glasses have not received much attention until now, so the circle of investigators devoting themselves to studies in this field is rather narrow. As was mentioned by Almeida,[256] this trend is probably due to the fact that the efforts of investigators are mostly concentrated at the problem of obtaining ultralow optical loss in the transparency range of these glasses.

Most of the available sources on the fundamental IR spectra of these glasses present the qualitative transmission spectra registered with the KBr and/or polyethylene pellet technique (see, for example, reviews in References 254 through 256); only a few of the sources comprise the IR reflection spectra.[58,257-261] Those studies that extend to the calculation of optical constants of halide glasses are quite scarce.[58,257-259]

According to available data,[254-256] two major types of halide glasses, such as fluoroberyllate and heavy metal fluoride glasses, were the subject matter of most studies. The IR and Raman spectra of vitreous BeF_2 were studied rather widely. Beryllium forms the BeF_4 tetrahedra connected with each other by the bridging F atoms. Crystalline BeF_2 is the structural analog of SiO_2, thus exhibiting the same principal crystalline modifications, including those like quartz and cristobalite. A brief review of early data on the IR and Raman spectra of vitreous BeF_2 and alkali fluoroberyllate glasses is given by Wong and Angell[160] and by Baldwin, Almeida, and Mackenzie.[254] Four IR bands at about 910, 770, 420, and 248 cm^{-1} were reported for the qualitative absorption spectrum, the strongest, 770 cm^{-1}, being attributed to the asymmetric Be-F stretching vibration. More recently, Galeener, Leadbetter, and Stringfellow[58] studied both the IR reflection and Raman spectra of vitreous BeF_2 up to 1500 cm^{-1} and presented plots of its optical conductivity, $\rho(\omega) = \omega\varepsilon''(\omega)$, and of the energy loss function, $\mathrm{Im}[-\hat{\varepsilon}^{-1}(\omega)]$, vs. frequency calculated with the Kramers-Kronig transform. According to Galeener et al.,[58] the experimental IR $R(\omega)$ spectrum and also the corresponding $\omega\varepsilon''(\omega)$ and $\mathrm{Im}[-\hat{\varepsilon}^{-1}(\omega)]$ spectra of vitreous BeF_2 are qualitatively similar to those of vitreous SiO_2, but observed spectrum features are shifted appreciably to lower frequencies as compared to those for SiO_2. Such a shift is evidently the consequence of a lowered Be-F force constant due to a substantially higher degree of ionicity of this bond as compared to that of Si-O bond. The magnitude of this shift is much greater for the principal BeFBe asymmetric stretching mode (the frequency of the conductivity maximum being of about 750 cm^{-1} against about 1060 cm^{-1} for SiOSi) than for the band related, judging by analogy to the spectrum of vitreous SiO_2, to the FBeF bending vibration (of about 400 cm^{-1} against 450 cm^{-1} for OSiO). Moreover, a band similar to the weak band at about 800 cm^{-1} in the spectra of SiO_2 (i.e., the one that might be related to the symmetric stretching vibration of the BeFBe bridge) is practically not observed for BeF_2. In spite of this difference, the general pattern of the $\omega\varepsilon''(\omega)$ and $\mathrm{Im}[-\hat{\varepsilon}^{-1}(\omega)]$ spectra of vitreous BeF_2 is indeed rather similar to that for SiO_2. On the contrary, the model calculations conducted earlier by Bell and Dean[62,63,67] resulted in rather substantial differences between BeF_2 and SiO_2 in the pattern of the vibrational density of states, including the relative role of high- and low-frequency modes (probably due to a dissimilar mass ratio for the Be-F and Si-O pairs).

As is the case for vitreous SiO_2, the nature of the highest frequency band observed in the absorption spectrum of vitreous BeF_2 at about 910 cm^{-1} and forming a shoulder in the $R(\omega)$ spectrum deserves a special comment. Galeener et al.[58] tend, as was already mentioned above, to relate this band in the Raman and, implicitly, in the IR spectra of tetrahedral glasses (including BeF_2) to the longitudinal optical phonon corresponding to the principal transverse asymmetric stretching mode; criticism to such an approach was given in Sections II.A.1 and II.A.4. Following the analogy of vitreous SiO_2, this band should probably be assigned to the second component of the asymmetric stretching vibration of the BeFBe bridge.

According to Reference 254, an increase in the alkali fluoride content of BeF_2-based glasses results in lowering the frequency of the asymmetric stretching vibration of the BeFBe bridge, the magnitude of this shift increasing with the atomic number of alkali metal. No data on the optical constants of binary and more complex fluoroberyllate glasses are available.

The notion of heavy metal fluoride glasses covers a broad variety of glass types, the principal are being based on ZrF_4 as the glassforming component. In addition to fluorozirconate glasses, such glass types as fluorohafnate, ThF_4-based, fluorogallate, fluoroindate, and ZnF_2-CdF_2-based ones were developed. Though aluminium is not a heavy metal, fluoroaluminate glasses are usually also considered together with such glasses because AlF_3 is a popular component of complex heavy metal fluoride glass compositions. The ionic components of heavy metal fluoride glasses are usually alkaline earth and/or rare earth metal fluorides.

The fundamental vibrational frequencies of heavy metal fluoride glasses are naturally much lower than those of not only oxide glasses but also of fluoroberyllate glasses;[255-261] neither of the frequencies exceeds 650 cm^{-1}, thus providing a substantially extended transparency range in the infrared (up to wavelengths of about 7 to 9 μ depending on the particular glass composition and the choice of limiting transmittance value). The IR spectra of actual heavy metal fluoride glasses are rather simple in structure and comprise, as a rule, only two bands in the 350 to 570 and 200 to 280 cm^{-1} ranges traceable to the internal asymmetric and symmetric vibrations of various fragments of anionic structure, respectively. In addition, the lowest frequency band related to the external cage-like vibrations of Me^+ and Me^{2+} cations with respect to their fluorine surrounding can be observed, thus being similar in nature to the external Me^+O^- or $Me^{2+}O^-$ vibrations discussed above for oxide glasses (see Section II.A). This band is not always observable in the experimental IR spectra of heavy metal fluoride glasses (see, for example, References 257, 258, and 261), which is probably due to overlapping with the band related to the internal symmetric stretching vibrations. Due to the substantially smaller atomic mass of aluminium, usage of AlF_3 as a component of heavy metal fluoride glasses leads to the appearance of one more band at about 620 to 640 cm^{-1}, which results in an unfavorable high-frequency shift of the transparency range limit.[258] For purely fluoroaluminate glasses, the IR spectrum in the 200 to 700 cm^{-1} range, naturally, again is reduced to two bands,[252] their frequency locations being about 620 to 640 and 360 to 400 cm^{-1}.

In view of the simple structure of the IR and Raman spectra of heavy metal fluoride glasses, it is tempting for investigators to interpret these spectra in terms of Sen and Thorpe's central force model (see Section II.A.1). Starting with this model, Almeida[255,256] developed a rather specific approach that presupposes distinguishing between the stretching vibrations of the bridging F atoms accompanied by displacements of Zr or similar network-forming atoms (in Almeida's notation, AS(C) or SS(C) for the asymmetric or symmetric vibrations, respectively) and the same vibrations involving no motion of Zr or other network-forming atoms (AS and SS, respectively). According to this approach, the IR spectra of fluorozirconate and similar heavy metal fluoride glasses in the 200 to 700 cm^{-1} range are assumed to be dominated by the AS(C) and SS(C) vibrations resulting in the bands at about 490 to 540 and 230 to 270 cm^{-1}, respectively, whereas the Raman spectra of these glasses are admittedly dominated by the AS and SS vibrations. A different pattern is assumed for fluoroindate glasses. Their IR spectra are interpreted by Almeida[261] almost completely in terms of the vibrations of InF_6^{3-} octahedra that are formed by F^- ions being predominantly nonbridging in nature (F_{nb}). Almeida attributed two IR bands at about 460 to 490 and 220 to 260 cm^{-1} to the asymmetric stretching vibrations of InF_6^{3-} octahedra (AS(F_{nb}), in Almeida's notation) and the bending $F_{nb}InF_{nb}$ vibration, respectively. No distinct band due to the vibrations involving Ba^{2+} cations was found in the IR spectra of fluoroindate glasses;[261] however, the reflectivity at frequencies lower than 160 cm^{-1} is high enough to allow the presence of one more unresolved band. The reconstruction of spectra of optical constants would probably reveal this band unambiguously, as is the case for similar low-frequency $Ba^{2+}O^-$ bands in the spectra of borate glasses (see Section II.A).

As mentioned above, there are a few sources presenting the spectra of optical constants of heavy metal fluoride glasses and/or band parameter values. Bendow, Drexhage et al.[257] presented the plots of $\varepsilon''(\omega)$ and $\mathrm{Im}[-\hat{\varepsilon}^{-1}(\omega)]$ vs. frequency for HBT glass of a batch composition of $33.75BaF_2 \cdot 57.5HfF_4 \cdot 8.75ThF_4$ (mol%), the $\varepsilon''(\omega)$ values being reconstructed with the classical dispersion analysis from the $R(\omega)$ spectrum. The corresponding computer fit also presented in Reference 257 is not quite satisfactory, which is natural in view of that said in Chapter 2, Section III.A concerning the inadequacy of the classical $\hat{\varepsilon}(\omega)$ model used in the standard dispersion analysis with respect to the spectra of glasses. This inadequacy, as was specified above in Chapter 3, Section IV.B, does not wholly rule out the possibility of using the classical dispersion analysis for estimating the spectra of optical constants of glasses but predetermines increased errors in the optical constants, the magnitudes of the errors depending on the quality of best fit obtained. The spectrum under study in Reference 257 was found to comprise two oscillators with frequencies of 200 and 425 cm^{-1}. The specific feature of the $\varepsilon''(\omega)$ spectrum of HBT glass distinguishing it from those of all glasses discussed above is the ratio of the low- and high-frequency $\varepsilon''(\omega)$ peak magnitudes being as high as of about 16.0/6.0, i.e., practically opposite to the ratio characteristic of vitreous BeF_2.[58] As for the data of Reference 261, no distinct band due to the vibrations involving Ba^{2+} cations was found. According to Reference 136, the frequency of the principal oscillator in the IR spectrum of crystalline BaF_2 is 184 cm^{-1}, so a spectrum feature related to the $Ba^{2+}F^-$ vibration can overlap greatly with the $\varepsilon''(\omega)$ peak at about 200 cm^{-1}, thus contributing to the extremely high intensity of this peak. Furthermore, Bendow, Banerjee et al.[258] also conducted the classical dispersion analysis of the $R(\omega)$ spectra for ternary fluorozirconate glass and three more complex ThF_4-based glasses, two of them containing AlF_3. No information on the quality of computer fits, unlike the preceding study of these authors,[257] is reported in Reference 258. Data presented in Reference 258 comprise both the plots of $\alpha(\omega)$ spectra in the broad frequency range and the numerical values of band parameters. Using the band parameter and ε_∞ values given in Reference 258, it is easy to reconstruct the

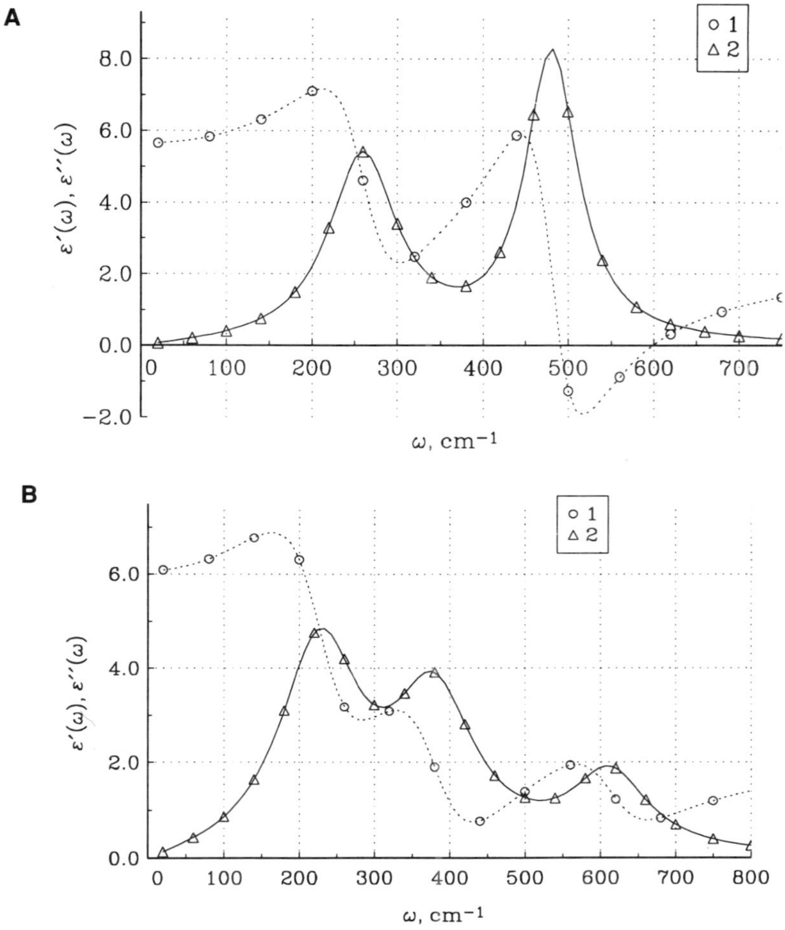

Figure 15 Model spectra of optical constants of some heavy metal fluoride glasses reconstructed with the dispersion analysis program based on Model (50) from band parameters and ε_∞ values presented in Reference 258. (A) ZrBL glass of $33BaF_2 \cdot 62ZrF_4 \cdot 5LaF_3$ batch composition, mol%; 1 is $\varepsilon'(\omega)$ and 2 is $\varepsilon''(\omega)$. (B) BYAT glass of $20BaF_2 \cdot 29YF_3 \cdot 29AlF_3 \cdot 22ThF_4$ batch composition, mol%; 1 is $\varepsilon'(\omega)$ and 2 is $\varepsilon''(\omega)$.

spectrum of any optical constants with some available dispersion analysis program.[m] The author has made such a reconstruction to illustrate the general pattern of the $\varepsilon'(\omega)$ and $\varepsilon''(\omega)$ spectra of heavy metal fluoride glasses (see Figure 15). The spectra shown in Figure 15 are calculated from the data of Reference 258 with the author's dispersion analysis program;[75,76] in this calculation, all the σ_j values entering into Model (50) are taken to be zero to reduce this model to the classical case (see

[m] It should be kept in mind, however, that the accuracy of the optical constants reconstructed in such a way remains unspecified. There are two reasons for this: (1) the insufficient adequacy of the classical model as such, and (2) the lack of information concerning the quality of computer fits obtained in Reference 258 (see also Chapter 3, Section IV.B).

Chapter 3, Section IV.B). As follows from Figure 15, the lower frequency $\varepsilon''(\omega)$ peak in the spectrum of ZrBL glass, unlike that in the spectrum of HBT glass reported in Reference 257, is less intense than the higher frequency peak related to the asymmetric stretching vibrations of the Zr-F-Zr bridge. In the spectrum of BYAT glass, however, the intensity of the $\varepsilon''(\omega)$ peak at about 380 cm^{-1} related to the asymmetric stretching vibrations of the Hf-F-Hf bridge is appreciably lowered at the expense of the emergence of the highest frequency peak at 615 cm^{-1} having no counterparts in the spectrum of ZrBL glass and related, as was already mentioned above, to the AlF_6^{3+} vibrations.

Among studies providing data on the spectra of optical constants of heavy metal fluoride glasses, that of Almeida[259] should be also mentioned. Almeida[259] presented the plots of TO and LO responses (i.e., of $\omega\varepsilon''(\omega)$ and $\mathrm{Im}[-\hat{\varepsilon}^{-1}(\omega)]$, respectively) vs. frequency for ZB and ZBLA glasses of $33.3BaF_2 \cdot 66.7ZrF_4$ and $35BaF_2 \cdot 3LaF_3 \cdot 5AlF_3 \cdot 57ZrF_4$ batch compositions (mol%); the $R(\omega)$ spectra registered at different incidence angles were used for these calculations.

There are some qualitative IR spectroscopic data on other halide glasses such as vitreous $ZnCl_2$,[160] $ZnCl_2$- and $ZnBr_2$-based glasses,[255] and also some others; no information on the spectra of optical constants of these glasses is available.

2. Chalcogenide Glasses

As is the case for other glass types, studies into the spectroscopy of the fundamental IR excitations in chalcogenide glasses, including the early well-known study of Hilton, Jones, and Brau,[262] are in most cases qualitative in character. However, the amount of data on the spectra of optical constants and/or on the IR band parameters for chalcogenide glasses[73,74,101,263-272] is not as small as that for halide glasses. As mentioned above, the range of the fundamental IR excitations of chalcogenide glasses is the lowest among inorganic glasses as a whole;[n] this, in accordance with Formula (61), is due to the greater atomic masses of chalcogens as compared to those of oxygen or fluorine and also to lower force constants in chalcogenides as compared to those in oxides. The high-frequency limit for the fundamental bands in the IR spectra of typical chalcogenide glasses containing no elements of first through third rows of the Periodic Table except sulfur varies from about 350 to 400 cm^{-1} (values observed for sulfide glasses) to about 190 to 200 cm^{-1} (those characteristic of telluride glasses).[262-271] Such a low frequency of the fundamental IR bands is the main reason responsible for the substantial extension of the transparency range of chalcogenide glasses as compared to the rest of the glass types. Another reason contributing to this effect is the rather low intensities of both the fundamental IR bands and, hence, of multiphonon absorption due to the predominantly covalent nature of chemical bonds. As a result, the transparency range of chalcogenide glasses covers the medium IR range, the long-wavelength transparency limit varying from 10.5 to 11 μ for most sulfide glasses to 18 to 20 μ for telluride. This predetermined the concentration of spectroscopists' efforts mainly on the problem of impurity absorption in the IR transparency range of chalcogenide glasses (see, for example, Reference 273). Studies extended to the calculations of optical constants or absorption coefficient values and of the IR band parameters of chalcogenide glasses are devoted mostly to vitreous selenium;[264,271] stoichiometric

[n] The only marked exception is phosphorus sulfide glasses, for which the fundamental frequencies in the 500 to 700 cm^{-1} range (that are close to those observed for heavy metal fluoride glasses) were reported in Reference 262.

vitreous chalcogenides of arsenic and germanium, such as As_2S_3,[101,263-268,272] As_2Se_3,[74,101,263,269,272] GeS_2,[264-266,270] and $GeSe_2$;[265] and also to pseudobinary As_2S_3-As_2Se_3,[101,264] As_2S_3-GeS_2,[265] GeS_2-$GeSe_2$,[265] As_2Se_3-$GeSe_2$,[264] As_2S_3-Sb_2S_3,[268] and As_2Se_3-Sb_2Se_3[269] glasses. There are also data on the optical constants spectra of binary (non-stoichiometric) As-S[267] and Ge-S[270] glasses. All sources mentioned present only plots of spectra of optical constants in the range of the fundamental excitations; no numerical values of optical constants for chalcogenide glasses are available. The $n(\omega)$ and $\kappa(\omega)$ spectra are shown in References 263 and 267; $\alpha(\omega)$ spectra in References 101, 264, and 271; $\varepsilon'(\omega)$ and/or $\varepsilon''(\omega)$ spectra in References 74 and 263 through 270; and $\mathrm{Im}[-\hat{\varepsilon}^{-1}(\omega)]$ spectra in References 264, 268 and 269. Moreover, the spectra of optical conductivity $\rho(\omega) = \omega\varepsilon''(\omega)$ in the IR range are reported for vitreous As_2Se_3 in References 73 and 74 and for $Tl_2Se \cdot As_2Te_3$ in Reference 73. In addition, plots of absorption coefficient $\alpha(\omega)$ spectra at frequencies lower than the fundamental ones for vitreous As_2S_3, As_2Se_3, and non-stoichiometric As-S and As-Se glasses are presented in References 181, 274, and 275. The numerical values of band parameters of chalcogenide glasses are given in References 73, 74, 268, 269, and 272.

Naturally, the magnitude of the constant dipole moment of the Se-Se bonds in elemental selenium is zero, so it is the induced dipole moment that is responsible for the occurrence of the fundamental IR spectrum of both vitreous and crystalline selenium. Available data on the IR spectrum of vitreous selenium[73,264,271,276] reveal at least four bands that are grouped into two pairs in the 220 to 260 and 90 to 150 cm^{-1} ranges well separated from one another, these ranges commonly being related to the stretching and bending vibrations, respectively. In addition, other IR bands depending on the thermal prehistory of a sample were reported.[277] As is known, the structure of vitreous selenium comprises both the long polymeric chains, such as those characteristic of the trigonal crystalline modification and the Se_8 rings similar to those characteristic of the monoclinic crystalline modification. According to Reference 278, bands at 95, 120, and 254 cm^{-1} are due to the IR active vibrations of a Se_8 ring, and a band at 135 cm^{-1} is due to the IR active vibration of a long polymeric chain, whereas a shoulder in the IR spectrum at about 225 cm^{-1} corresponds to the predominantly Raman active mode of this chain. The theoretical calculations of Lucovsky and Martin[279] conducted in the molecular approach using the triatomic structural fragment of the polymeric chain presuppose the occurrence of three bands, the calculated frequencies being of 251, 239, and 123 cm^{-1}.

The IR spectra of vitreous As_2S_3 and As_2Se_3 are quite similar to one another in character, the only difference being the shift of As_2Se_3 spectrum to lower frequencies. The most pronounced features of these spectra are a strong band at about 300 to 330 or 215 to 240 cm^{-1} and a weak low-frequency band at about 150 to 165 or 95 to 105 cm^{-1} for As_2S_3 or As_2Se_3, respectively.[74,101,262-269] The spectra of mixed As_2S_3-As_2Se_3 glasses comprise both the 300 to 330 and 215 to 240 cm^{-1} bands,[101,264,272] thus being a good example of two-mode behavior similar to that characteristic of solid solutions, such as $GaAs_xP_{1-x}$. Also, additional weak bands (including a high-frequency satellite of the principal $R(\omega)$ peak seen clearly in the experimental spectra obtained in References 265 and 269) were found, the number of these bands and their assignment varying from study to study (see below).

In general, the 300 to 450 or 200 to 300 cm^{-1} frequency ranges in the spectra of As_2S_3 or As_2Se_3, respectively, are commonly related to the stretching vibrations, whereas the bending ones are assumed to be located at frequencies lower than 250 cm^{-1} for As_2S_3 or lower than 200 cm^{-1} for As_2Se_3. Lucovsky and Martin[279] conducted the calculation of the fundamental vibrational frequencies inherent in vitreous As_2S_3 and As_2Se_3 in the molecular approximation, interpreting the spectra under study as the totality of the independent normal modes related to both the AsS_3 or $AsSe_3$ "molecule" and the As-S-As or As-Se-As bridge. The expediency of taking into account the "molecular" and "bridging" modes independently of each other was assumed to follow from weak "intermolecular" coupling. Therefore, seven normal modes, four of them belonging to the AsS_3 or $AsSe_3$ molecule and three to the As-S-As or As-Se-As bridge, were calculated and compared to the experimental data on both the IR and Raman spectra of these glasses. Based on these calculations, the major IR active modes at 310 or 216 cm^{-1} in the IR spectra of vitreous As_2S_3 or As_2Se_3 were related to the asymmetric stretching vibrations and the Raman modes at 344 or 227 cm^{-1} to the symmetric stretching vibrations of the pyramidal AsS_3 or $AsSe_3$ molecules, respectively. As is seen, this approach differs strikingly from the above approach applied most commonly now to the spectra of vitreous oxides (see Section II) in which a spectrum of an oxide is described with only a single kind of simple structural groups. As mentioned in Section II.A.1, the IR spectrum of vitreous SiO_2 is described quite satisfactorily in terms of vibrations related to the Si-O-Si bridge, with no vibrations belonging to the SiO_4 tetrahedron being involved (see Section II.A.1). On the contrary, when treating the IR spectrum of a vitreous oxide in terms of vibrations related to some polyhedron, such as BO_3 for vitreous B_2O_3 (see Section II.A.3), no vibrations of the B-O-B bridges are considered. In spite of this contradiction, the model of Lucovsky and Martin[279] was often cited until now when considering the spectra of chalcogenide glasses.

There are several sources discussing band frequencies and other band parameters for these glasses. In an early study,[272] Felty, Lucovsky, and Myers fitted the $R(\omega)$ spectra of As_2S_3-As_2Se_3 glasses, including pure As_2S_3 and As_2Se_3, in the 150 to 550 cm^{-1} range with the classical dispersion analysis; the quality of the fits at the frequencies of $R(\omega)$ maxima and minima was not too satisfactory. Only a single oscillator corresponding to the principal band was used for treating the spectra of each individual chalcogenide and two oscillators were used for the case of mixed glasses. The oscillator frequencies were found to be about 300 and 217 cm^{-1} for As_2S_3 and As_2Se_3, respectively. For mixed glasses, the oscillator frequencies varied with glass composition only slightly in the narrow 300 to 325 and 217 to 230 cm^{-1} limits for the As-S and As-Se modes, respectively. Because a too narrow frequency range was studied, there is no sense in discussing the data of Reference 272 in terms of band intensity. As was mentioned above, Taylor, Bishop, and Mitchell[74] calculated the spectra of $\varepsilon'(\omega)$ and optical conductivity $\rho(\omega) = \omega\varepsilon''(\omega)$ of vitreous As_2Se_3 from the $R(\omega)$ spectrum with the Kramers-Kronig transform and interpreted band shapes in the conductivity spectrum in terms of the Gaussian distribution of the Lorentzian oscillators (see Chapter 2, Section II.B). The results obtained for vitreous As_2Se_3, along with those for some other glasses (see below), are cited also in Reference 73. The Gaussian bandwidths were estimated crudely from band shapes in the vicinity of band maxima, and damping coefficients were estimated from the shapes of band wings. Four bands were used in Reference 74 for describing the IR spectrum of vitreous As_2Se_3; band parameter values are given in Table 15. More recently, Onari et al. studied in detail the IR spectra of As_2S_3-As_2Se_3,[101] As_2S_3-Sb_2S_3,[268] and As_2Se_3-Sb_2Se_3[269] glasses; the experimental $R(\omega)$ spectra were treated with both the Kramers-Kronig transform and the classical dispersion analysis. In addition to the $\varepsilon'(\omega)$ and $\varepsilon''(\omega)$ spectra[268,269] or

Table 15 Comparison Between Band Parameter Values for Vitreous As_2Se_3 Obtained in References 74 and 269

Type of vibration	Kind of parameter	Band parameters according to Ref. 269	Band parameters according to Ref. 74
Bond-stretching	ω_j, cm^{-1}	268	—
	$\Delta\varepsilon_{0j}$	0.085	
	γ_j, cm^{-1}	24	
	ω_j, cm^{-1}	248	—
	$\Delta\varepsilon_{0j}$	0.085	
	γ_j, cm^{-1}	24	
	ω_j, cm^{-1}	218	237
	$\Delta\varepsilon_{0j}$	1.35	0.79
	γ_j, cm^{-1}	24	≤5
	σ_j, cm^{-1}	—	23
Bond-bending	ω_j, cm^{-1}	—	156
	$\Delta\varepsilon_{0j}$		0.03
	γ_j, cm^{-1}		?
	σ_j, cm^{-1}		12
	ω_j, cm^{-1}	100	102
	$\Delta\varepsilon_{0j}$	0.23	0.57
	γ_j, cm^{-1}	40	?
	σ_j, cm^{-1}	—	19
	ω_j, cm^{-1}	—	(45)
	$\Delta\varepsilon_{0j}$		(0.7)
	γ_j, cm^{-1}		?
	σ_j, cm^{-1}		(28)
High-frequency dielectric constant, ε_∞, used in calculations		7.9	7.3
"Static" dielectric constant, ε_0, calculated via the contributions of the IR fundamentals and ε_∞		9.6	9.4

$\alpha(\omega)$ spectra[101] obtained with the Kramers-Kronig transform, the computer fits to the experimental $R(\omega)$ spectra obtained with the classical dispersion analysis are shown in all three sources mentioned. Band parameter values are published for As_2S_3-Sb_3[268] and As_2Se_3-Sb_2Se_3[269] glasses, the IR spectra of vitreous As_2S_3 and As_2Se_3 being represented by four oscillators. For As_2Se_3, band parameter magnitudes and corresponding ε_∞ values according to Reference 269 are given in Table 15.

It is of interest to compare band parameter values for vitreous As_2Se_3 given in References 74 and 269. For the bond-stretching frequency range, Kato et al.[269] used both the principal 218 cm^{-1} oscillator and two weak oscillators with frequencies of 248 and 268 cm^{-1} (see Table 15), the last one corresponding to a weak but distinct feature in the $R(\omega)$ spectrum. On the contrary, Taylor, Bishop, and Mitchell,[74] based on their own $R(\omega)$ spectrum that does not contain a weak high-frequency feature, assessed the frequency of the principal oscillator to be as large as of 237 cm^{-1} and used no additional weak bond-stretching oscillators at all. For the bond-bending frequency range, Kato et al.[269] took into account only a single oscillator at 100 cm^{-1}, whereas Taylor et al.[74] used as much as three oscillators. As a result, the estimates of band intensities obtained in these two studies differ crucially from each other. This is illustrated in Table 15, which presents band intensities in terms of oscillator strengths $\Delta\varepsilon_0$ being the contributions of the bands to the "static" dielectric constant ε_0. As is seen from Table 15, the $\Delta\varepsilon_0$ value for the principal band in the 215 to 240 cm^{-1} range reported by Kato et al.[269] (1.35) is about 1.7 times higher than that presented by Taylor, Bishop, and Mitchell[74] (0.79). Conversely, the total intensity of bond-bending oscillators according to Reference 74 (about 1.3) is almost six times higher than that of a single oscillator used by Kato et al.[269] (0.23). Thus, the data of References 74 and 269 on individual band intensities are practically incompatible with one another.

A well-known proof of the IR band intensities is the comparison between the static dielectric constant ε_0 calculated according to the sum rule given by Formula (18) and the $\varepsilon'(\omega)$ value recorded experimentally in the UHF range (see also Chapter 3, Section IV.B). The ε_0 values estimated from the data of References 74 and 269 are also presented in Table 15. The experimental $\varepsilon'(\omega)$ magnitude of about 9.7 in the UHF range is given both in Reference 74 (for the linear frequency of 36 GHz) and in Reference 280 (within the 10 to 150 GHz range). When assessing the results of comparison, the initial ε_∞ magnitudes used in the calculation of the ε_0 values should be taken into account. Typically, the ε_∞ magnitudes should correspond approximately to the square of the refractive index in the transparency range because the refractive index dispersion of the electronic origin from the transparency range to the UHF can be neglected as compared to the error of $\varepsilon'(\omega)$ measurement in the UHF. Therefore, in accordance with the data of Reference 281 on the refractive indices of vitreous As_2Se_3 in the 8 to 12 μ range, the ε_∞ value should be of about 7.7 to 7.9. However, the ε_∞ value fit in Reference 74 is as low as 7.3. Therefore, although the ε_0 value of 9.4 reported in Reference 74 is in a reasonable agreement with the experimental $\varepsilon'(\omega)$ value of 9.7, the total contribution of the IR fundamentals given in Reference 74 (2.1) is probably overestimated. Conversely, the data of Reference 269 are obtained with the reasonable ε_∞ value of 7.9 and result in the ε_0 of 9.6, which is in quite good agreement with the above experimental $\varepsilon'(\omega)$ value. Therefore the total contribution of the IR fundamentals obtained in Reference 269 (1.7) should be closer to the actual one than that found in Reference 74. However, the 80 to 500 cm^{-1} range studied in Reference 269 does not cover the frequency of the oscillator at about 45 cm^{-1} found in Reference 74. If this band is not an artifact, the resultant balance between band intensities can differ from that reported in Reference 269.

On the whole, differences between data on band parameters presented in References 74 and 269 are too great to be explained only by differences in the assumed number of oscillators and band shapes; probably, they are caused mainly by differences in the $R(\omega)$ spectra. The small-scale plots of the initial $R(\omega)$ spectra given in References 74 and 269 cannot be used for their quantitative comparison. These plots show, however, that a peak $R(\omega)$ value for the principal band is only of about 0.3 according to Reference 74 and of about 0.4 according to Reference 269, the difference being substantial. For a more detailed comparison, computer fits to the $R(\omega)$ spectra of vitreous As_2Se_3 corresponding to band parameters presented in References 74 and 269 and also respective $\varepsilon''(\omega)$ spectra were

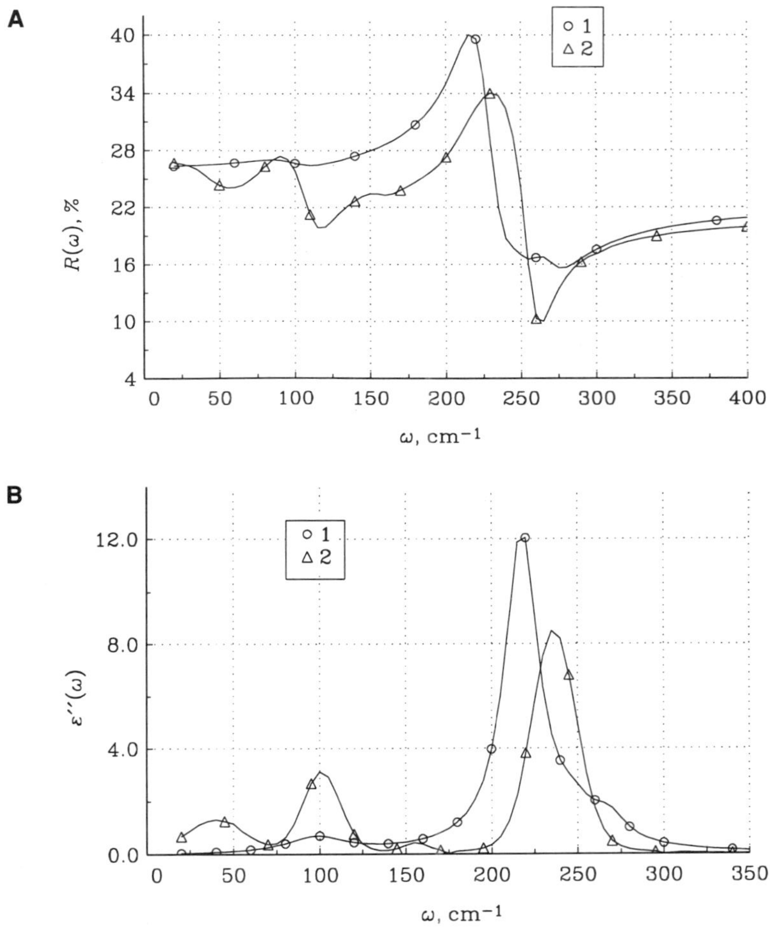

Figure 16 Comparison of the model IR spectra of vitreous As_2Se_3 reconstructed with the dispersion analysis program based on Model (50) from band parameters and ε_∞ values available in literature. (A) The model reflection spectra (i.e., best fits) according to the data of (1) Reference 269 and (2) Reference 74. (B) The model $\varepsilon''(\omega)$ spectra according to the data of (1) Reference 269 and (2) Reference 74.

reconstructed with our dispersion analysis program.[75,76] (See Figure 16A, B).[o] As is seen from Figure 16A, computer fits of Kato et al.[269] and of Taylor et al.[74] differ from one another not only in the peak $R(\omega)$ values for the principal band but also in peak location and in peak shape, not mentioning differences in the low-frequency range. In general, such differences are too great to be explained only by the errors of fits and therefore should be considered to represent approximately the

[o] When computing the fit of Reference 74, available data on the Gaussian bandwidths, Γ_j, (that are connected to the standard deviations σ_j entering into Model (50) by the relation $\Gamma_j = 2\sqrt{2\ln 2}\sigma_j$) were taken into account, and Lorentzian widths γ_j estimated crudely in Reference 74 to be no more than 5 cm^{-1} were taken to be of 5 cm^{-1}. For the fit of Reference 269, the σ_j values were taken to be zero.

actual differences in the initial experimental $R(\omega)$ spectra used in these two studies. Hence it follows that Taylor et al.[74] greatly overestimated the width of the principal band because of the excessively smoothed character of their initial $R(\omega)$ spectrum; as a result, their conclusion mentioned in Chapter 2, Section II.B that band shapes in the IR spectrum of vitreous As_2Se_3 are predominantly Gaussian does not seem very convincing (see also Chapter 2, Section III.B). On the contrary, the principal $\varepsilon''(\omega)$ peak corresponding to the data of Kato et al.[269] (see Figure 16B) can be too sharp, because the classical dispersion analysis does not take into account the possibility of Gaussian broadening at all. Thus, band intensities and band shapes in the IR spectrum of vitreous As_2Se_3 need further verification. Probably, the actual $\varepsilon''(\omega)$ peak shape lies somewhere in between those shown in Figure 16B.

Along with the spectra of As_2Se_3, those of some other chalcogenide glasses, such as elemental selenium (see above), $Tl_2Se \cdot As_2Se_3$, $Tl_2Se \cdot As_2Se_3$, and $As_{55}Te_{45}$, were analyzed by Mitchell, Bishop, and Taylor[73] in the same way as described above for As_2Se_3. For thallium-containing glasses, three bands with frequencies in the 246 to 255, 189 to 213, and 80 to 86 cm^{-1} ranges, respectively, were assumed. Only two bands at about 195 and 70 cm^{-1} were found for $As_{55}Te_{45}$ glass. Following the above approach of Reference 74 for As_2Se_3, Mitchell, Bishop, and Taylor[73] concluded that the IR band shapes in the spectra of chalcogenide glasses with the chain-like structures (such as selenium, $Tl_2Se \cdot As_2Se_3$, and $Tl_2Se \cdot As_2Se_3$) and layer-type structures (such as As_2Se_3 and $As_{55}Te_{45}$) are predominantly Gaussian. In view of criticism expressed above to such a conclusion for As_2Se_3, the extent of the Gaussian band broadening in the IR spectra of chalcogenide glasses should be considered to be unspecified, although the existence of such broadening is beyond doubt (see Chapter 2, Section III.B).

For vitreous GeS_2 and $GeSe_2$, the plots of $\varepsilon''(\omega)$ spectra obtained with the Kramers-Kronig transform are given in References 264 to 266 and 270; in Reference 270 the plot of the $\mathrm{Im}[-\hat{\varepsilon}^{-1}(\omega)]$ spectrum of vitreous GeS_2 is also shown. As is the case for As_2S_3 and As_2Se_3, the IR spectra of vitreous GeS_2 and $GeSe_2$ are quite similar in general character. A strong and sharp peak related to the principal bond-stretching Ge-S or Ge-Se mode is observed in the $\varepsilon''(\omega)$ spectra of GeS_2 and $GeSe_2$ glasses, this principal peak being accompanied by two weak features or shoulders located at its lower- and higher frequency wings. The principal peak frequencies are of about 365 to 370 and 250 cm^{-1} for GeS_2 and $GeSe_2$, respectively; according to Reference 270, a weak lower frequency feature is observed in the GeS_2 spectrum at about 330 cm^{-1} and a weak higher frequency shoulder at about 410 cm^{-1}. In addition to the principal peak and its satellites, a weak and broad low-frequency maximum related to the bond-bending vibrations is observed at about 160 and 100 cm^{-1} for GeS_2 and $GeSe_2$, respectively. For treating the spectrum of vitreous GeS_2, Lucovsky et al.[270] used not only the Kramers-Kronig transform but also the classical dispersion analysis. Data presented in Reference 270 comprise the plot of the computer fit to the experimental $R(\omega)$ spectrum and a plot illustrating a comparison between the $\varepsilon''(\omega)$ spectra obtained with the Kramers-Kronig transform and with the classical dispersion analysis. The quality of this fit is not too good which is only natural in view of the shortcomings of the classical dispersion analysis when applying to glasses (see Chapter 3, Section IV.A). As a result, the $\varepsilon''(\omega)$ values obtained with the dispersion analysis at frequencies corresponding either to the range in between two bands (200 to 300 cm^{-1}) or to the range of the higher frequency wing of the 370 cm^{-1} peak are appreciably greater than those found with the Kramers-Kronig transform.

The plots of $\varepsilon''(\omega)$ spectra for some pseudobinary As_2S_3-GeS_2, GeS_2-$GeSe_2$, and As_2Se_3-$GeSe_2$ glasses are shown in References 264 and 265. For all these spectra, the two-mode behavior was found, caused either by mass difference between S and Se atoms (as is the case for As_2S_3-As_2Se_3 and GeS_2-$GeSe_2$ glasses) or by the difference in the polyhedron symmetry (as is the case for As_2S_3-GeS_2 and As_2Se_3-$GeSe_2$ glasses). The plots of $\varepsilon''(\omega)$ spectra for non-stoichiometric $Ge_{1-x}S_x$ glasses, x varying in the 0.55 to 0.90 range, are presented in Reference 270. According to Reference 270, the structure of the $\varepsilon''(\omega)$ spectrum characteristic of GeS_2 remains intact with an increase in x up to 0.9; on the contrary, an increase in Ge content above that in pure GeS_2 results in appreciable smearing of the spectrum features.

Available IR spectroscopic data for other chalcogenide glasses are confined to raw experimental spectra (such as those cited in References 252 and 262); in addition, crude estimates of band frequencies can be available in some cases. For example, Hilton, Jones, and Brau[262] also presented, along with the qualitative IR transmission spectra, the plots of the reflection spectra for Ge_2S_3, $Ge_{15}P_{15}Te_{70}$, and $Si_{10}As_{10}Te_{80}$ glasses and estimated the inherent frequencies of the most intense fundamental IR bands in these spectra from data on the $R(\omega)$ peak magnitudes and peak frequencies according to Moss.[3] For various chalcohalide glasses, only qualitative IR transmission spectra are available (see, for example, References 283 through 285).

III. OPTICAL CONSTANTS OF GLASSES IN THE RANGE OF ELECTRONIC EXCITATIONS

The specific features of optical spectra of the fundamental electronic excitations distinguishing them from the IR spectra are reviewed in Chapter 2, Section II.C. When presenting the experimental spectroscopic quantities and optical constants in this range, they are usually expressed, to avoid numbers that are too large, as functions of wavelength λ or energy E (in eV) rather than of frequency or wavenumber; this form of data presentation is chosen for this section, too. Energies and wavelengths can be easily transformed into one another using a simple relation $E(\text{eV}) \simeq 1.239/\lambda$, λ being expressed in microns. In general, data on the optical constants of inorganic glasses in the range of the fundamental electronic excitations are scarce. For oxide and halide glasses whose spectra of the fundamental electronic excitations are located in the VUV range, serious difficulties in the methodology of spectroscopic experiments in this range are among the major reasons responsible for this situation.

A. VITREOUS SILICA

Among the VUV spectra of inorganic glasses, that of vitreous silica has been studied most thoroughly. The structure of the experimental $R(E)$ spectrum of vitreous silica and a relationship between it and the spectrum of α-quartz were the topic of study for many years;[86-88,286] a brief review of the basic notions used when interpreting this spectrum is given in Chapter 2, Section II.C. New information on the $R(E)$ spectrum of silica keeps appearing; for example, Bosio et al.[287] revealed recently a shoulder at about 24 eV and related it to the edge formed by plasma excitations with energy of 25 eV. The plots of the spectra of optical constants of vitreous silica in the energy range up to 21 eV obtained from the near-normal incidence reflection spectra with the Kramers-Kronig transform are presented in References 86, 88, and 92; in addition, experimental points corresponding to the

$\kappa(E)$ values of low expansion commercial silica glass sample at several wavelengths are plotted by Osantovsky.[288] Vitreous silica also remains until now the only glass for which literature sources report the numerical values of optical constants in the VUV range.[159,289] Lamy[289] reconstructed the $n(\lambda)$ and $\kappa(\lambda)$ spectra of vitreous silica from the near-normal incidence reflection spectrum with the Kramers-Kronig transform in the 1000 to 1500 Å range; two sharp maxima characteristics of this range were found to be located at about 1215 and 1080 Å (which is similar to estimates given in References 86 and 92). Recently,[159] the authors calculated the $n(\lambda)$ and $\kappa(\lambda)$ spectra of vitreous silica with the dispersion analysis version based on Model (50) in the broad wavelength range from 580 to 1500 Å covering all four distinct maxima in its VUV spectrum. An expediency of using this dispersion analysis version for treating the VUV spectra of glasses is substantiated in Chapter 3, Section IV.B. Good best fit to the experimental $R(\lambda)$ spectrum was obtained ($Q = 1.2 \cdot 10^{-3}$). The $n(\lambda)$ and $\kappa(\lambda)$ values according to References 159 and 289 are compared in Table A10 of the Appendix.

The pattern of the $n(\lambda)$ and $\kappa(\lambda)$ spectra of vitreous silica in the VUV range is illustrated by Figure 17 using the data of References 159 and 289. As was mentioned in Chapter 2, Section II.C, this pattern is rather similar to that characteristic of α-quartz, minor differences consisting in the negligibly small long-wavelength shift of peak positions (about 24 Å for the first peak) and in insignificant peak broadening. There is no single acknowledged view on the nature of four $\kappa(\lambda)$ peaks in the VUV spectrum of silica (see Chapter 2, Section II.C). Approaches published thus far differ in many aspects, including the number of peaks attributed to the exciton-like transitions; the extreme case is the approach of Laughlin[81] that assigns to such transitions all four peaks at 1215, 1080, 900, and 710 Å shown in Figure 17A.

As is seen from Figure 17A, the $\kappa(\lambda)$ spectra of References 159 and 289 differ perceptibly from one another, the systematic differences being about 0.1 to 0.2. As was shown in Reference 159, this is caused only by appreciable differences in the initial $R(\lambda)$ spectra and not by computational errors inherent in these methods. When conducting the dispersion analysis of the $R(\lambda)$ spectrum mimicking the experimental spectrum of Reference 289 (this spectrum being reconstructed using the $n(\lambda)$ and $\kappa(\lambda)$ values of Reference 289), the resultant spectra of optical constants coincide excellently with those of Lamy[289] (see Figure 17B). This fact, in addition to a quite good best fit obtained with the dispersion analysis, demonstrates clearly that computational errors inherent in the data of Reference 159 are indeed negligible. Therefore, the choice between the data of References 159 and 289 on the $n(\lambda)$ and $\kappa(\lambda)$ values of vitreous silica in the VUV range should be determined wholly by the choice between the initial experimental $R(\lambda)$ spectra used in these studies. In this respect, the data of Reference 159 are probably preferable because the array of $R(\lambda)$ values used in Reference 159 is registered taking into account the effect of a polished surface[174] and, moreover, covers a substantially broader wavelength range. Also, as is shown in Reference 159, the corresponding oscillator parameters allow for calculating the $n(\lambda)$ values in the visible range accurate to within $1 \cdot 10^{-3}$, whereas an estimate reported by Lamy[289] is of about $7 \cdot 10^{-3}$. Therefore, the data of Reference 159 cited in the Appendix (Table A10) can be recommended as best representing the optical constants of vitreous silica in the VUV range. In accordance with the above in Chapter 2, Section II and Chapter 3, Section IV.B, the oscillator parameter values for the VUV spectrum of vitreous silica also given in Reference 159 should be considered only to be useful effective quantities rather than the actual parameters of individual electronic excitations.

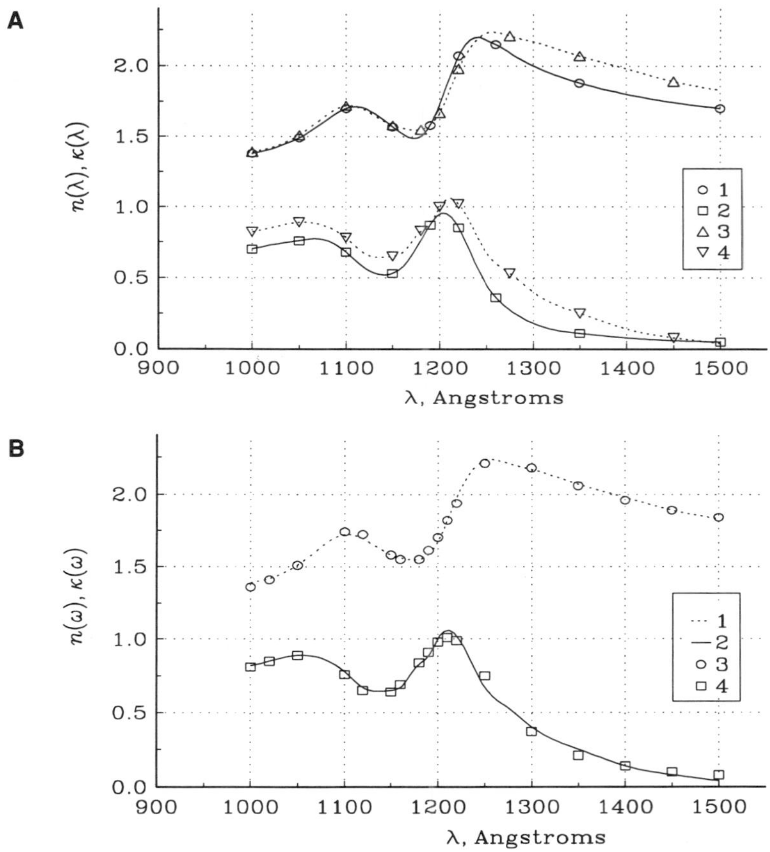

Figure 17 Spectra of optical constants of vitreous silica in the VUV range according to available literature data. (A) Comparison of the spectra plotted with the data of (1, 2) Reference 159 and (3, 4) Reference 289; (1, 3) *n(λ)*, and (2, 4) *κ(λ)* spectra. (B) Comparison between the spectra of optical constants (1, 2) calculated by Lamy[289] with the Kramers-Kronig transform and (3, 4) obtained with the dispersion analysis from the similar *R(ω)* spectrum[159] (this spectrum was reconstructed from *n(λ)* and *κ(λ)* numerical values of Reference 289); (1,3) *n*(λ), and (2,4) κ(λ) spectra.

B. VITREOUS GERMANIA

The plots of the $\varepsilon''(E)$ spectra of both vitreous germania and crystalline (hexagonal) GeO_2 in the 0.5 to 25 eV range reconstructed from the near-normal incidence reflection spectra with the Kramers-Kronig transform were published by Pajasova.[94] In spite of structural similarity between vitreous GeO_2 and SiO_2, the VUV spectrum of vitreous germania differs rather substantially from that of silica. This spectrum comprises two strong, well-separated $\varepsilon''(E)$ peaks at about 6.6 and 10.7 eV, peak $\varepsilon''(E)$ values being of about 2.0, and at least two additional poorly resolved features on the high-energy side of the 10.7 eV peak. In general, the VUV spectra of vitreous germania and of crystalline (hexagonal) GeO_2 are rather similar to one another in the location of two principal maxima and their relative intensities. However, differences between these spectra are not so negligible as is the case for vitreous silica and α-quartz. The spectrum of hexagonal GeO_2 contains additional broad features on both sides of a quite sharp 6.6-eV peak (these features being smeared completely in the spectrum of vitreous germania) that make the spectrum as a whole look more similar to the spectra of typical semiconductors than to the spectrum of α-quartz. No specific interpretation of these spectra in terms of notions mentioned in Chapter 2, Section II.C was proposed in Reference 94.

C. OTHER GLASSES

For binary and more complex inorganic glasses, available data on the spectra of the fundamental electronic excitations are confined mostly (for oxide and chalcogenide glasses) or entirely (for halide glasses) to the raw reflection spectra.[92,93,95-99,290,291] As mentioned in Chapter 2, Section II.C, some of the available experimental VUV spectra of binary and more complex oxide glasses are probably not too reliable due to a difference in chemical composition between a surface layer formed when polishing a sample and a bulk material. This concerns, first, silicate glasses. The VUV reflection spectra of silicate glasses registered from sample surfaces polished in the usual way, i.e., using water,[92,95,99,291] comprise strong peaks similar in location to those at 10.2 and/or 11.5 eV in the spectrum of vitreous silica. As was shown in References 97 and 98, the VUV spectra of sodium silicate glasses of Na_2O content greater than 28 mol% registered with all precautions against the formation of above surface layers have little in common with the spectrum of silica. For example, the VUV spectrum of $33.4Na_2O \cdot 66.6SiO_2$ glass (by analysis) registered from a surface broken in vacuum shows no signs of silica-like peaks at 10.2 and 11.5 eV; only a strong peak at 8.5 to 8.7 eV and weaker ones at about 9.3 and 10.8 eV are observed in the 8.0 to 12.5 eV range.[97] The disappearance of the 10.2 eV peak and emergence of the 8.5 and 9.3 eV peaks in the spectra of alkali silicate glasses were also observed earlier by Sigel;[95] however, unlike the data of Reference 97, Sigel also mentioned the presence of the silica-like 11.5 eV peak in the spectrum of $Na_2O \cdot 2SiO_2$ glass. The extent of influence of sample preparation on the VUV spectra of glasses other than silicate has not been specially studied until now, although polishing the samples of borate glasses to promote spectrum feature broadening was mentioned in Reference 290. The possibility of two-photon excitations occurring is also among the factors capable of influencing the VUV spectra of glasses, because such excitations can increase the intensity of high-energy features above the actual one. The role of such excitations can be especially substantial when using powerful synchrotron VUV radiation. As a result, the VUV spectrum of a glass registered with different VUV radiation sources can differ greatly in the relative intensity of its higher- and lower-energy sections although exhibiting a similar general pattern (compare, for example, the reflection spectrum of vitreous silica from References 86, 286, and 287 with that from References 99 and 291).

The question of whether an experimental VUV spectrum of interest is reliable did not stimulate the development of quantitative spectrum treatment. Studies extending calculations to optical constants of binary and more complex oxide glasses in the VUV range are quite scarce and present only plots of the spectra.

Ellis et al. published the $\varepsilon''(E)$ spectra in the 2.5 to 25 eV range calculated from the $R(E)$ spectra with the Kramers-Kronig transform for binary $30Na_2O \cdot 70SiO_2$ glass and commercial soda-lime-silica-based float glass[92] and also for three potassium-containing lead silicate glasses,[93] the SiO_2 content of these glasses ranging from 64.8 to 8.8 mol%. The spectra of both sodium silicate glass and float glass presented in Reference 92 look like the smeared spectrum of vitreous silica. In accordance with the above, such spectra do not correspond to the actual spectra of glasses under study and should be ascribed to the hydrated surface layers. The spectra of lead silicate glasses shown in Reference 93 are characterized by a more complex structure; weak silica-like peaks near 10 and 11.5 eV are seen only in the spectrum of most SiO_2-rich glass.

An example of a study presenting numerical values of optical constants in the VUV range for oxide glass-based materials is that of Osantovsky[288] for Cer-Vit (lithium aluminosilicate glass ceramics). Osantovsky[288] calculated $n(E)$ and $\kappa(E)$ values of this commercial material for nine wavelengths ranging from 256 to 1048 Å from the reflectivity of polished samples measured at several different incidence angles.

The range of the fundamental electronic excitations for chalcogenide glasses is shifted, as compared to that for oxide glasses, to much lower energies: the location of the fundamental absorption edge of chalcogenide glasses ranges from about 2.5 (for sulfide glasses) to about 0.6 eV (for glasses containing elements of the sixth row of the Periodic Table), and the upper frequency limit for the optical excitations of the valence electrons is usually no more than 10 to 12 eV. Therefore, the optical spectra of the electronic excitations of chalcogenide glasses are much more accessible to study than those of oxide and halide glasses. As a result, there are available data on the optical constants in this range for chalcogenide glasses (see, for example, References 294 through 296).

The $R(E)$ spectra of vitreous As_2S_3 and As_2Se_3 in the 1 to 12 eV range reported in References 292 and 293 are similar in structure and comprise a broad strong maximum, a shoulder at its higher energy wing, and a weak highest energy maximum. The reflection spectra of the corresponding crystalline materials are similar in general pattern but reveal an additional structure. There are some differences between the data of Belle, Kolomietsz, and Pavlov[292] and those of Zallen et al.[293] in the peak $R(E)$ values and in the relative intensities of the above shoulder. Using the reflection spectra of Reference 293, Drews et al.[294] calculated the $\alpha(E)$ and $\varepsilon''(E)$ spectra of vitreous and crystalline As_2S_3 and As_2Se_3. The $\varepsilon''(E)$ features in the spectra of glasses were found to be appreciably smeared and shifted by about 1 eV to higher energies as compared to those in the spectra of corresponding crystals. Velicky, Zavetova, and Pajasova[295] registered the $R(E)$ spectrum of vitreous As_2Te_3, calculated its $\varepsilon''(E)$ spectrum with the Kramers-Kronig transform, and compared these spectra with those reported earlier for As_2S_3 and As_2Se_3.[292-294] The $\varepsilon''(E)$ spectra of all three glasses were shown in Reference 295 to be similar in general pattern, the lower energy principal peak and higher energy weak peak being separated by an interval of about 4 eV. In the As_2S_3-As_2Se_3-As_2Te_3 sequence, the principal $\varepsilon''(E)$ peak shifts appreciably to lower energies from about 5 eV to about 3 eV, the intensity of the peak for As_2Te_3 being much higher than that for the two other glasses.

Aspnes et al.[296] published plots of $\varepsilon'(E)$ and $\varepsilon''(E)$ spectra of $GeSe_2$ glass in the 2 to 6 eV range. It was shown that the spectrum of this glass is similar to the overlapped spectra of corresponding crystal for two different orientations of light beam polarization with respect to the crystallographic axes.

As for chalcogenide glasses of other compositions, no data on their spectra of optical constants in the energy range of electronic excitations are yet available.

Chapter 5

Refractive Index Dispersion of Glasses in the Transparency Range: Applied Aspects

As mentioned in Chapter 1, Section I, the term "optical constants" spectroscopically refers to either $n(\omega)$ and $\kappa(\omega)$ or $\varepsilon'(\omega)$ and $\varepsilon''(\omega)$. At the same time, there exists a practice in applied optics (especially in the Russian literature) to use this term while referring to refractive indices at some fixed wavelengths within the transparency range and some characteristics derived from these indices, such as the *main dispersion, Abbe number, partial dispersions,* and *relative partial dispersions* that are specified below in Section I.A. To avoid misunderstanding, these quantities are referred to in this book as *optical characteristics.*

Prior to discussing the optical characteristics of inorganic glasses in the transparency range, some comments on the location of this range should be made. As is well known, the transparency range of dielectrics and ionic conductors, including inorganic glasses, is located in between two frequency ranges of strong absorption, one due to fundamental vibrational excitations and the other to fundamental electronic excitations. Major factors causing substantial differences in the frequency ranges of the fundamental excitations of every kind that are observed for various glass-forming compounds and, hence, for glass types based on these compounds are outlined briefly in Chapter 4, Section I. Naturally, the same factors are also responsible for the variations in the location of the transparency range that extends, for most non-metallic solids (such as oxides and halides, including glasses), from the near UV to the near IR or, for chalcogenide glasses, from the visible or near IR to the medium IR. The absorption edge formed by the electronic excitations directly determines the short-wavelength limit of the transparency range. However, a situation with the long-wavelength limit of the transparency range needs an additional comment, because it is not the high-frequency wing of the nearest fundamental IR band that causes an increase in intrinsic absorption with a decrease in frequency. As is known, the long-wavelength limit of the transparency range for vitreous silica and silicate glasses is no more than 4.0 to 4.5 μ; for germania and germanate glasses, about 5 to 5.5 μ; for tellurite glasses, about 6 μ; for fluorozirconate glasses, about 7 to 9 μ; and for chalcogenide glasses, about 12 to 18 μ, the magnitudes cited being much closer to the visible than the wings of the nearest fundamentals. The additional absorption responsible for these transparency limits is determined by the multiphonon excitations, such as overtones and combination modes of fundamentals, the number, frequencies, and intensities of these multiphonon bands being determined by the fundamental vibrational frequencies. In the case of a rather intense multiband fundamental spectrum, there arises a great number of overlapped multiphonon bands with successively vanishing intensities that form a more or less smooth absorption "tail" described phenomenologically by the exponential function. Such a description of a multiphonon tail was reported, for example, for heavy-metal-fluoride glasses[297] and for vitreous silica.[88] In the case of simple fundamental spectra of low intensity, such as those of chalcogenide glasses,

Table 1 Wavelengths and Letter Symbols of Spectral Lines of Chemical Elements Used in Standard Light Sources

Wavelength, nm	Designation	Element
365.0146	*i*	Hg
404.6561	*h*	Hg
435.8343	*g*	Hg
479.9914	*F′*	Cd
486.1327	*F*	H
546.0740	*e*	Hg
587.5618	*d*	He
589.2938	*D*	Na[a]
643.8469	*C′*	Cd
656.2725	*C*	H
706.5188	*r*	He
852.110	*s*	Cs
1013.98	*t*	Hg

[a] The location of the center of a close doublet is given; in some studies, the 589.262 nm component of the doublet is also used.

weak distinct bands related to particular second-order excitations (and, sometimes, to the third-order) are observed (see, for example, References 258 and 260).

I. REFRACTIVE INDEX DISPERSION IN TERMS OF OPTICAL CHARACTERISTICS USED IN APPLIED OPTICS

A. BASIC NOTATIONS AND TERMINOLOGY

The principal optical characteristics of glasses (and other optical materials) in applied optics used as a basis for classifying optical glasses are the refractive index n_{γ_1} at a certain fixed wavelength λ_1 (usually somewhere in the middle of a range of interest) and several simple functions of some other n_{λ_m} indices. To denote the respective particular wavelengths, the well-known letter symbols of standard spectral lines are commonly used; the correspondence between these symbols and wavelengths for the most common lines is given by Table 1. The derived functions are the *main,* or *principal,* dispersion, $n_{\lambda_2} - n_{\lambda_3}$, λ_2 and λ_3 usually being chosen near the opposite limits of a range of interest), and the so-called Abbe number (or *constringence*),

$$\nu_{\lambda_1} = \frac{n_{\lambda_1} - 1}{n_{\lambda_2} - n_{\lambda_3}}. \tag{62}$$

From the viewpoint of planning an optical system, the Abbe number is the principal characteristic of the refractive index dispersion of a glass or other optical material. The achromatization of a two-component lens, i.e., correcting its chromatic aberration for two wavelengths, can be achieved by using, for the components, materials differing greatly in Abbe number. Optical glass catalogs contain, as reference data, the magnitudes of the above three characteristics for various combinations of spectral lines or wavelengths. One of these combinations is usually considered to be preferable for a given catalog; it is the optical characteristics for this combination that are used when classifying optical glasses by different types (see Section I.B) and are presented in test certificates covering standard measurements for a particular glass melt. Companies producing commercial optical glasses use optical characteristics for different preferable combinations of spectral lines. Initially, the wavelengths of *D, F,* and *C* lines were taken commonly for λ_1, λ_2, and λ_3 when referring to the visible; correspondingly, the Abbe number was chosen to be $v_D = (n_D-1)/(n_F-n_C)$. Now, n_D and v_D are no longer used as preferable optical characteristics; *D* line is not a single line but a close doublet, which results in an n_D measurement that is less accurate when compared to the refractive indices for other standard lines. For optical glass catalogs of Schott, Hoya, and Corning France, the principal optical characteristics are chosen to be n_d, $n_F - n_C$, and v_d; also used are n_e, $n_{F'} - n_{C'}$, and $v_e = (n_e-1)/(n_{F'}-n_{C'})$. The catalog of Russian manufacturers now uses n_e, $n_{F'} - n_{C'}$, and v_e as the principal optical characteristics. For characterizing glasses used in the IR range, the corresponding special combinations of optical characteristics should be chosen (see Section I.C).

For characterizing the refractive index dispersion of optical materials in more detail, additional optical characteristics are used, such as partial dispersions, $n_{\lambda_4} - n_{\lambda_5}$, and relative partial dispersions,

$$P_{\lambda_4,\lambda_5} = \frac{n_{\lambda_4} - n_{\lambda_5}}{n_{\lambda_2} - n_{\lambda_3}}, \tag{63}$$

$n_{\lambda_2} - n_{\lambda_3}$ being the main dispersion.[a] Correction of the so-called secondary spectrum (i.e., achromatization for more than two wavelengths) can be obtained when using, for two of the components of a complex lens, materials that differ substantially in Abbe number but have similar relative partial dispersions in the wavelength range of interest. To match these requirements, special optical glasses (see Section IV.A) have been developed. Initially, the most frequently used were relative partial dispersions for an interval that covers blue visible lines, such as $P_{g,F}$, $P_{F,d}$, $P_{F,e}$, and also for an interval that covers visible red lines, such as $P_{d,C}$, $P_{C,r}$ and others similar. Later, relative partial dispersions additionally involving the *i* line in the near UV, such as $P_{i,g}$, some lines in the near IR, such as $P_{C,s}$ and $P_{C,t}$, were put into practice. Along with the relative partial dispersions mentioned above, those corresponding to the main dispersion $n_{F'} - n_{C'}$, such as $P'_{g,F'}$, $P'_{F',e}$, etc are used broadly.

Along with the optical characteristics mentioned above, one additional specific optical function such as *the material dispersion,*

$$M = \frac{\lambda}{c} \cdot \frac{d^2 n}{d\lambda^2}, \tag{64}$$

is used in waveguide optics,[298-300] c being the light velocity. In view of the form of the $n(\lambda)$ curve in the region of normal dispersion (see Chapter 1, Section I.C), M decreases rapidly with wavelength. Due to the negative sign of the contribution from IR excitations, $\Delta\varepsilon'_{ir}(\lambda)$, to the total magnitude of the real dielectric constant

[a] This presupposes that n_{λ_4} and/or n_{λ_5} are not equal to n_{λ_2} and/or n_{λ_3}, respectively.

$\varepsilon'(\lambda)$ in the transparency range, the $n(\lambda)$ curve for all materials except elemental substances has an inflection point (usually somewhere in the near IR). As is known, an inflection point corresponds to the zero magnitude of the second derivative, so the so-called crossover wavelength λ_c, in which M changes sign, corresponds to the inflection point of the $n(\lambda)$ curve. Information on the location of λ_c for a material is vital for designing low dispersion optical fibers. Data on the material dispersion of a heavy metal fluoride glass (such glasses being, as is known, promising materials for optical fibers) are presented, for example, in Reference 301.

B. ABBE DIAGRAMS AND THE PHENOMENOLOGICAL CLASSIFICATION OF GLASSES

The principal refractive index n_{λ_1} and the corresponding main dispersion of a material are naturally not independent quantities but are interconnected. The general character of the resultant relationship between these quantities for all optical glasses (and other optical materials) can be understood crudely in terms of the simplest Sellmeier dispersion formula (30). The magnitudes of three wavelengths used for determining the Abbe number of a material (62) usually correspond to the sequence $\lambda_2 < \lambda_1 < \lambda_3$, the magnitudes of respective refractive indices following the reverse sequence due to the normal dispersion in the transparency range. The contribution of the electronic excitations, $\Delta\varepsilon'_{el}(\lambda)$, to the total $\varepsilon'(\lambda)$ values of oxide and halide glasses in the visible range is several orders of magnitude greater than that of the IR excitations, so it is the behavior of $\Delta\varepsilon'_{el}(\lambda)$ that determines the general pattern. The closer the inherent wavelengths of the electronic excitations λ_j of a material to particular wavelengths lying in the visible range, the less the differences $\lambda^2 - \lambda_j^2$ entering into Formula (30) and, hence, the greater the refractive indices at these wavelengths. For the above sequence of wavelengths, the magnitude of n_{λ_2} increases more rapidly than that of n_{λ_1} and, even more so, than n_{λ_3}. For virtually the most simple case of a single excitation, this can be qualitatively illustrated using Chapter 1, Figure 1. Imagine that the entire $n(\omega)$ maximum approaches some fixed frequency lying to the left of the $n(\omega)$ maximum frequency. Therefore, the main dispersion also increases with an increase in inherent wavelengths of electronic excitations. Within the transparency range, an inequality $n_{\lambda_1} - 1 \gg n_{\lambda_2} - n_{\lambda_3}$ always holds, so a relative change in the denominator of (62) due to an increase in the inherent wavelengths of electronic excitations is always greater than a relative change in the numerator. As a result, the Abbe number determined by (62) always decreases with an increase in inherent wavelengths of the electronic excitations, i.e., in the direction opposite that of the refractive index n_{λ_1}. Actual variations in the optical characteristics of glasses with glass composition are somewhat more complicated due to the role of the intensities of electronic excitations, variations in these intensities substantially influencing the absolute values of refractive indices but having almost no effect on Abbe numbers (see Section IV.B).

The n_{λ_1} vs. ν_{λ_1} diagram illustrating the above tendency in the interconnection of these quantities for the totality of optical glasses and/or other optical materials is known as the *Abbe diagram*. Initially, the term Abbe diagram referred to the n_D–ν_D or to the quite similar n_d–ν_d diagram; later, this term was extended to other kinds of the diagram such as n_e–ν_e, etc. On the whole, the totality of inorganic glasses extends from glasses with minimum refractive indices and maximum Abbe numbers to glasses with maximum refractive indices and minimum Abbe numbers, thus covering a rather broad area in Abbe diagrams. Abbe diagrams are usually divided into individual subareas occupied by optical glasses of particular types. Such

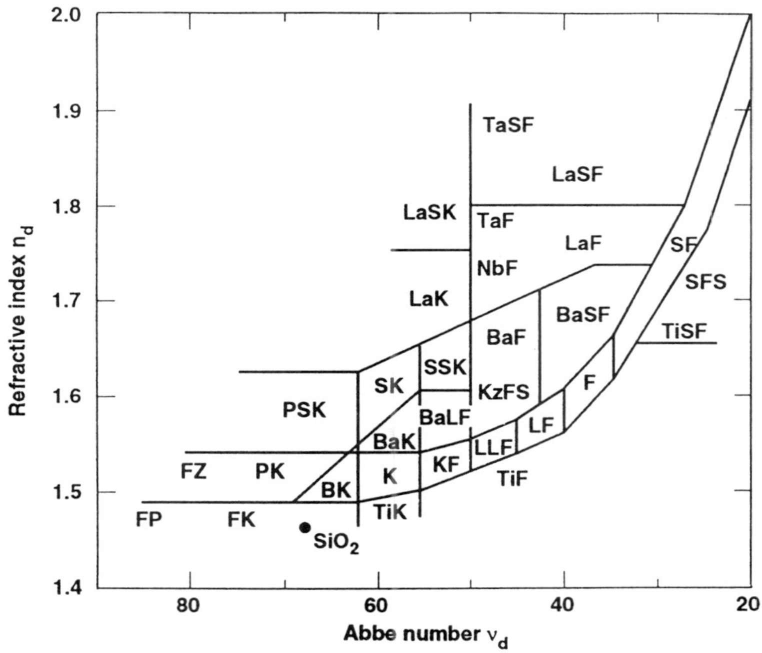

Figure 1 An example of the Abbe diagram for Schott glasses.

diagrams are presented in all optical glass catalogs and in general are quite well known (see, for example, Figure 1). An area occupied by optical glasses in an Abbe diagram increases successively in the course of developing newer glass types. In general, the optical characteristics of commercial optical glasses of standard types vary now from those of fluorine-rich glasses (such as fluorophosphate and some other ones with n_d in the 1.40 to 1.45 range and v_d, at the greatest, in the 90 to 95 range) to those of lead-oxide-based glasses (with n_d up to 1.95 to 2.06 and v_d down to 20.0 to 16.5). Glasses more specific in composition can appreciably exceed these limits. In a recent review paper, Feltz, Burckhardt, Voigt, and Linke[302] compared areas occupied by optical glasses at Abbe diagram n_e–v_e in 1884 and 1937 with the current area. The lowest refractive indices (down to about 1.3) and greatest Abbe numbers (above 100) within this area correspond to fluoroberyllate glasses. The upper limit of refractive indices of glasses within this area is shown at the level of about 2.1, Abbe numbers of these glasses extending to low magnitudes of about 5 to 10. However, the maximum refractive index among commercial optical glasses is known to correspond to a Russian tellurite glass STF 3 with n_e of 2.18626 and v_e of 16.89 that is not included in the diagram of Reference 302. Simple glasses synthesized in the laboratory can span even in greater areas of Abbe diagrams. The limiting case of the lowest refractive index is vitreous beryllium fluoride with n_D of 1.27475 and v_D of 106.9.[303] The greatest refractive indices ($n_D \geq 2.2$) are known to be characteristic of glasses based on oxides such as Bi_2O_3, TiO_2, WO_3, etc. Values of n_D up to about 2.31 and of v_D down to about 14.1 were reported for quaternary TeO_2-WO_3-Bi_2O_3-based glasses.[304] Also, the location of chalcogenide glass such as vitreous GeS_2 is shown at the n_e–v_e diagram of Reference 302, n_e

being about 2.15 and v_e about 10. GeS_2 glass has a pronounced yellow tint; probably, other glasses with $v_e = 5$ to 10 occupying the right upper corner of this diagram are also colored.

Optical glasses are commonly divided into two large species: as a rule, all glasses with high Abbe numbers (in most cases, those with $v_{\lambda_1} \gtrsim 55$) are referred to as *crowns* and glasses with low Abbe numbers (in most cases, those with $v_{\lambda_1} \lesssim 50$) as *flints*. In between $50 \leq v_{\lambda_1} \leq 55$, the kind of notation depends, moreover, on n_{λ_1} values; glasses with $n_{\lambda_1} \gtrsim 1.6$ are also related to crowns and those with $n_{\lambda_1} \lesssim 1.6$ to flints.[b] Crowns are denoted by a capital letter "K" in Schott and Russian notation or a "C" in Hoya and Corning France notation, whereas flints everywhere are denoted by "F". Subdivisions distinguish between *light* glasses (those with the lowest refractive indices within a type) denoted by "L", "usual" glasses of the type accompanied by no special notation, *dense* glasses, and even *extra dense* glasses (those with high and the highest refractive indices within the type) denoted by "S" and "SS" in Schott notation (from German "Schwer" and "Schwerst"), "D" and "ED" in Hoya and Corning France notation, and "T" and "ST" in Russian transliterated notation, respectively. Another subdivision is based on the presence of particular components in glass chemical composition; such subdivisions are denoted by the symbols of corresponding chemical elements. As a result, rather complicated designations of particular glass types such as BaLK (barium light crowns), LaSF (lanthanum dense flints), or TaFD (tantalum dense flints) are in use. The designations of similar glass types used by different manufacturers can differ appreciably from each other. For example, not only LaC but also TaC glasses of Hoya fall into the LaK type of Schott; both TaFD and, partially, NbFD glasses of Hoya fall into the LaSF type of Schott. To some extent, "chemical" designations have now lost their initial literal sense, because many glass components are interchangeable with respect to their influence on glass optical characteristics. As a result, attributing a glass to a particular chemical subdivision can have little in common with its actual composition. For example, compositions of some of glasses referred to as "dense barium (or baryte) flints" comprise no barium oxide at all. There are, however, glass types with optical characteristics that cannot be obtained with other components except for particular ones. This concerns, first of all, fluorophosphate and purely fluoride glasses (particular designations varying with manufacturer but always comprising "fluorine"). Also, glasses with optical characteristics related to some other types, such as TiF, LaK (LaC), and TaFD, can hardly be obtained without titanium oxide, lanthanum oxide (or, at least, other lanthanide oxides), and tantalum oxide plus lanthanum oxide, respectively.

Along with these traditional designations of glass types based on the above subdivisions, the kind of designation based on six-digit code numbers is also widely used, these code numbers describing, in a rounded-off form, the refractive index (with unity omitted) and Abbe number of a glass. For example, Schott glass FK54 also has the designation of 437907 which means n_d of about 1.437 and v_d of about 90.7. In the six-digit numbers used in the Hoya and Corning France catalogs, figures related to the refractive index and to the Abbe number are separated by a dash.

[b] There can be additional exceptions; for example, there are Russian optical glasses with v_{λ_1} ranging from 45 to 50 that are related to extra dense crowns (STK) and glasses with v_{λ_1} ranging from 50 to 55 and n_{λ_1} of about 1.50 to 1.52 that are denoted as "crown-flints" (KF).

C. ABBE-LIKE DIAGRAMS FOR FREQUENCY RANGES OTHER THAN THE STANDARD

The choice of spectral lines or wavelengths for characterizing glasses by their refractive indices and Abbe numbers depends on the wavelength range of interest. For characterizing glasses with the transparency range entirely shifted from the visible to the infrared (as is the case for most chalcogenide glasses) the usual optical characteristics are not suitable and should be replaced by others corresponding to wavelengths greater than 1 μ. Another specific feature of optical characteristics of chalcogenide glasses is their great refractive indices in the transparency range. For glasses of basic chalcogenide systems, the refractive indices vary from 2.3 to 2.9,[281,305] this range extending from 3.1 to 3.3 for glasses containing the elements of the sixth row such as Tl, Pb, and Te[305] (see Table 2) and down to 1.9 to 1.6 for some chalcohalide glasses.[306] Available data on the characteristics of the refractive index dispersion of chalcogenide glasses are rather limited.[281,282,305,307]

The Abbe-like diagram presenting the optical characteristics of simple and commercial chalcogenide glasses in the near IR was published first by Aio, Efimov, and Kokorina.[305] This diagram uses the refractive index $n_{2.0}$ for n_{λ_1} and "Abbe-type number" $\nu_{2.0} = (n_{2.0} - 1)/(n_{1.8} - n_{2.2})$ for ν_{λ_1}, subscript indices denoting wavelengths in microns. The $\nu_{2.0}$ vs. glass composition plots were presented in Reference 305 for As-Se, Ge-Se, Ge-As-Se, and Ge-As-S-Se glasses, the $\nu_{2.0}$ values ranging from about 135 to about 215. Chalcogenide glasses were shown to reveal, at the $n_{2.0}$–$\nu_{2.0}$ diagram, the usual tendency of an increase in the refractive index with a decrease in Abbe number specified above for the location of oxide optical glasses at Abbe diagrams related to the visible. On the contrary, an area occupied at the $n_{2.0}$–$\nu_{2.0}$ diagram by oxide optical glasses extends, according to Reference 305, in the perpendicular direction, i.e., from glasses with high $\nu_{2.0}$ and high refractive indices to glasses with low $\nu_{2.0}$ and low refractive indices. This can be attributed to the much greater influence of the IR excitations on the refractive index dispersion of oxide glasses in this range as compared to that occurred in the visible (see also Section IV.B). Later, a similar $n_{2.0}$–$\nu_{2.0}$ diagram comprising some additional data on other materials was presented by Feltz et al.[302] and Gan Fuxi.[252] According to Reference 302, vitreous GeS_2 with $n_{2.0}$ of about 1.95 to 2.0 and $\nu_{2.0}$ of about 260 comes rather far outside the limits shown for chalcogenide glasses in Reference 305.

Further studies of the refractive index dispersion of chalcogenide glasses were devoted to the 3.0 to 5.0 and 8.0 to 12.0 μ ranges corresponding to the "windows" of atmospheric transparency. Data on the Abbe-type numbers of chalcogenide glasses for these ranges, such as $\nu_{4.0} = (n_{4.0} - 1)/(n_{3.0} - n_{5.0})$ and, especially, $\nu_{10.0} = (n_{10.0} - 1)/(n_{8.0} - n_{12.0})$, are scarce due to the shortage of experimental $n(\lambda)$ values; those presented for simple glasses in References 281 and 307 are illustrated by Table 3. The optical characteristics of commercial chalcogenide glasses intended for optical applications (see, for example, References 173 and 302) practically do not go outside the limits determined by the data of Tables 2 and 3.

II. REFRACTIVE INDEX INTERPOLATION AND EXTRAPOLATION

The problem of how to interpolate accurately the refractive index of various optical materials, including glasses, was studied quite intensely in the applied optics

Table 2 Examples of Refractive Indices of Various Chalcogenide Glasses at Several Wavelengths According to References 281, 282, 305, and 307

Glass composition, atomic %				Refractive indices at wavelengths (μ)				
Ge	As	Se	Other	3.0	5.0	8.0	12.0	Ref.
—	40	60	—	—	—	2.7840	2.7728	281
20	—	80	—	—	—	2.4071	2.3973	
10	20	70	—	—	—	2.4649	2.4526	
10	40	50	—	—	—	2.6108	2.6016	
30	10	60	—	—	—	2.4408	2.4271	
30	20	50	—	—	—	2.5690	2.5560	
—	20	80	—	2.5814	2.5745[a]	2.5697[b]	2.5657[c]	305
20	—	80	—	2.4190	2.4138[a]	2.4074[b]	2.4002[c]	
20	20	60	—	2.5168	2.5109[a]	2.5031[b]	2.4941[c]	
12.5	20	67.5	—	2.5713	2.5650[a]	2.5578[b]	2.5500[c]	
20	20	30	S 30	2.3501	2.3439[a]	2.3329[b]	—	
—	30	50	Te 20	2.9324	2.9211[a]	2.9126[b]	2.9055[c]	
17	20	58	Sn 5	2.6114	2.6045[a]	2.5962[b]	2.5870[c]	
25	20	50	Pb 5	2.7185	2.7100[a]	2.7010[b]	2.6910[c]	
12.5	—	67.5	Sb 20	2.7645	2.7547[a]	2.7450[b]	2.7336[c]	
—	30	60	Tl 10	3.0018	2.9899[a]	2.9790[b]	2.9690[c]	
—	30	45	Tl 15 and Te 10	3.2369	3.2219[a]	3.2089[b]	—	
30	13	57	—	2.4936	2.4859	2.4784	2.4650	282
30	13	47	Te 10	2.6118	2.6024	2.5952	2.5829	
30	13	37	Te 20	2.7412	2.7305	2.7229	2.7117	
30	13	27	Te 30	2.8818	2.8688	2.8610	2.8509	
—	—	50	Te 30 and Br 20	2.7030	2.6932	2.6863	2.6748[c]	307
—	30	40	Te 20 and I 10	2.8435	2.8357	2.8267	2.8194[c]	
—	40	30	Te 20 and I 10	2.9035	2.8918	2.8807	2.8702[c]	
—	60	10	Te 20 and I 10	3.0489	3.0382	3.0262	3.0162[c]	

[a] Data for the wavelength of 4.6 μ.
[b] Data for the wavelength of 7.8 μ.
[c] Data for the wavelength of 11.0 μ.

beginning from the 1870s. One general reason for an interest in this problem was the necessity to have at one's disposal the refractive index values of the commercial optical materials for wavelengths lying in between those of the spectral lines commonly used in the standard experimental methods. An interest was stimulated additionally by the development of laser optics, because laser radiation wavelengths usually do not coincide with the locations of spectral lines of standard light sources.

Table 3 Abbe-Type Numbers and Respective Refractive Indices of Chalcogenide Glasses According to References 281, 282, and 307

Glass composition, atomic %				3 to 5 μ range		8 to 12 μ range		
					Abbe-type number		Abbe-type number	
Ge	As	Se	Other	$n_{4.0}$	$\upsilon_{4.0}$	$n_{10.0}$	$\upsilon_{10.0}$	Ref.
—	40	60	—	—	—	2.7789	159	281
20	—	80	—	—	—	2.4027	143	
10	20	70	—	—	—	2.4594	119	
10	30	60	—	—	—	2.6201	135	
10	40	50	—	—	—	2.6067	176	
20	10	70	—	—	—	2.5583	156	
30	10	60	—	—	—	2.4347	104	
30	15	55	—	—	—	2.4914	113	
30	20	50	—	—	—	2.5633	120	
33	12	55	—	—	—	2.4942	111	
30	13	57	—	2.4887	165	2.4724	110	282
30	13	47	Te 10	2.6057	171	2.5897	129	
30	13	37	Te 20	2.7342	162	2.7178	154	
30	13	27	Te 30	2.8732	144	2.8563	185	
—	—	50	Te 30 and Br 20	2.6990	173[a]	2.6799	146[b]	307
—	30	40	Te 20 and I 10	2.8385	236[a]	2.8222	250[b]	
—	40	30	Te 20 and I 10	2.8979	162[a]	2.8741	179[b]	
—	60	10	Te 20 and I 10	3.0436	191[a]	3.0203	202[b]	

[a] Abbe-type number of $\nu_{3\text{–}5} = [(n_{3.0} + n_{5.0})/2 - 1]/(n_{3.0} - n_{5.0})$ kind.
[b] Abbe-type number of $\nu_{8\text{–}11} = [(n_{8.0} + n_{11.0})/2 - 1]/(n_{8.0} - n_{11.0})$ kind.

The solution of the problem of interpolation usually was sought in the form of a particular dispersion formula providing the explicit analytical presentation of dependence of the refractive index on wavelength within the transparency range. Early well-known examples are the interpolation formulas of Cauchy (see, for example, Reference 1) and Herzberger;[308] interpolation with other standard types of such formulas is reviewed in detail in Sections II.A.1 and II.A.2. The parameters of an interpolation formula for a particular glass are calculated by fitting to the experimental $n(\lambda)$ values for certain wavelengths. The number of these wavelengths in various studies ranges from the minimum determined by the number of unknowns in the interpolation formula (as is the case for most of studies) to the maximum equal to the total number of wavelengths accessible for $n(\lambda)$ measurements in the range of interest (as is the case for the well-known Malitson's study[309]). Two versions of a refined approach to the precise interpolation and extrapolation of the refractive indices of glasses, one of these versions additionally involving the model presentation of $n(\lambda)$ and $\kappa(\lambda)$ dispersion with the spectroscopically determined parameters of the VUV excitations, are described in Section II.B.

A. INTERPOLATION WITH THE STANDARD TYPES OF DISPERSION FORMULAS

1. Interpolation with the Sellmeier-, Lorentz-Lorenz-, and Gladstone-Dale-Type Dispersion Formulas

A great number of studies devoted to both vitreous and crystalline optical materials use interpolation formulas based on the above Sellmeier dispersion formulas (30) or (31), that is, the simplified (that presupposing the absence of damping of an excitation) version of the classical analytical model for the real part of the dielectric constant (see Chapter 1, Section II.C.1). A comprehensive review of papers on this topic published before 1977 is given in Reference 22. On the whole, the dispersion formulas of this type described in literature usually comprise from one to five terms.

Studies based on the single-term formulas[28,29,310,311] and two-term formulas[31,312] are mostly not intended for obtaining the accurate $n(\lambda)$ values as such but rather for revealing the general correlations between particular parameters of the dispersion formulas and the approximate magnitudes of optical characteristics of glasses. Conclusions made in these studies are discussed below in Section III.B.

Among studies using the Sellmeier type of interpolation formulas for obtaining refractive indices, that of Malitson[309] devoted to vitreous silica is probably the most well-known. Malitson approximated comprehensive data on the refractive indices of three different commercial samples of vitreous silica[c] measured at 60 wavelengths covering the entire 214 to 3700 nm range with the three-term Sellmeier interpolation formula (30),

$$n^2(\lambda)-1=\frac{0.6961663}{1-\left(68.4043/\lambda^2\right)}+\frac{0.4079426}{1-\left(116.2414/\lambda^2\right)}+\frac{0.8974794}{1-\left(9896.161/\lambda^2\right)}, \quad \mathbf{(65)}$$

wavelengths being given in nanometers. No appreciable systematic deviations from the averaged curve determined by Formula (65) were observed, so the spread of particular $n(\lambda)$ values around this curve was assumed in Reference 309 to be caused mostly by random experimental errors. Malitson's calculated data often are considered to be the best presentation of the actual refractive index dispersion for vitreous silica.

The Sellmeier-type interpolation formulas remain to be used as a tool for calculating the refractive indices. An interest in the Sellmeier formulas was further stimulated by studies of the material dispersion of glasses; formulas comprising no more than three terms were usually applied to these studies (see, for example, References 299 and 300), because there is no need for precise values of refractive indices in this case. Izumitani and Hirota beginning with the single-term formulas[310,311] recently began to use five-term interpolation formulas of (30) type[291] for analyzing factors influencing the optical characteristics of glasses (see below Section III.B). In a recent monograph, Gan Fuxi[252] published two kinds of the Sellmeier-type interpolation formulas. First, the parameters of three-term Sellmeier interpolation formula of (30) type,

$$n^2(\lambda)=A+\frac{\beta\lambda^2}{\lambda^2-\lambda_1^2}+\frac{c\lambda^2}{\lambda^2-\lambda_2^2}, \quad \mathbf{(66)}$$

[c] The Corning code 7940, Dynasil, and General Electric type 151 samples were used.

are presented for several oxide and chalcogenide glasses, such as vitreous silica, $50ZnO \cdot 50B_2O_3$, arsenic sulfide, $12.5Ge \cdot 20Sb \cdot 67.5Se$, $30As \cdot 50Se \cdot 20Te$, and some other glasses. Second, the "partial" parameter values of two-term interpolation formula of (31) type are given in Reference 252 for 39 fluorides being the components of fluoride glasses.

As follows from the totality of studies into the problem of the refractive index interpolation with the Sellmeier-type formulas, the accuracy of the interpolation procedure cannot be improved unlimitedly by means of an increase in the number of terms of the dispersion formula. Naturally, single-term formulas whose inherent frequencies serve as the "centers of gravity" of the entire fundamental spectrum are not capable of providing accurate interpolated $n(\lambda)$ values acceptable for the applied purposes and can only be used for some general considerations (such as those conducted by Wemple and DiDomenico in References 27 through 30 and Hirota and Izumitani in References 310 and 311); the results of this kind are discussed below in Section III.B. Two-term formulas (such as those used by Wemple in Reference 31 and by Bohl[312]) can satisfactorily approximate the $n(\lambda)$ values only in narrow wavelength ranges. Those interpolation formulas that comprise from three to five terms can reproduce rather accurately the refractive indices of most optical crystals and glasses in relatively broad ranges, the errors in $n(\lambda)$ in the visible being, in general, from about $5 \cdot 10^{-6}$ to (1 to 3) $\cdot 10^{-5}$. There are, however, examples showing that an increase in the number of terms from three to four or five does not necessarily promote a corresponding increase in the accuracy of the interpolation procedure. For example, Stozharov and Frolov[313] used both three- and four-term Sellmeier formulas of (30) type for approximating the refractive indices of thirteen Russian commercial glasses of various types; the error of refractive index interpolation with both the three- and four-term interpolation formulas remained practically intact being of about $1 \cdot 10^{-5}$ in the visible and $1 \cdot 10^{-4}$ in the infrared. The behavior of interpolation errors depending on the number of terms in the formula of (31) type was studied in Reference 314 using the array of refractive indices of ternary BaO-La_2O_3-B_2O_3 glass as an example. It was shown that an increase in the number of terms from three to six did not provide a perceptible decrease in the interpolation errors. Reference 314 also demonstrates the regular character of error distribution along the frequency axis typical of calculation procedures involving the minimum number of experimental $n(\lambda)$ values. Rigorous estimations of error origin and magnitudes for the case of interpolation with the Sellmeier-type interpolation formulas are not available in literature.

It should be specially noted that the parameters of a particular Sellmeier-type interpolation formula found by fitting to the experimental $n(\lambda)$ data have, as a rule, rather little in common with the parameters of actual features of fundamental spectra. Vivid examples are the data of References 312 and 313 for the totality of optical glasses. Both the second inherent wavelength in Bohl's two-term dispersion formula[312] and the fourth one in Stozharov and Frolov's four-term formula[313] fall within the transparency range, these wavelengths varying with glass composition from approximately 300 to 1700 nm and from 450 to 1140 nm, respectively. Therefore, some authors[22,291,314,315] choose the oscillator frequencies based on additional independent information on the frequency locations of $R(\omega)$ maxima in the experimental VUV spectra; however, this does not promote an appreciable increase in the accuracy of the interpolation of refractive indices.

From the viewpoint of Chapter 3, Section III concerning the specific features of the inverse spectroscopic problem, such a poor correspondence between the Sellmeier formula parameters and the spectroscopic evidence is natural. First, the number of inherent wavelengths is usually less than that of actual intense features in the fundamental VUV spectrum (not mentioning the IR one). Second, the totality of experimental $n(\lambda)$ values as such simply cannot contain enough information for reconstructing the actual parameters of individual features of the fundamental spectra. This is due to the remoteness of the transparency range from the ranges of these spectra that makes subtle contributions to the refractive index dispersion to be practically indistinguishable against the experimental errors. Therefore the inverse spectroscopic problem becomes, in this case, undoubtedly incorrect. This also is the reason why, as mentioned above, a successive increase in the number of terms of Sellmeier dispersion formulas cannot provide a corresponding increase in the accuracy of interpolation. However, all of the above does not prevent the Sellmeier-type formulas (30) and (31) from being quite appropriate tools for conducting those empirical calculations that do not require great accuracy.

All the reservations specified above for the Sellmeier-type formulas also hold for any other types of interpolation formulas, such as the Lorentz-Lorenz formula (35), etc. The Lorentz-Lorenz formula (35) was used for the interpolation of the refractive indices of glasses but much more rarely than the Sellmeier one; examples are the early studies of Engl[316] and of Florinskaya.[317] Engl's study was devoted to vitreous silica and that of Florinskaya to three commercial flint glasses, both of the studies using the two-term interpolation formulas and, hence, being confined only to the visible. The errors of interpolation obtained in both studies were about the same as those characteristic of the interpolation formulas of (30) and (31) types lacking the Lorentz-Lorenz correction.

Some early studies also used the Gladstone-Dale-type interpolation formulas, i.e., those involving the $n(\lambda) - 1$ difference as the left-hand side of interpolation formula, similar to the case for the Gladstone-Dale refractivity (37). Many of these studies, such as those of Young and Finn,[318] Huggins,[319] Huggins, Sun, and Davis,[320] and Ramachandran,[321] attempted to combine, in a single formula, the description of both the frequency dispersion of the refractive index and the variation in the refractive index magnitude with glass composition. Therefore, these sources will be discussed below together with others devoted to the problem of interrelation between the refractive index dispersion and glass composition (see Section III.A.2).

Also rather widely known is the early approach of Herzberger.[308] Using the Gladstone-Dale-type left-hand side of the interpolation formula, Herzberger had chosen for silicate glasses, a modified formula lacking any theoretical background

$$n - 1 = a + b\lambda^2 + \frac{c}{\lambda^2 - 0.035} + \frac{d}{\left(\lambda^2 - 0.035\right)^2}, \tag{67}$$

a, b, c, and d being constants. According to Herzberger, this formula, when applied to the visible region, can provide $n(\lambda)$ values accurate to within $5 \cdot 10^{-6}$, i.e., those characterized by the calculation error almost coinciding with the experimental error.

An attempt to further develop usage of Gladstone-Dale-type interpolation formulas was made by Polukhin and Krymova.[322] They proposed, for the commercial optical glasses, an empirical approach to the interpolation and extrapolation of the refractive index and its dispersion based on the Gladstone-Dale specific refraction (see Chapter 1, Section II.C.2) and the so-called specific dispersions,

$$\Delta r_{\lambda_4 \lambda_5} = \frac{n_{\lambda_4} - n_{\lambda_5}}{\rho} . \tag{68}$$

According to Polukhin and Krymova, the plots of these quantities vs. ω^2 or vs. log ω can be tentatively treated as the broken lines with two or more approximately linear sections, those in the 365 to 435 and 487 to 768 nm wavelength ranges being characteristic of all kinds of plots mentioned and those in the range of wavelengths greater than 852 nm being observed only for the plots of $\Delta r(\omega)$ vs. ω^2 or vs. log ω. The authors of Reference 322 believe that the procedure of interpolation along such linear sections plotted using two experimental $n(\omega)$ values might provide a good accuracy of obtained refractive indices. This belief is supported by the calculated data on the refractive indices of four commercial glasses accurate to within $1 \cdot 10^{-4}$ in the 435 to 706 nm range.

2. Interpolation with the Polynomial Dispersion Formulas

Those most widely used are the polynomial interpolation formulas of the general form

$$n(\lambda) = \sum_{m=-m_1}^{m_2} C_m \lambda^{-2m} , \tag{69}$$

C_m being constants that lack any definite physical meaning. Such formulas often are treated as the power series expansions of the Sellmeier dispersion formula (30). However, as was shown in Reference 7, this type of interpolation formula also can be derived directly from the Kramers-Kronig relations and, therefore, provides the most general (i.e., freed from restrictions imposed by any particular analytical model of the dielectric constant) presentation of the $n(\lambda)$ dispersion in the transparency range (see Section II.B). Modern optical glass catalogs include, for each particular glass, such polynomial interpolation formulas using m values from –1 to +4. In most cases, the estimates of tolerance limits for the refractive index values calculated with the interpolation formulas of this kind are taken to be from $\pm 1 \cdot 10^{-5}$ to $\pm 5 \cdot 10^{-6}$. The Schott Optical Glass catalog provides the most rigid tolerance limits of these formulas assumed to be $\pm 5 \cdot 10^{-6}$ for the entire 365 to 1014 nm range and only $\pm 3 \cdot 10^{-6}$ for the narrower 400 to 750 nm range. For heavy-metal fluoride glasses, five-term polynomial formulas with m from –2 to +2 were reported;[301,323] in Reference 301 such a formula was used for calculating the material dispersion of these glasses.

The origin of errors in approximating the $n(\lambda)$ dispersion with the polynomial interpolation formulas of (69) type was analyzed theoretically in Reference 7. The optimum values of m_1 and m_2 limits in (69) were shown to depend on a variety of factors such as the transparency range length, the refractive index magnitude and its dispersion in this range, the specific features of the fundamental excitations, the relative locations of reference wavelengths, and also, which is of utmost importance, the $n(\lambda)$ experimental errors at the reference wavelengths. As a result, the number of reference wavelengths used when treating a particular array of experimental data for a glass should correspond to some optimum value. Therefore, an unlimited increase in the number of reference wavelengths can be of no use or even lead to an increase in the interpolation error.

B. PRECISE INTERPOLATION AND EXTRAPOLATION OF THE REFRACTIVE INDEX VALUES: RECENT REFINEMENTS

Until recently, the problem of refractive index interpolation was studied extensively, whereas the problem of how to extrapolate precisely the refractive indices of optical glasses to wavelengths lying far outside the range of measurements was not even raised seriously. At the same time, solving the problem of extrapolation into the ultraviolet is of interest for photolithography and into the infrared for laser optics.

Two versions of a refined approach to the problem of precise interpolation and extrapolation of refractive indices of vitreous silica were developed recently by Efimov and Makarova.[7] In both versions, the above theoretical conclusions concerning the optimum ways to construct an interpolation polynomial of the (69) type were taken into account. The first version, like the standard interpolation procedures, uses only the polynomial constructed on the basis of the experimental reference $n(\lambda)$ values, i.e., involves no additional information. In contrast, the second version intended specifically for silica glass presupposes the construction of a correcting function that involves, in addition to usual experimental $n(\lambda)_{DA}$ reference values, model $n(\lambda)$ values calculated with the dispersion analysis data on the parameters of the fundamental VUV spectrum of silica glass borrowed from Reference 152 (see Section III.A). The dispersion analysis data of Reference 159 as such provide the refractive indices, $n(\lambda)_{DA}$, accurate to within $1 \cdot 10^{-3}$ in the broad wavelength range,[d] the error of these indices varying monotonically with wavelength. In the second version of the extrapolation procedure, this model $n(\lambda)$ dependence was taken to be the starting one. The experimental reference $n(\lambda)$ values were not used for calculating the coefficients of the polynomial determining the refractive indices themselves but of the correcting function, $f(\lambda)$, chosen in such a way that $n(\lambda)_{DA} + f(\lambda) = n(\lambda)$. The $f(\lambda)$ values are 3 to 5 orders lower than the refractive indices themselves; therefore, the accuracy of the resultant procedure, especially for the case of extrapolation to wavelengths lying far outside the range of $n(\lambda)$ measurements, is substantially higher than that of the standard extrapolation procedure (see below).

Based on these two versions of the extrapolation procedure, two independent computational programs were written. Both programs provide the possibility of varying the number of reference $n(\lambda)$ values and the locations of reference wavelengths on the wavelength axis. Bearing in mind that precise refractive index measurements are laborious, the possibility of using a limited number of the reference $n(\lambda)$ values was presupposed by these programs. Four reference $n(\lambda)$ values were shown to be sufficient for providing quite precise calculated refractive indices of vitreous silica. With the reference wavelengths chosen in the visible, these two programs can be used both for interpolating the refractive indices of vitreous silica between the reference wavelengths and for extrapolating the refractive indices from the visible into the ultraviolet.[e] For testing the procedures developed, the array of $n(\lambda)$ values calculated with Malitson's formula (65) was used in Reference 159 as the starting data. The accuracy of interpolation and extrapolation

---

[d] In view of the rather large experimental error of the initial $R(\omega)$ values in the VUV range (at best, 1%), such an accuracy seems to be quite satisfactory.

[e] The possibility of extrapolating into the IR range was not tested in Reference 7; however, there is no obstacle preventing the extension of the procedure into the IR range, too.

Table 4 Interpolated and Extrapolated Refractive Indices of Vitreous Silica Based on Model Reference $n(\lambda)$ Values Chosen from the Data of Malitson[309]

Wavelength, nm	$n(\lambda)$ according to Formula (62)	The $n(\lambda)$ errors, $\cdot 10^6$, obtained when calculating with two combinations of reference wavelengths: First program (with no additional information used)		Second program (that involving the data of Ref. 159)	
		h-g-F′-C′	*h-g-e-C′*	*h-g-F′-C′*	*h-g-e-C′*
214.438	1.5337225	–2154.8	–2241.0	+12.8	+20.6
226.747	1.5227505	–1298.9	–1361.2	+11.3	+17.0
237.833	1.5147295	–832.1	–878.7	+9.2	+13.5
248.272	1.5083985	–550.1	–585.6	+7.3	+10.6
269.885	1.4980475	–234.9	–254.9	+4.2	+6.0
280.347	1.4940392	–155.0	–170.1	+3.1	+4.5
302.150	1.4871940	–63.8	–71.9	+1.6	+2.3
330.259	1.4805390	–18.5	–21.8	+0.6	+0.9
346.620	1.4774676	–8.2	–10.0	+0.3	+0.4
365.015	1.4745391	–2.8	–3.6	+0.1	+0.2
377.572	1.4727998	–1.2	–1.6	+0.05	+0.08
479.991	1.4635027	—	–0.3	—	+0.02
486.133	1.4631265	+0.04	–0.3	–0.003	+0.03
508.582	1.4618630	+0.2	–0.2	–0.02	+0.02
546.074	1.4600781	+0.6	—	–0.05	—
587.561	1.4584637	+0.7	+0.2	–0.07	–0.03
656.272	1.4563666	–0.3	–0.2	+0.04	+0.03
667.815	1.4560668	–0.7	–0.3	+0.09	+0.06
706.520	1.4551455	–2.3	–1.2	+0.4	+0.3

obtainable with these two programs is illustrated by Table 4 that repeats, with some refinements, the results of Reference 7. When calculating data of Table 4, the array of $n(\lambda)$ values obtained with Malitson's Formula (65) was used as a substitute for the experimental $n(\lambda)$ values. Interpolated and extrapolated refractive index values placed into Table 4 were obtained with two different combinations of the reference wavelengths denoted by the sequence of symbols of corresponding spectral lines. The accuracy of interpolation and extrapolation illustrated by Table 4 directly characterizes the methodological errors of the procedures developed, because the initial array of $n(\lambda)$ values calculated with Malitson's formula (65) is in essence the model array and, hence, does not comprise any experimental errors.

According to the data of Table 4, both programs are capable of providing the interpolated refractive indices of vitreous silica in the visible accurate to within $1 \cdot 10^{-6}$, i.e., with substantially better accuracy than the interpolation formulas presented in the optical glass catalogs. However, the precision of extrapolation provided with the second program (that involving the VUV dispersion analysis data) is substantially higher than that of the first program involving no additional information. The refractive indices extrapolated with the second program into the UV range down to 330 nm are characterized by the methodological errors not exceeding $1 \cdot 10^{-6}$; as is seen from Table 4, an increase in this error up to $1 \cdot 10^{-5}$ is observed only at wavelengths as short as 245 to 250 nm. The methodological error characteristic of the first program increases with a decrease in wavelength much more rapidly; the error of $1 \cdot 10^{-5}$ is reached at wavelengths of about 345 to

Table 5 Examples of the Extrapolated Refractive Indices for the Commercial KS4V and KU1 Glasses Obtained with Various Combinations of the Reference Wavelengths According to Reference 7

	Data obtained by extrapolation with the second program using various combinations of reference wavelengths			
	KS4V glass sample		KU1 glass sample	
Wavelenth, nm	Averaged refractive index value	Spread over combinations, $\cdot 10^6$	Averaged refractive index value	Spread over combinations, $\cdot 10^6$
214.438	1.534196	±43	1.533913	±31
248.272	1.508697	±17	1.508486	±13
269.885	1.498290	±10	1.498103	±7
302.150	1.487391	±4	1.487225	±3
330.259	1.480716	±2	1.480560	±1
365.015	1.474704	<±1	1.474554	<±1

350 nm. Therefore, it is the second of the programs that is preferable for the extrapolation of the refractive indices of silica glass.

When interpolating only, the errors of refractive indices calculated in Reference 7 for the samples of commercial silica glasses remain lower than the limits of the experimental errors. Much more seriously, the experimental errors of the reference $n(\lambda)$ values influence the extrapolation errors. According to the data of the numerical experiments described in Reference 7, it is the random kind of experimental errors that causes great errors of extrapolated refractive indices for actual glass samples. Fortunately, the total experimental error of the precise $n(\lambda)$ measurements usually is due to the systematic error contributions rather than the contributions of random errors. Additionally, the random errors can be diminished to an acceptable limit by using the reference $n(\lambda)$ values from data subjected initially to a statistical treatment, i.e., taking the averages for several runs. As is seen from Table 5 illustrating the data of Reference 7, using the reference $n(\lambda)$ values averaged over four or five runs for each particular sample of two Russian commercial KS4V and KU1 glasses provides a rather small spread in the extrapolated refractive indices over different combinations of reference wavelengths down to 248 nm (the second of the programs being used). Taking the magnitudes of the spread as the crude estimates of the total extrapolation errors, Reference 7 concludes that this procedure is capable of providing, for actual glass samples, extrapolated refractive indices accurate to within $1 \cdot 10^{-5}$ for wavelengths down to 270 nm and to within $2 \cdot 10^{-5}$ for wavelengths down to 240 to 250 nm. Therefore the extrapolated refractive indices obtained for particular commercial silica glass melts can be used directly as starting data for designing optical systems for the UV range. Such a potentiality can be of special interest for the 240 to 300 nm range, because the standard refractive index measurements usually do not extend to wavelengths shorter than 300 nm. Moreover, the interpolation and extrapolation procedure developed in Reference 7 can be used for checking the accuracy of precise $n(\lambda)$ measurements throughout the entire transparency range. For glasses other than silica glass, only the first of the two extrapolation programs of Reference 7 (i.e., that characterized by more substantial methodological errors at wavelengths lower than 330 nm) can be used because reliable quantitative data on their VUV spectra (see Chapter 4, Section III) are not available.

III. REFRACTIVE INDEX DISPERSION VS. GLASS COMPOSITION

Studies devoted to the problem of interrelation between the refractive index dispersion and glass composition can be divided into two groups. The first group comprises a great number of works intended to develop empirical calculation methods for obtaining the numerical values of the refractive indices themselves and/or some characteristics of the refractive index dispersion based on glass composition with accuracy that might meet the requirements of applied research. A search for such methods between the 1930s and 1950s was among the most tempting goals pursued by investigators in the field of optical glasses; some of the developed methods have acquired an appreciable applied significance. Studies of this kind are discussed in Section III.A. The second group of studies is intended to discover the general correlations between the nature of principal glass components and the numerical values of parameters in the simplest dispersion formulas, such as the Sellmeier-type, single- or two-term. Such studies are discussed below in Section III.B.

A. APPLIED CALCULATION METHODS

1. Principal Limitations

The starting point of all empirical methods developed for calculating the glass refractive index and its dispersion based on glass composition was the additivity of the molar refractivities (see Chapter 1, Section II.C.2). This fact enabled investigators to believe that some appropriate function of the refractive index (such as that of Drude, Lorentz-Lorenz, or Gladstone-Dale) or the refractive index value as such can be presented as a sum of contributions related to particular glass components (either initial oxides or binary compounds formed when synthesizing a glass), the specific increments being assumed to be constants and, hence, the magnitudes of the contributions being taken to be proportional to the concentrations of the components. This idea, however, has substantial shortcomings of a principal nature that are evident from both the general considerations and those in Chapter 2. The most evident shortcoming is as follows: the properties of particular compounds, including initial oxides, are used as constant increments in such calculations, whereas chemical reactions occur between these compounds when melting and the glass network is, in most cases, a highly polymerized structure. To avoid this contradiction, the notion of "oxide property in glass" presupposing the increment values to differ from the actual properties of the oxides was introduced (see below Section III.A.3). This approach, however, cannot evade the fact that both the distribution of the electron density around an atom and the electrical, dynamical, and kinematic interactions of any given structural fragment with others depend greatly on the nature of particular neighboring fragments (see also above Chapter 2, Sections II.B and II.C), hence the inevitable variations of all optical characteristics with glass composition. Moreover, the analytical forms of all $\hat{\varepsilon}(\omega)$ models discussed in Chapters 1 and 2 clearly reveal that there can be no direct proportionality between the refractive index and the concentration of any component, because it is the real part of the complex dielectric constant (that, for the transparency range, equals approximately to the square of the refractive index) and not the refractive index as such that is proportional to the number of oscillators per unit volume. As a result, an acceptable accuracy of calculations with such methods can be achieved either by limiting the applicability of particular increment values by rather narrow composition ranges or by means of approximating these values with some function of glass composition. In general, such methods should be treated only as useful tools for applied calculations that lack any perceptible physical significance.

2. Early Approaches

The development of various empirical methods for calculating the refractive index of silicate glasses in the transparency range based on their composition was started rather widely in the 1920s to 1930s. Most of the studies used a certain unified calculational scheme applied to both the refractive index and density of the glasses under study. When calculating the refractive index, the notion of refractivity taking into account the interconnection between the refractive index and density (see Chapter 1, Section II.C.2) was commonly used. Later, other optical and physical characteristics such as the main and partial dispersions and also the expansion coefficient were considered. The early approaches to the problem under discussion were reviewed, for example, in Reference 324. In most cases, these approaches describe the refractive index dispersion (if any) solely in terms of the main dispersion, thus involving no analytical description of the phenomenon. There were, however, a few approaches aimed directly at obtaining the composition-dependent dispersion formulas, i.e., at combining the description of both the refractive index magnitude and the frequency dispersion of the refractive index in terms of glass composition. Among studies of this kind, those of Young and Finn,[318] Huggins,[319,320] and Ramachandran[321] should be mentioned. All of these studies used the Gladstone-Dale refractivity presenting the latter as the function of wavelength and glass composition. Young and Finn[318] used the simplest two-term formula comprising no particular frequency values,

$$\frac{n(\lambda)-1}{\rho}=\sum_{m=1}^{M}\left(\alpha_m+\beta_m\cdot\frac{1}{\lambda^2}\right)\cdot P_m, \tag{70}$$

ρ being the glass density, P_m being the molar fraction of m-th oxide, and α_m and β_m being the constant increments characteristic of the oxide. The α_m and β_m values were found for 47 various oxides; no estimations of $n(\lambda)$ calculation errors was given in Reference 318.

Huggins[319] and then Huggins, Sun, and Davis[320] used a formula of the following general type

$$n(\lambda)-1=\sum_{m=1}^{M}a_m(\lambda)\cdot\frac{N_c}{V_0} \tag{71}$$

where N_c is the number of the electropositive element atoms per oxygen atom, V_0 is the oxygen molar volume, and a_m is the function of increments d_m and g_m characteristic of m-th oxide. The analytical form of the a_m function was chosen to be

$$a_m(\lambda)=\frac{d_m}{g_m-\dfrac{1}{\lambda^2}} \tag{72}$$

in Reference 319 and

$$a_m(\lambda) = \frac{d_m}{g_m - \frac{1}{\lambda^2}} - Kd_m\lambda^2 \tag{73}$$

in Reference 320, K being a constant. As is seen from Formulas (72) and (73), the g_m increments play the role of the $1/\lambda_m^2$ ratios in some Sellmeier-type formulas. The values of d_m and g_m increments for 31 oxides were found; the error of calculating n_D and $n_F - n_C$ with Formula (73) for ternary Na_2O-CaO-SiO_2 glasses was estimated to be $\pm 4 \cdot 10^{-4}$ and $\pm 5 \cdot 10^{-5}$, respectively. For multicomponent glasses, however, much greater errors were obtained.

Ramachandran,[321] in his study of thermo-optic behavior of glasses, also used a dispersion formula quite similar to that of Huggins, Sun, and Davis:[320]

$$\frac{n(\lambda)-1}{\rho} = \sum_{m=1}^{M} f_m \left[b_m \frac{c_m}{\frac{1}{\lambda_m^2} - \frac{1}{\lambda^2}} - K_m\lambda^2 \right], \tag{74}$$

f_m being the weight fraction of m-th oxide and λ_m being the fixed wavelength characteristic of m-th oxide.

3. Appen and Demkina Calculation Methods: Initial Versions and Later Refinements

Among such methods, those of Appen[325] and of Demkina[326] are probably the most widely known. The studies resulting in these methods were started as early as in the 1940s to 1950s;[327,328] however, both methods remain to be used among Russian researchers of optical glasses. The methods of Appen and Demkina, like most early methods for calculating the refractive indices of glasses based on their composition, do not involve any explicit analytical description of the refractive index dispersion, thus being confined to the optical characteristics such as the refractive index n_d or n_e, main and partial dispersions. Naturally, glass density is also calculated with both methods; some other glass properties can be estimated with Appen's method.

Appen[319] developed a method for calculating the density, n_D, $n_F - n_C$, and the linear expansion coefficient of silicate and borosilicate glasses based on a simple relation,

$$n = \frac{\sum_{m=1}^{M} n_m a_m}{\sum_{m=1}^{M} a_m}, \tag{75}$$

n_m being the increment of the refractive index (or other quantity mentioned) for m-th oxide and a_m being the molar fraction of the oxide. In other words, the calculations were carried out in terms of the notion of partial molar quantities. As is known, this notion can be valid, strictly speaking, only for properties being dependent on the total amount of a substance and, hence, should not be extended to the refractive index and other properties such as those studied by Appen. The plots of $n(\lambda)$ (and, hence, of $n_F - n_C$) vs. glass molar composition are not linear; therefore, the n_m and $(n_F - n_C)_m$ increments used in Appen's method were either considered to be composition-dependent (for oxides such as SiO_2, B_2O_3, TiO_2, PbO, and CdO) or were taken to be the averages over an entire composition range under consideration (for the rest of the oxides). Nonetheless, Appen's method is used rather widely. In general, the calculations with Appen's method result in smaller errors of n_m and $(n_F - n_C)_m$ as compared to errors obtained with the method of Huggins.

Demkina[326] developed her method based on the assumption that there exists a possibility of presenting glass properties such as the density, refractive index, and also main and partial dispersions as true linear functions of a particular glass composition. According to Demkina, such a possibility can be tentatively provided when expressing glass compositions in terms of the volume fractions (instead of the molar or weight ones) of oxides. Because the chemical nature of initial oxides does not remain intact when forming a glass, the volume fractions of oxides become the virtual quantities and, hence, cannot be calculated directly from their weight or molar fractions. Therefore, Demkina introduced the so-called "structural" coefficients, s_m, of oxides in glass, thus obtaining the calculation formula of the following form,

$$n = \frac{\sum_{m=1}^{M} \frac{w_m}{s_m} n_m}{\sum_{m=1}^{M} \frac{w_m}{s_m}}, \tag{76}$$

n_m being the increment of the refractive index (or other above property) of m-th oxide and w_m being the weight fraction of the oxide. Formula (76) implies that the totality of s_m values for the constituent oxides should promote transferring from the weight fractions of the oxides in a glass to the volume fractions. For glass-forming oxides, the n_m and $(n_F - n_C)_m$ values were assumed to be equal to the experimental n_D and $n_F - n_C$ values of corresponding single-component glasses, whereas s_m values were chosen voluntarily to be equal to the molar masses of the oxides; for the rest of oxides, the n_m, $(n_F - n_C)_m$, and s_m values were calculated by fitting Formula (76) to the experimental data. As in the case of Appen's method, the n_m and $(n_F - n_C)_m$ values can be considered to be constants only within rather limited composition ranges. Therefore, refining the arrays of the increments for various glass types lasted for years. For some oxides, more than one combination of the n_m, $(n_F - n_C)_m$, and s_m values turned out to be necessary: two substantially different combinations of these increments were assumed for oxides such as SiO_2, B_2O_3, Bi_2O_3, and Tl_2O and even three combinations for PbO. According to Demkina's estimations, the minimum calculation errors in n_D and $n_F - n_C$ obtainable with this method for multicomponent silicate and borosilicate optical glasses are of about $\pm(1 \text{ to } 2) \cdot 10^{-3}$ and $\pm(5 \text{ to } 10) \cdot 10^{-5}$, respectively. Recently, Demkina's method was extended to lanthanum borate glasses.[329]

The accuracy of data obtained with the methods of Appen and Demkina[319,326] and with two earlier methods, including that of Huggins et al.[320] was compared by

Gan Fuxi[330] for a broad variety of glass types. According to Gan Fuxi, the methods of Appen and Demkina provide a satisfactory accuracy of n_D and $n_F - n_C$ for various types of multicomponent silicate and borosilicate optical glasses, the magnitude of errors being practically similar, whereas errors typical of the other two methods are much greater than those. For more simple glasses of laboratory synthesis, however, the accuracy of data obtained with both methods of Appen and Demkina was found to be substantially less compared to that obtainable for commercial glasses. Also, Gan Fuxi[330] attempted to supplement and refine Appen's method by means of ascribing particular increment values to rather narrow composition ranges chosen in a complicated way: a ratio between the total ionic oxide content (different ionic oxides being taken with different coefficients ranging from 1.0 to 0.3) and B_2O_3 content was involved. According to the author's estimations, the errors in n_D and $n_F - n_C$ obtainable with this method for multicomponent silicate and borosilicate optical glasses are about $\pm 1.6 \cdot 10^{-3}$ and $\pm 1.0 \cdot 10^{-4}$, respectively. Gan Fuxi also attempted to extend such empirical calculations to commercial lanthanum borate glasses. However, the errors of data for such glasses were rather large ($\pm 5.6 \cdot 10^{-3}$ and $\pm 4.0 \cdot 10^{-4}$ for n_D and $n_F - n_C$, respectively). Somewhat better accuracy (the errors in n_D and $n_F - n_C$ being within the $\pm(2 \text{ to } 4) \cdot 10^{-3}$ and $\pm(2 \text{ to } 4) \cdot 10^{-4}$ limits, respectively) was obtained recently in terms of Demkina's method.[329]

As is seen from that above, none of the elaborate empirical methods can be applied to glasses other than silicate, borosilicate, and, at best, lanthanum borate; the accumulation of data sufficient for extending the methods to other glass types such as phosphate, germanate, etc. would take years. Luckily, when varying the chemical composition of any voluntarily chosen initial glass with the optical characteristics being known from experiment, the resultant variations in these characteristics can be crudely estimated in much simpler ways than those specified above, as is especially the case when variations in glass composition can be confined to the successive equimolar replacement of the constituent oxides by their analogs according to the Periodic Table (SiO_2 by GeO_2, etc.). Such analogs mostly (though not always) form the isostructural polyhedra in the glass network or play similar roles in the depolymerization of these networks, which often causes the corresponding property vs. composition dependences to look more like the approximately linear ones than is the case for other kinds of variations in glass compositions. This allows one to obtain experimentally, in the linear approximation, the increments of the optical characteristics per unit molar fraction of a substituted component; these increments then can be used directly for estimating changes in the optical characteristics of other glasses that should result from similar substitutions. Errors originated from the crudely approximated nature of such a procedure are compensated to some extent by the fact that increments used in this case are not characteristics of particular components but differences between corresponding pairs of these characteristics. Naturally, such estimates are too crude for any applied purposes; however, they can serve for prognosis of tendencies characteristics of refractive index variations with glass composition in glassforming systems that are not yet studied experimentally.

4. Concluding Remarks

The usefulness of the additive calculational methods discussed in Section III.A.3 for applications is beyond doubt. It should be noted, however, that the physical meaning of rather good accuracy of data on glass properties obtainable with some of these methods should not be overestimated. Neither of the mentioned approaches can evade the fact that the linear approximation being the basis of such empirical calculations meets principal limitations mentioned briefly above in Section III.A.1. Moreover, each new increment value is calculated by subtracting contributions assumed to be known from the experimental magnitude of glass property. Hence, errors in the estimates of these contributions will be compensated by the error in the new increment value of the opposite sign. Evidently, it is this mutual compensation of errors introduced by constant increments that can explain the satisfactory accuracy of the additive calculations of glass properties in limited concentration ranges. Such a compensation can take place irrespective of whether the scheme of the particular glass structure presupposed by the calculation procedure is only simplified or wholly erroneous. Therefore various additive calculations for some particular glass differing crucially in the underlying model of its structure can result in the comparable accuracy of obtained property magnitudes. Bearing this in mind, one should not take too seriously any assumptions of glass structure having no justifications except the satisfactory accuracy of the additive description of glass properties based on these assumptions.

Bearing in mind that all of the methods for calculating the optical characteristics in the transparency range based on glass composition lack a rigorous physical background, there is probably no use in trying to discriminate between them from the viewpoint of their physical significance. Therefore, the criterion for choosing between different methods of this kind can be only pragmatic, such as that of providing an acceptable accuracy in the easiest way.

B. GENERAL CORRELATIONS IN TERMS OF WEMPLE-DIDOMENICO, HIROTA-IZUMITANI, AND SIMILAR APPROACHES

The simplest single- and two-term Sellmeier-type dispersion formulas have been used widely as a basis when searching for general correlations between the optical characteristics of glasses and their chemical compositions or principal physical properties.[29,31,291,310-312] Using the single-term formula (40) (which is, in essence, the Sellmeier formula (31) rewritten for energies — see Chapter 1, Section II.C.3) Wemple[29] discussed the variations in the parameters of this formula with glass composition for Na_2O-SiO_2 glasses, various ternary and quarternary Me_2O-MeO-SiO_2 glasses, and also three Schott glasses. The parameters of Formula (40) are only the effective characteristics of the entire fundamental spectrum of electronic excitations taken as a whole and therefore, as was stressed in Chapter 1, Section II.C.3, have much less physical significance than that ascribed to them by Wemple himself. For further discussion, it is worth recalling that these parameters are (1) the energy of the single classical oscillator E_0 that plays the role of center of gravity of the entire fundamental spectrum and (2) the so-called dispersion energy E_d, which is proportional to the oscillator strength of this effective oscillator (for the connection between E_d and the usual parameters of the single-term Sellmeier formula, see Formula (42)). According to Reference 29, E_0 values decrease when adding other oxides to SiO_2, the effect becoming more pronounced in the CaO-BaO-Na_2O-PbO sequence. The E_d values of all glasses, except those rich in PbO and Schott lanthanum glasses, were found to vary only weakly in the ±10% limits around an average value close to that characteristic of vitreous silica. Within the above limits, E_d increases with the addition of BaO, CaO, and PbO (the presence of La_2O_3 promoting much greater E_d values) and decreases with the addition of Na_2O. Assuming E_d to be constant within large groups of structurally similar substances (see Chapter 1, Section II.C.3), Wemple took the above variations in E_d of glasses to be meaningful and interpreted them to result from the higher coordination numbers of Ba^{2+}, Ca^{2+}, Pb^{2+}, and La^{3+} cations in respective crystalline oxides as compared to those of Si^{4+} and Me^{+}. However, there is no guarantee that the coordination numbers of Ba^{2+}, Ca^{2+}, Pb^{2+}, and La^{3+} cations in glasses remain the same as those in individual crystalline oxides. Moreover, a decrease in E_d with the addition of Na_2O contradicts this explanation, because the coordination number of Na^{+} is by no means smaller than that of silicon atom.

In addition, Wemple[29] expressed Abbe numbers v_D of these glasses in terms of E_0 and E_d.

$$v_D = 0.72\mathrm{E}_0^2\left[n_D \big/ \left(n_D + 1\right) \right] \tag{77}$$

and

$$\nu_D = 0.72\mathrm{E}_d^2\left(n_\infty^2 - 1\right)^2\left[n_D \big/ \left(n_D + 1\right) \right] \tag{78}$$

Due to the narrow limits of possible variations in the $n_D/(n_D + 1)$ ratio, the Abbe values, in accordance with (77), were assumed to depend predominantly on E_0. Based on (78), Wemple described the location of the majority of optical glasses to lie in between two v_D vs. n_D curves corresponding to the limiting E_d values of 14.3 (that found for vitreous silica) and 25.0. Naturally, such a presentation of an Abbe diagram does not provide new information because E_0 and E_d are calculated from the same refractive indices that are used directly for plotting the points of glasses at the diagram.

Moreover, DiDomenico and then Wemple used their approach to discuss material dispersion M[28,31] and crossover wavelength λ_c values[31] of various materials, including glasses. In Reference 31, Wemple equipped (40) with the additional asymptotic term E_l^2/E^2 for the vibrational contributions and expressed M and λ_c in terms of the relations

$$M \sim f^*\ \beta z_a \big/ \left(E_0^3 \lambda_0^3\right) \tag{79}$$

and

$$\lambda_c \sim \left[f^* m^* d^3 \big/ \left(E_0^3 z_a \right) \right]^{1/4}, \tag{80}$$

m^* being the average reduced mass of atomic pairs and the rest of the parameters being the same as those in (41). Using (80), the lowest expected λ_c value of about 1.05 μ was calculated in Reference 31 for BeF_2. Because it is sufficient to have at one's disposal the λ_c values accurate to within the second decimal figure, such calculations probably can serve, in spite of the crudely approximate character of the Wemple-DiDomenico approach, as the useful empirical prognoses of λ_c.

Studies of the optical characteristics of inorganic glasses with various dispersion formulas was conducted for years by Hirota and Izumitani.[291,310,311] In References 310 and 311, a simple formula for $1/(n_\lambda^2 - 1)$ (being the reciprocal of the single-term Sellmeier formula (30)) was used. With this formula, Hirota and Izumitani treated the refractive indices of about 200 commercial glasses in the 365 to 1014 nm range[310] and the n_F, n_d, and n_C values of various ternary $10BaO \cdot 5M_{m/n}O \cdot 85B_2O_3$ glasses[311] (*M* being elements of Ia-VIa and IIb-Vb groups of the Periodic Table). Parameters under discussion were the inherent wavelength of a single effective oscillator λ_0, the number of the oscillators per unit volume *N*, and the oscillator strength f_0. According to References 310 and 311, n_d and the Abbe number of glasses are approximately the linear functions of the $Nf_0\lambda_0^2$ product and of the $1/\lambda_0^2$ ratio, respectively; at the same time, the Abbe number does not depend on either the number of oscillators per unit volume, *N*, or on the oscillator strength f_0. Oscillator strengths f_0 were found in Reference 310 to be large for glasses of the BK, SK, LaK, LaF, TaK, TaF, and TaFD types and small for various flints and dense flints; for any given glass type, the product $Nf_0\lambda_0$ is approximately a constant. In Reference 311, a classification of the above elements M is given based on the behavior of λ_0 and f_0. The first kind of elements comprising most of those studied in Reference 311 provides small λ_0 and large f_0 values of glasses which is attributed to the absorption by nonbridging oxygens. The second kind of elements, such as Tl, Pb, and Bi, provides, on the contrary, large λ_0 and small f_0 values of glasses which are attributed to the absorption due to the transitions from the $6s^2$ electronic shell of the elements. For the third kind of elements such as Ti, Zr, Hf, Nb, Ta, and W, an intermediate character of λ_0 and f_0 relationship was found. An attempt was also made to connect λ_0 and f_0 to the Z/a ratio, where *Z* is the formal electrostatic charge and *a* is the corresponding ionic radius. The complicated character of λ_0 vs. Z/a dependences shown in Reference 311 probably results from the fact that the Z/a ratio loses any physical significance when applied to elements with a substantial covalency of chemical bonds.

In Reference 291, Izumitani et al. calculated the parameters of the five-term Sellmeier formula of (30) type,

$$n^2 - 1 = \frac{KN_0f_0}{1/\lambda_0^2 - 1/\lambda^2} + \frac{KN_1f_1}{1/\lambda_1^2 - 1/\lambda^2} + \frac{KN_2f_2}{1/\lambda_2^2 - 1/\lambda^2} + \frac{KN_3f_3}{1/\lambda_3^2 - 1/\lambda^2} - \frac{KN_4f_4}{1/\lambda_4^2 - 1/\lambda^2}, \quad (81)$$

K being a constant, for vitreous silica and some commercial glasses. Four of the oscillators in this formula were related to the VUV[f] or near UV ranges and the fifth to the IR range. The inherent wavelengths of the first and second VUV oscillators, λ_0 and λ_1, were chosen based on the locations of the most intense farthest-from-the-visible and nearest-to-the-visible features (for vitreous SiO_2, 68 and 122 nm) in the experimentally measured VUV reflection spectra of glasses, these oscillators being attributed to the inner-shell electronic transitions and to transitions of the

[f] Izumitani et al.[291] distinguish between the VUV range (5 to 10 eV) and the extreme ultraviolet range, EUV (10 to 35 eV). For simplicity, the author in this book uses the standard notation of the entire range of energies higher than 5 eV as the VUV range.

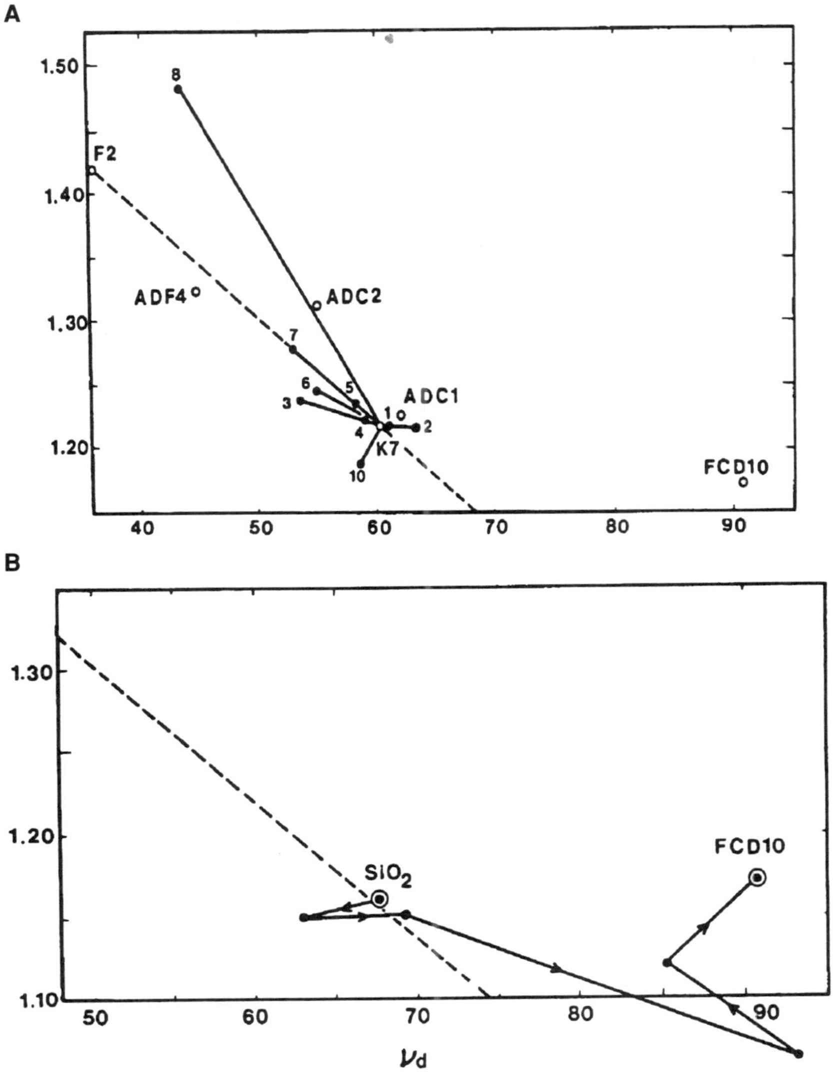

Figure 2 The data of numerical experiments by Izumitani et al. based on the Sellmeier-type formula (81). (A) The displacements of an initial point of Hoya's K7 glass at the $P_{i,g}$–υ_d diagram when increasing (1) λ_0 by 10 nm, (2) KN_0f_0 by 20.0, (3) λ_1 by 10 nm, (4) KN_1f_1 by 10.0, (5) λ_2 by 10 nm, (6) KN_2f_2 by 2.0, and (7) and (8) KN_3f_3 by 1.0 for λ_3 of 207 and 251 nm, and also KN_4f_4 by 0.004. The dashed line denotes the "normal line." (From Izumitani, T. et al., *J. Non-Cryst. Solids,* 86, 361, 1986. With permission.) (B) The difference in the location of vitreous silica and aluminium fluoride glass FCD10 at the $P_{i,g}$—υ_d diagram presented in terms of successive steps by the parameters of the Sellmeier-type formula (81). Steps are: for λ_0 from 68.4 to 54.5 nm, for KN_0f_0 from 148.21 to 185.83, for λ_1 from 116 nm to 97.2 nm, for KN_1f_1 from 30.52 to 58.29, for λ_4 from 9.3 to 12.0 μ, and for KN_4f_4 from 0.00915 to 0.00453. The dashed line denotes the normal line. (From Izumitani, T. et al., *J. Non-Cryst. Solids,* 86, 361, 1986. With permission.)

valence electrons of bridging oxygen atoms, respectively.[g] The inherent wavelength λ_2 of the third oscillator was chosen to be somewhere in the 142 to 179 nm range depending on glass composition and that of the fourth oscillator (if any), λ_3, to be at about 208 nm; these oscillators were attributed to the electronic transitions of the nonbridging oxygens and to the $6s^2$ electronic transitions in heavy cations such as Pb^{2+}, respectively. The λ_4 value related to the IR oscillator was, for silicate glasses, 9.8 μ. All the oscillator strengths, KN_jf_j, were found by fitting the experimental refractive indices of glasses with Formula (81).

The $P_{i,g}$–v_d relationship (concerning such relationships, see Section IV.A) as a function of variations in oscillator parameters of Hoya optical glass K7 and vitreous silica was studied in Reference 291 with a kind of numerical experiment. Making one-step changes in the parameters of Formula (81) for K7 glass chosen as the initial point, Izumitani et al. modeled the effect of an increase in each of these parameters on an Abbe number and the relative partial dispersion $P_{i,g}$ of this glass. In the $P_{i,g}$ vs. v_d plot presented in Reference 291, the resultant displacements of the initial point were approximated by straight lines. As is seen from Figure 2A borrowed from Reference 291, the Abbe number decreased and $P_{i,g}$ increased with an increase in all λ_j (lines 3 and 5 in Figure 2A) and KN_jf_j (lines 4 and 6 to 8 in Figure 2A), except those for the farthest-from-the-visible and the IR oscillators. When increasing λ_0 and KN_0f_0, the Abbe number, on the contrary, weakly increased[h] (lines 1 and 2 in Figure 2A). An increase in KN_4f_4 was found to be the only factor promoting a decrease in $P_{i,g}$. In addition, the difference in the location of vitreous silica and aluminium fluoride glass FCD10 was presented to be the result of successive steps by some of the parameters of Formula (81) (see Figure 2B). Based on Figure 2B, Izumitani et al. concluded that the higher Abbe number of FCD10 glass was due to shorter λ_1 and greater KN_0f_0 values, whereas the higher $P_{i,g}$ value of FCD10 glass was due mostly to the low oscillator strength of the IR oscillator.

Bohl[312] used the two-term Sellmeier formula (30) for treating the refractive indices of commercial optical glasses of the Schott (Jena) catalog and for interpreting their optical characteristics. The inherent wavelength of the first oscillator was found to be located in the VUV range; an approximately linear dependence between the values of this wavelength and Abbe numbers v_e of optical glasses was found, fluorine-containing crowns deviating to some extent from the common dependence. Also, an approximately linear dependence between the oscillator strengths of the first oscillator and the refractive indices of glasses was observed. The second inherent wavelength considered as an average for several excitations of different origin was found to vary in broad limits from the near UV to the near IR range; a not too distinct curvilinear relationship between the values of this wavelength and Abbe numbers was observed for most glasses, the points of fluorine-containing crowns and B_2O_3-containing glasses falling completely out of this regularity.

The interconnection between the parameters of the two-term Sellmeier formula and relative partial dispersions $P_{g,F'}$ of optical glasses was also considered in Reference 312. The shortest values of inherent wavelength of the first oscillator obtained for fluorine-containing crowns were assumed to be the main factor

---

[g] As is seen from Chapter 2, Section II.C, such attributing contradicts the acknowledged interpretation of the VUV spectra of oxides such as SiO_2 (see also Reference 81).

[h] This effect was not confirmed in later studies, such as that described in Reference 121.

responsible for the positive deviations of their $P_{g,F'}$ values from the "normal line" (for the meaning of these notions, see Section IV.A). The roles of both the oscillator strength of this oscillator and variations in the second inherent wavelength influenced by the location of the IR excitations were assumed to be much less substantial. On the contrary, the negative deviations of $P_{g,F'}$ values of B_2O_3-containing glasses were attributed to the influence of the location of the IR excitations. For dense flints, the roles of the second inherent wavelength and of corresponding oscillator strength were found to predominate in providing the positive deviations of their $P_{g,F'}$ values. In addition, based on the single-term Sellmeier formula of (30) type, Bohl showed the curvilinear character of the relationship between $\Delta P_{g,F'}$ and the inherent wavelength of an effective oscillator, λ_0, the wavelength range of negative $\Delta P_{g,F'}$ values corresponding approximately to $100 < \lambda < 220$ nm and the ranges of positive $\Delta P_{g,F'}$ values corresponding to $\lambda_0 < 100$ nm and to $\lambda_0 > 220$ nm. The role of variations in the effective oscillator strength, A_0, was found to be negligible.

As seen from the above discussion, the conclusions of Wemple,[29] Hirota and Izumitani,[310,311] and Bohl[312] concerning the behavior of Abbe numbers of glasses derived with the single- or two-term Sellmeier-type formulas agree, in general, with each other; the role of inherent wavelengths (or E_0, which is equivalent) is considered to predominate and the role of oscillator strengths is assumed to be much less substantial. However, the pattern becomes less simple with an increase in the number of terms in the dispersion formula. According to the data of Reference 291, the influence of variations in the λ_0 and λ_1 is weaker than that of variations in KN_3f_3 (see Figure 1A). There are also differences in explaining the behavior of relative partial dispersions; these differences are touched upon in Section IV.B.

IV. BEHAVIOR OF RELATIVE PARTIAL DISPERSIONS

A. EMPIRICAL REGULARITIES: NORMAL LINE AND ABNORMAL GLASSES

As was found initially by Abbe and then confirmed by many investigators, most optical glasses, when plotting their relative partial dispersions, P, vs. corresponding Abbe numbers, ν, are approximately located along certain straight lines. Such an interrelation between relative partial dispersions and Abbe numbers described by the empirical linear equation,

$$P_{\lambda_4,\lambda_5} = A_{\lambda_4,\lambda_5} + B_{\lambda_4,\lambda_5} \cdot \nu_{\lambda_1} , \tag{82}$$

is known as Abbe's rule; the corresponding straight lines are referred to as *normal lines*. The values of parameters A and B for different spectral ranges are given in Table 6.

Normal glasses, i.e., those satisfying Abbe's rule (82) do not allow for correcting the so-called secondary spectrum of a complex lens, because the magnitude of secondary spectrum, Δs, for a pair of normal glasses,

$$\Delta s = f' \cdot \frac{P_1 - P_2}{\nu_1 - \nu_2} \tag{83}$$

Table 6 Coefficients in Abbe's Empirical Relationship (82) for Different Spectral Ranges

Symbols of spectral lines for λ_4, λ_5		A	B
Schott catalog,	*i,g*	1.7241	–0.008382
main dispersion $n_F - n_C$	*g,F*	0.6438	–0.001682
	F,e	0.4884	–0.000526
	C,s	0.4029	+0.002331
	C,t	0.5450	+0.004743
Russian catalog,	*i,F′*	2.2580	–0.009327
main dispersion $n_{F'} - n_{C'}$	*g,F′*	0.5704	–0.001483
	F′,e	0.5429	–0.000570
	F′,r	1.1793	+0.000721

(f' being the focal length of a lens made from these glasses, P_1, ν_1, P_2, and ν_2 being the relative partial dispersions and Abbe numbers of the first and second glass, respectively), is a constant determined by the slope tangent of the normal line, $(P_1 - P_2)/(\nu_1 - \nu_2)$. Therefore, planning an optical system achromatized for more than two wavelengths is possible only if at least one glass is used that deviates appreciably from Abbe's normal line (82). Such glasses, the number of which in optical glass catalogs is usually rather limited, are often denoted as *abnormal.*

As a measure of the deviation of particular glasses from the normal line at a P_{λ_4,λ_5} vs. ν_{λ_1} plot, either the ordinate difference, $\Delta P_{\lambda_4,\lambda_5}$, or the abscissa difference, $\Delta\nu_{\lambda_1}$, can be used. In most cases, $\Delta P_{\lambda_4,\lambda_5}$ is commonly used as a parameter characterizing a glass from the viewpoint of its deviation from the normal line. Naturally, a boundary between the normal and abnormal glasses is somewhat conditional, because an optical glass catalog containing about 200 to 300 particular glass types presents a practically continuous sequence of $\Delta P_{\lambda_4,\lambda_5}$ values. As a criterion for the location of such a conditional boundary, not only $\Delta P_{\lambda_4,\lambda_5}$ but also $\Delta\nu_{\lambda_1}$ values are used; glasses with $\Delta\nu_{\lambda_1} \leq -3.0$ and with $\Delta\nu_{\lambda_1} \geq +3.0$ are commonly taken to be abnormal.

Two general kinds of abnormal glasses should be distinguished. Glasses characterized by the negative deviations of their relative partial dispersions from the normal line are referred to as *short,* whereas glasses characterized by the positive deviations from the normal line are referred to as *long*. Attributing a particular glass to short, normal, or long glasses depends on the wavelength range of interest. For example, Schott glasses PK51, FK51, FK52, and FK54 are typical long crowns from the viewpoint of their $\Delta P_{g,F}$ values; they are, however, short glasses when considering their $\Delta P_{C,s}$ values. Short glasses with respect to $\Delta P_{g,F}$ are usually flints and dense flints with ν_d values from about 35 to 55 and lanthanum crowns with ν_d of about 50 to 55. Long glasses with respect to $\Delta P_{g,F}$ are either crowns, especially fluorophosphate and purely fluoride crowns, or long dense flints. Historical terms (German in origin) for short flints and long crowns (with respect to a situation for the blue interval of the visible) are *kurtzflints* and *langkrons,* respectively; designations based on these terms are used for some Schott glasses, such as KzF and KzFS types and also LgSK2 glass. At the same time, many Schott glasses denoted in a usual way are also abnormal glasses (LaK8, LaKN16, and LaSFN31 being, for example, short ones, whereas SF56 to SF59 and TiSF1 are long glasses). Designations for abnormal glasses from the Hoya catalog are ADC and ADF (abnormal dispersion crowns and flints) and for those of the Russian catalog are OK and OF ("osobye," i.e., special, crowns and flints), respectively. Studies of abnormal glasses were mainly confined to a search for empirical regularities connecting relative partial dispersions with glass composition.

B. THE NATURE OF THE REGULARITIES AS FOLLOWS FROM NUMERICAL MODELING

Until recently, deviations of abnormal glasses from Abbe's rule (82) were commonly assumed to result from some not too clearly understood specific features of their structure and/or VUV spectra which distinguished them from the rest of the glasses. Correspondingly, all glasses lacking such specific features were considered to obey Abbe's rule. For example, Feltz et al.[302] cite a view that "as a premise for apochromatic correction, glasses of distinct short-range order have absolutely to be combined. Therefore, glasses belonging to isostructural series such as SiO_2, GeO_2, GeS_2... are precluded from apochromats." It was admitted tentatively for some kinds of Abbe-type diagrams (see Reference 302) that a nonlinear term of the $C\nu_{\lambda_1}^2$ type, in addition to the linear terms of (82), can present in the expression for the P_{λ_4,λ_5} vs. ν_{λ_1} relationship, but this idea was not widely used. Further insight into the pattern of relationships between relative partial dispersions and Abbe numbers of glasses was gained from numerical modeling that turned out to be an especially useful tool for studying the empirical regularities such as Abbe's rule (82); (see Section IV.B.1).

As mentioned in Section III.B, such modeling was first attempted in the above study[312] by Bohl based on the single-term Sellmeier formula of (30) type for the $P_{g,F'}$–ν_e diagram. Modeling was also conducted by Izumitani et al.[291] As is seen from the discussion in Section III.B, the conclusions of Bohl[312] and of Izumitani et al.[291] concerning the influence of oscillator strengths on relative partial dispersions of glasses contradict each other. Bohl considers the effect of variations in oscillator strengths on the magnitudes of deviations from the normal line to be negligible, whereas Izumitani et al. assume the effect of the IR oscillator strength to be substantial. In general, variations in the relative role of different parameters of Sellmeier-like dispersion formulas with the number of terms in the formulas are inevitable, because this number is always smaller than the number of individual features in the experimental fundamental spectra of glasses. Therefore, one should not put too much weight on views developed on the basis of data obtained with these formulas (especially the single- or two-term ones and with respect to subtle characteristics such as relative partial dispersions).

1. Numerical Modeling: Recent Findings

Recently, the problem of origin of Abbe's rule (82) was studied in detail with numerical modeling by Efimov[78,121] based on the analytical model (50) for the complex dielectric constant of glasses. The behavior of relative partial dispersions and of Abbe numbers was modeled with the calculation of these characteristics for artificial (that is, virtual) optical materials determined by several arbitrarily chosen oscillators. This guaranteed that each of the oscillators under consideration was indeed the individual one (and not the effective presentation of a series of individual excitations, as is the case when conducting the numerical experiment with data for actual glasses). Three VUV oscillators were used; the first, second, and third of the oscillators being considered the model representations of the farthest-to-the-visible, intermediate-in-location, and nearest-to-the-visible excitations, respectively. For the initial artificial material, the inherent wavelengths of these VUV

Table 7 Parameters of Oscillators Chosen in Reference 121 for Determining the Optical Characteristics of an Initial Artificial Material

Kind of parameter	Parameters of oscillators nos. 1	2	3	4
λ_j, nm	75	105	135	9250
$\Delta\varepsilon_{0j}$	0.35	0.30	0.35	0.6 and 0.8
γ_j/ω_j	0.09	0.07	0.05	0.015
σ_j/ω_j	0.08	0.05	0.02	0.013
High-frequency dielectric constant, ε_∞: 1.21				

oscillators were chosen to be 75, 105, and 135 nm, thus covering the entire range of the electronic excitations of typical oxide glasses; the rest of the parameters were assumed to be close to those found in Reference 159 for the strong features of the VUV spectrum of vitreous silica (see Table 7). Various cases of $P = P(\nu)$ relationships were studied in Reference 121 by means of calculating the changes in the optical characteristics of the artificial optical materials resulting from successive many-step variations in each of oscillator parameters. At first, a $P = P(\nu)$ relationship predetermined only by the behavior of the electronic excitations was considered; then, an additional IR oscillator was introduced for revealing the role of the IR excitations.

According to Reference 121, the magnitude of the Abbe number for the ensemble of only electronic oscillators is determined almost wholly by the location of the center of gravity of the entire VUV spectrum which, in general, depends on the inherent wavelengths and strengths of all oscillators. However, the role of the inherent wavelength of the nearest-to-the-visible (i.e., the longest-wavelength) excitation greatly predominates. Variations in the location of this excitation from one artificial material to another in reasonable limits (97.5 to 240 nm) result in varying both Abbe numbers and relative partial dispersions in broad limits that practically cover the entire ranges of these characteristics for optical glasses (for example, $22 < \nu_d < 94$). Compared to this effect, the influence of oscillator strength redistribution between the nearest-to-visible and the shorter wavelength oscillators (or, more so, the effects of variations in both the total oscillator strength of electronic excitations and in the rest of their parameters) are rather weak. All of the above holds for P vs. ν diagrams using any relative partial dispersions for the visible. For the case of the $P_{g,F}$ vs. ν_d plot, this is illustrated by Figure 3. Figure 3 shows that all such $P_{g,F} = P_{g,F}(\nu_d)$ dependences, unlike those of Izumitani et al.[291] for P_{ig} shown in Figure 2A, are not straight lines but lines of higher order. Only a small section of entire $P_{g,F} = P_{g,F}(\nu_d)$ dependence determined by the above variations in the location of the nearest-to-visible oscillator is covered by Figure 3. On the contrary, the sections of $P_{g,F} = P_{g,F}(\nu_d)$ curves corresponding to reasonable variations in the rest of the oscillator parameters are located in a rather small area $\nu_d \cong 61 \pm 4$ and $P_{g,F} \cong 0.5635 \pm 0.003$ (end points of these curves shown in Figure 3 for illustrating their shapes correspond to such oscillator parameter values that lie far outside the physically acceptable limits). Moreover, variations in the inherent wavelength of the nearest-to-the-visible electronic oscillator substantially influence the role of variations in other parameters of electronic excitations, thus promoting a rapid decrease in this role with an increase in ν_d. As a result, the location of the "electronic" $P = P(\nu)$ dependence as a whole in the P–ν coordinates

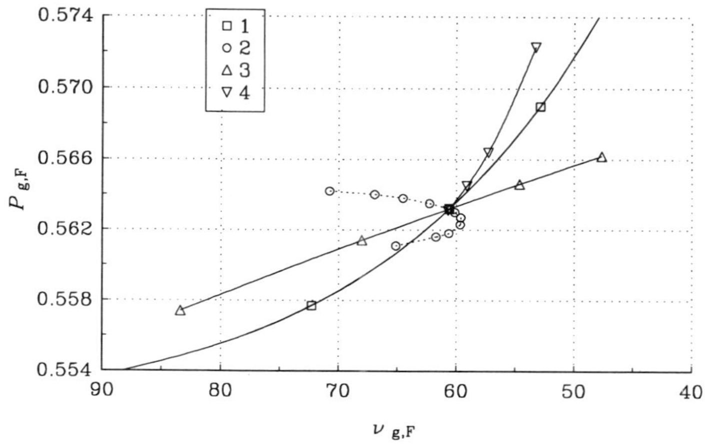

Figure 3 The model $P_{g,F} = P_{g,F}(\upsilon_d)$ dependences corresponding to successive variations in different parameters of the electronic oscillators such as (1) inherent wavelength of the third oscillator (a small part of full curve), (2) the total oscillator strength ($0.4 \leq \Delta\varepsilon_{0el} \leq 8.0$), (3) oscillator strength distribution between the first and third oscillators (the ratios between their $\Delta\varepsilon_{0j}$ values ranging from 0.6:0.1 to 0.1:0.6), and (4) the standard deviation of the third oscillator distribution ($0.001 \leq \sigma_3/\omega_3 \leq 0.14$).

for a variety of artificial materials can vary only rather weakly. This can be clearly seen from Figure 4; note that a difference between the two electronic $P_{g,F} = P_{g,F}(v_d)$ curves 1 and 2 corresponding to substantially different distributions of oscillator strengths over the VUV wavelength range practically vanishes in the region of $v_d >$ 60 to 65. Thus, the location of the electronic $P = P(v)$ dependence is rather insensitive to any possible changes in electronic oscillator parameters and in the region of high Abbe numbers becomes practically fixed. Hence it follows that an essentially similar electronic $P_{g,F} = P_{g,F}(v_d)$ dependence determined by the effect of varying the inherent wavelength of the nearest-to-visible electronic excitation is an underlying factor predetermining the general pattern of the $P_{g,F} = P_{g,F}(v_d)$ relationship for optical glasses, too. This dependence, however, as is seen from Figure 4, lies above the normal line, which reveals the existence of one more factor substantially influencing the magnitudes of relative partial dispersions.

When additionally including the IR excitation with an oscillator strength of the same order as the total oscillator strength of the IR spectrum of $33.4Na_2O \cdot 66.6SiO_2$ glass ($\Delta\varepsilon_0 \cong 0.6$) or of vitreous silica ($\Delta\varepsilon_0 \cong 0.8$), the calculated $P_{g,F}$ values greatly decrease. As a result, the location of the $P = P(v)$ curve as a whole shifts to the region of the normal line (see Figure 4, curves 4 and 5). This agrees well with the above assumption of Izumitani et al.[291] made for the particular case of fluoroaluminate FCD10 glass that a decrease in the IR intensity increases $P_{i,g}$ (see Section III.B). Such a substantial influence of IR excitations on the relative partial dispersions as that demonstrated by Figure 4 might seem rather unexpected because the contribution of the electronic excitations, $\Delta\varepsilon_{el}(\lambda)$, to the real dielectric constant $\varepsilon'(\lambda)$ at wavelengths of interest exceeds the contribution of the IR excitations, $\Delta\varepsilon'_{ir}(\lambda)$, by several orders of magnitude. However, the locations of the two ranges of fundamental excitations with respect to the visible estimated in terms of differences $\omega^2 - \omega_j^2$ also should be taken into account. As is shown in Reference 121,

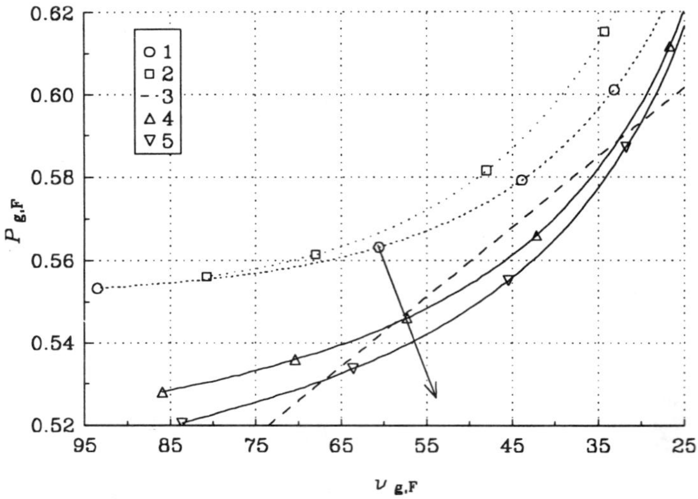

Figure 4 Variations in the location of a model $P_{g,F} = P_{g,F}(\upsilon_d)$ curve with respect to the normal line. (1) The location of the basic "electronic" $P_{g,F} = P_{g,F}(\upsilon_d)$ curve determined by successive variations in the inherent wavelength of the third oscillator for the case of oscillator strengths given by Table 7; (2) the same for the case of oscillator strengths of the first, second, and third oscillators equal to 0.45, 0.30, and 0.25, respectively; (3) the normal line; and (4) and (5) the $P_{g,F} = P_{g,F}(\upsilon_d)$ curves originated from curve 1 due to the introduction of the IR oscillator with oscillator strengths $\Delta\varepsilon_{0ir}$ of 0.6 and 0.8, respectively. An arrow shows the direction of displacement of a point from the initial $P_{g,F} = P_{g,F}(\upsilon_d)$ curve when successively increasing the IR oscillator intensity.

the relationship $(\omega^2 - \omega_{ir}^2)/(\omega_{el}^2 - \omega^2) \ll 1$ always holds; as a result, the frequency dispersion of the $\Delta\varepsilon'_{ir}(\omega)$ value in the visible is not negligible compared to that of $\Delta\varepsilon'_{el}(\omega)$. It is this factor that is responsible for such an appreciable influence of the IR excitations on the relative partial dispersions of optical materials.

As is seen from examples given by curves 4 and 5 in Figure 4, the model $P = P(\nu)$ dependences taking into account the role of the IR excitations remain to be of the same general form as the initial electronic curve, i.e., unlike the prediction of Abbe's rule, they are not straight lines. Their curvatures are substantial in the range of ν values corresponding to those obtained for actual glasses. The sign of curvature of a particular $P = P(\nu)$ curve can be both positive (as is the case for curves 4 and 5 in Figure 4) and negative; this depends on relative location of wavelengths determining the main and partial dispersions (see Section IV.B.2).

For comparison with experiment, the $P = P(\nu_D)$ model curves corresponding to two different intensities of the IR oscillator (the same as those used for calculating curves 4 and 5 of Figure 4) were presented in Reference 121 for the $P_{g,F}$ vs. ν_D and $P_{D,C}$ vs. ν_D diagrams. The area between these two curves was assumed to be the model representation of the most probable location of oxide glasses with high glassforming oxide content. As was shown in References 78 and 121, most Russian abnormal glasses are indeed located either within this area or, at least, rather close to its boundaries. Figure 5 demonstrates the same for Schott abnormal glasses with respect to the $P_{g,F} = P_{g,F}(\nu_d)$ diagram, two model curves 2 and 3 corresponding to curves 4 and 5 of Figure 4. As is seen from Figure 5, the area between the model $P_{g,F} = P_{g,F}(\nu_d)$ curves 2 and 3 covers the location of most of the abnormal Schott

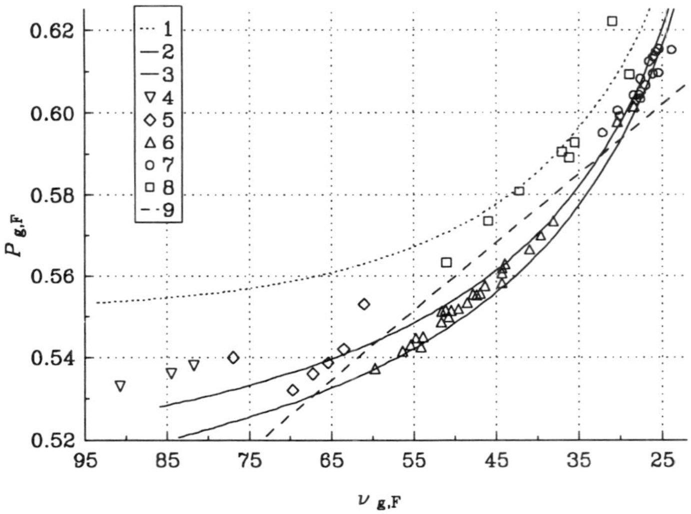

Figure 5 The location of Schott abnormal glasses characterized by $\Delta P_{g,F}$ absolute values greater than 0.0050 with respect to the model $P_{g,F} = P_{g,F}(\upsilon_d)$ curves. (1) the basic electronic $P_{g,F} = P_{g,F}(\upsilon_d)$ curve (the same as curve 1 in Figure 4); (2) and (3) the $P_{g,F} = P_{g,F}(\upsilon_d)$ curves corresponding to the oscillator strengths of the IR oscillator of 0.6 and 0.8 (the same as curves 4 and 5 in Figure 4), respectively; (4) glasses of FK type; (5) glasses of PK and PSK types and also LgSK2 glass; (6) glasses of LaK, KzF, KzSF, LaF, and LaSF types; (7) glasses of SF type; (8) glasses of TiF and TiSF types and also FN11 glass; and (9) the normal line.

glasses, such as practically all PK (except PK 51), PSK, LaK, KzF, KzFS, LaF, and LaSF and also most of SF glasses. There are very few marked exceptions from this pattern; such exceptions are the FK glass type (with the addition of particular PK51 and LgSK2 glasses) and also TiF and TiSF glass types (with the addition of FN11 glass), these cases being discussed below in Section IV.B.3. Thus, for most abnormal glasses, there exist “normal curves” rather than the “normal straight lines” assumed by Abbe’s rule.

Why then do normal glasses obey Abbe’s rule? As follows from Figure 4, the main reason for a scatter in P values of materials with a given v_d value is a scatter in the total intensities of their IR spectra. For actual glasses, these intensities are determined, first of all, by the glassforming oxide content, because the frequencies of vibrations of typical cations in their oxygen cages are located much farther from the visible than the stretching vibrations of structural groups making up glass networks (see Chapter 4). The SiO_2 and/or other glassforming oxide content of most normal glasses located in the central areas of P–v diagrams is much less than 67 mol%, thus resulting in lowered intensities of the fundamental IR spectra of these glasses and, hence, in increased $P_{g,F}$ values as compared to those corresponding to the upper of two $P_{g,F} = P_{g,F}(v_d)$ model curves[i] (Figure 4, curve 4 or Figure 5, curve 2). Therefore, such glasses are located randomly above this model $P_{g,F} = P_{g,F}(v_d)$ curve, thus masking the curvature of the upper boundary limiting the area

[i] As was mentioned above, the IR oscillator strength chosen for calculating this curve corresponds approximately to that found experimentally for $33.4Na_2O \cdot 66.6SiO_2$ glass.

occupied by the totality of optical glasses at the $P_{g,F}$–v_d diagram. As a result, an opportunity arises to approximate crudely the location of optical glasses as a whole at this diagram with the straight line. Hence it follows that it is normal glasses rather than most of the abnormal ones that deserve to be considered as deviating from the natural regularities.

2. Consequences of Curvilinear Character of *P*–*v* Diagrams: Natural Limitations Imposed on the Ranges of Relative Partial Dispersions Attainable for Glasses

The first principal consequence of the curvilinear character of the $P = P(v)$ relationship revealed by the data of Reference 121 is as follows: the deviation of most abnormal glasses (such as Schott glasses of PK, PSK, LaK, KzF, KzFS, LaF, and LaSF types) from the normal lines has an "algebraic" nature rather than a physically meaningful one. In other words, this deviation is explained solely by the mathematically predetermined curvilinear character of the $P = P(v)$ relationship which, in contrast to previous views, is not based on any specific features of glass composition and/or structure. This conclusion, however unexpected it might seem, is confirmed unequivocally by the fact that the location of most abnormal glasses is indeed predicted rather well by two $P = P(v)$ curves (Figure 5, curves 2 and 3) obtained solely with the above model calculations, calculations involving no information about the compositions and properties of particular glasses.

Further, the data of Reference 121 show that two possible types of $P = P(v)$ curves differing in the sign of curvature can exist and, hence, two respective types of P vs. v diagrams can be observed for actual glasses. The first type covers the cases when $\lambda_4 \leq \lambda_2$ and $\lambda_5 \leq \lambda_3$ (excluding, naturally, the case when both the inequalities become equalities). In these cases, the $P = P(v)$ curves forming the boundaries of area of the most probable location of glasses have a positive sign of curvature. For such diagrams, an increase in the intensity of the IR spectrum of a material diminishes its relative partial dispersion (see Figure 4); an example of these diagrams for optical glasses is $P_{g,F}$ vs. v_d illustrated by Figure 5. The second type covers the opposite cases when $\lambda_4 \geq \lambda_2$ and $\lambda_5 \geq \lambda_3$. In these cases, the $P = P(v)$ boundary curves have a negative curvature; along with this, an increase in the intensity of the IR spectrum of a material increases its relative partial dispersions. As an example of diagrams of the second type, $P_{D,C}$ vs. v_D is presented in Reference 121; an area between two model $P_{D,C} = P_{D,C}(v_D)$ curves plotted for the same IR oscillator strengths also satisfactorily represents the location of most abnormal glasses.

One more consequence of the curvilinear character of P vs. v diagrams imposes limitations on the ranges of relative partial dispersions attainable for inorganic glasses. This consequence is as follows: in the regions of very high and very low Abbe numbers, all glasses should be abnormal (i.e., no normal glasses can be obtained) and only one sign of anomaly with respect to the normal line can be observed depending on the type of P–v diagram (i.e., on the sign of curvature of $P = P(v)$ curve). This is due to the intersection of the $P = P(v)$ curve limiting the area of glasses (such as curve 3 in Figure 5) with the normal line in two points corresponding to different v values. In the case of the $P_{g,F}$–v_d diagram (and of other diagrams of the first type), all glasses located to the left or to the right from points in which the lower of the curves intersects with the normal line (see Figure 5) should inevitably be long crowns or long flints, respectively. Conversely, in the case of P vs. v diagrams of the second type, glasses located to the left or to the right from the intersection points should be short crowns and short flints. Judging by Figures 4 and 5, such intersection points for high-silica glasses (and for other glasses whose $\Delta\varepsilon'_{ir}(\lambda)$ values also are not greater than that of vitreous silica) should correspond to $v_d \cong 68$ and $v_d \cong 30$. Indeed, there are practically no normal optical glasses (not mentioning "short" glasses for the g,F interval or "long" glasses for the C,s interval) with $v_d > 68$ to 70 and $v_d < 28$ to 30. For simple glasses with the intense IR spectrum located nearer to the visible than that of SiO_2 (such as glasses close to pure B_2O_3) the intersection points can probably be separated by a somewhat broader interval, thus corresponding to $v_d \cong 78$ and $v_d \cong 22$.

Naturally, the maximum possible $\Delta\varepsilon'_{ir}(\lambda)$ values are, moreover, factors predetermining the maximum deviations of abnormal glasses from the normal line in between the intersection points. In general, there are not many opportunities to increase the attainable limit of $\Delta\varepsilon'_{ir}(\lambda)$ values of glasses by increasing their IR intensities or by diminishing the IR inherent wavelengths as compared to those characteristic of existing B_2O_3-based glasses. As a result, there are not many chances to enhance substantially the negative deviations of $P_{g,F}$ and like values for short glasses as compared to those attained until now. On the contrary, the IR spectra of polar hydrogen-containing compounds (mostly low-molecular liquids) can be located much nearer to the visible. As a result, such substances can have relative partial dispersions lying far outside the limits estimated above for glasses, thus being the extremely abnormal optical materials. For example, water having the IR spectrum extended up to about 3900 cm^{-1} is characterized by a $P_{g,F}$ value as low as about 0.518 for v_d of about 55.6.

3. Role of Specific Components: the Abnormal Behavior of Fluorine-Containing Glasses

As was mentioned above, there are also two kinds of abnormal glasses located substantially higher than the upper of two model $P = P(v)$ curves that do not follow the above pattern. As is seen from Figure 5, these are mostly either fluorine-containing (such as those of Schott FK type) or titanium-containing glasses (such as those of Schott TiF and TiSF types). The causes responsible for the behavior of relative partial dispersions of fluorine-containing glasses also can be interpreted rather unequivocally in terms of the approach developed in Reference 121.

Usually, the abnormal behavior of fluorine-containing crowns is ascribed to the specific role of fluorine. This view is supported by the fact that crystalline fluorides such as LiF, NaF, MgF_2, CaF_2, SrF_2, and BaF_2 show, along with very high Abbe numbers, even higher deviations of $P_{g,F}$ values from Abbe's rule than fluorine-containing crowns. However, as is seen from Figure 6, crystalline fluorides can substantially differ from each other in location with respect to the $P_{g,F} = P_{g,F}(v_d)$ normal curves 2 and 3 in Figure 5 resulting from the above model calculations. LiF, like high-silica glasses, falls within the area between these two normal curves; MgF_2 lies not too far from the upper of the curves. Thus, the abnormal behavior of LiF (and, to a lesser extent, of MgF_2) is caused by the curvature of the actual $P_{g,F} = P_{g,F}(v_d)$ dependence. No specific role of fluorine as such except providing quite high Abbe numbers is observed. On the contrary, NaF, CaF_2, SrF_2, and BaF_2 show great positive deviations from the model normal curves, thus being abnormal with respect to these curves, too (and not only with respect to the normal line). To a lesser extent, this is also valid for FK51, FK52, FK54, PK51, and LgSK2 Schott glasses also shown in

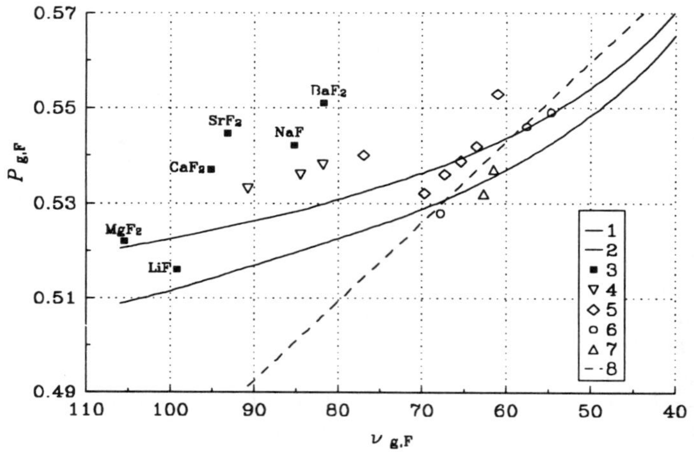

Figure 6 A section of $P_{g,F}$ vs. υ_d diagram for high Abbe numbers. (1) and (2) the model $P_{g,F} = P_{g,F}(\upsilon_d)$ curves (the same as curves 2 and 3 in Figure 5); (3) crystalline fluorides; (4) FK glasses; (5) PK and PSK glasses and also LgSK2 glass; (6) vitreous silica; $23.6Na_2O \cdot 76.4SiO_2$, and $33.4Na_2O \cdot 66.6SiO_2$ glasses; (7) $25.9BaO \cdot 74.1B_2O_3$ and $36.0 \cdot BaO \cdot 64.0B_2O_3$ glasses (mol% by analysis); and (8) the normal line.

Figure 6. Therefore the abnormal behavior of such fluorides and glasses cannot be reduced only to the curvature of the actual $P_{g,F} = P_{g,F}(v_d)$ dependence; some additional factors also participate. As was mentioned above in Section IV.B.1, the relative partial dispersion of a material with a given Abbe number exceeding 60 to 65 depends almost solely on the relative contribution of its IR spectrum, $|\Delta\varepsilon'_{ir}(\lambda)|/\varepsilon'(\lambda)$, to the total magnitude of the real dielectric constant $\varepsilon'(\lambda) = n^2(\lambda)$ at wavelengths of interest in the visible. Hence it follows that the extremely abnormal behavior of the above fluorine-containing materials can be caused only by substantially decreased magnitudes of this relative contribution as compared to oxide glasses. This is confirmed by the data of Table 8 taken from Reference 79. As is seen from Table 8 which presents data for D line, LiF is characterized by a $|\Delta\varepsilon'_{ir}(\lambda)|/\varepsilon'_D$ ratio close to that of $33.4Na_2O \cdot 66.6SiO_2$ glass, whereas CaF_2 and BaF_2 show greatly decreased values of this ratio (MgF_2 being in between these two extremities), which is in full agreement with the above pattern of location of these materials at the $P_{g,F}$–v_D diagram. Earlier, Izumitani et al.[291] assumed the lowered intensity of the effective IR oscillator to result in the increased $P_{i,g}$ dispersions of fluoroaluminate and fluorozirconate glasses (see also Section III.B). In general, the IR intensity is not the only factor determining the $|\Delta\varepsilon'_{ir}(\lambda)|/\varepsilon'(\lambda)$ ratio: a decreased magnitude of this ratio might also be caused by the location of strong features of the IR spectrum too far from the visible. Although CaF_2 shows, unlike LiF, a great positive deviation from the normal curves (see Figure 6), the principal IR frequencies of LiF and CaF_2 are rather close to one another (307 and 254 cm^{-1}, respectively[133,136]); clearly, a small difference in 50 cm^{-1} cannot result in the above difference in the behavior of relative partial dispersions of LiF and CaF_2. Therefore, only the substantially lowered IR oscillator strengths can be responsible for the deviations of relative partial dispersions of the above fluorine-containing optical glasses and crystalline fluorides from model normal curves. The above exceptions to this tendency illustrated by LiF and, to a lesser extent, MgF_2 demonstrate, however, that a lowered IR oscillator strength is not inherent in all

Table 8 Experimental Optical Characteristics and Properties of the IR Excitations for Some Simple Glasses and Fluoride Crystals

Material	Experimental optical characteristics		IR contribution to the ε' for the D line	
	ν_D	$P_{g,F}$	$\Delta\varepsilon'_{ir}$	$\Delta\varepsilon'_{ir}/\varepsilon'_D$
		Glasses		
SiO_2	67.8	0.528	0.00319	0.00150
$23.6Na_2O \cdot 76.4SiO_2$	57.6	0.546	0.00286	0.00127
$33.4Na_2O \cdot 66.6SiO_2$	54.7	0.549	0.00247	0.00109
$25.9BaO \cdot 74.1B_2O_3$	62.7	0.532	0.00449	0.00178
$36.0BaO \cdot 64.0B_2O_3$	61.5	0.537	0.00424	0.00160
		Crystals		
LiF	99.2	0.516	0.00229	0.00118
MgF_2[a]	105.5	0.522	0.00138	0.00073
CaF_2	95.1	0.537	0.00111	0.00054
BaF_2	81.7	0.551	0.00055	0.00025

[a] MgF_2 is the tetragonal crystal; data are given for the ordinary ray.

fluorine-based materials. Some other conditions (probably related to the structure of a material) should be obeyed, in addition to a high fluorine content, for promoting this effect.

Unlike the above case of fluorine-containing glasses, the substantially increased $P_{g,F}$ values of TiF and TiSF glasses cannot be explained only by the lowered IR intensities. As is seen from Figures 4 and 5, the displacement of points of the initial $P_{g,F} = P_{g,F}(\nu_d)$ model curve due to the introduction of the IR oscillator appreciably diminishes with a decrease in Abbe number. Moreover, some of these are located directly at the electronic $P_{g,F} = P_{g,F}(\nu_d)$ curve 1 of Figures 4 and 5 (as if there would be no IR excitations at all) and TiF6 glass even lies above this curve. No direct evidence concerning the reasons of such a situation is available; however, some qualitative assumptions can be expressed. According to Section IV.B.1, the location of the electronic $P_{g,F} = P_{g,F}(\nu_d)$ curve in the region of ν_d lower than 55 to 60 also depends somewhat on the parameters of the electronic excitations, too; the most appreciable role belongs to variations in the oscillator strength distribution between the nearest-to-the-visible and far-from-the-visible excitations (compare curves 1 and 2 of Figure 4). Bearing in mind the location of curve 2 of Figure 4, one may assume that the substantially increased $P_{g,F}$ values of titanium flints can be traceable to the much greater intensities of high-energy features of their VUV spectra as compared to the intensities of nearest-to-visible features. This factor combined with the low intensity of the IR spectrum might tentatively explain the observed location of TiF and TiSF glasses at the $P_{g,F}$–ν_d diagram.

As a concluding remark, I would like to stress once more that the analytical description of the complex dielectric constant and of related optical characteristics of inorganic glasses, when applied properly, allows one to extract a great amount of useful information unattainable by any other way from the standard experimental data and even provides reliable predictions of important optical properties. I hope that this book can serve, to some extent, as a demonstration of such opportunities.

Appendix: Numerical Data on the Optical Constants of Various Glasses in the Ranges of the Fundamental Excitations

Table A1 Optical Constants and Reflectivity of Vitreous Silica in the 900 to 1200 cm^{-1} Range According to References 79, 169, and 172

Frequency	Optical constants and reflectivity according to sources								
	Ref. 169		Ref. 172		Ref. 79				
(cm^{-1})	$n(\omega)$	$\kappa(\omega)$	$n(\omega)$	$\kappa(\omega)$	$R(\omega)$,%	$n(\omega)$	$\kappa(\omega)$	$\varepsilon'(\omega)$	$\varepsilon''(\omega)$
900	1.90	0.15	1.84	0.02	8.6	1.83	0.02	3.34	0.07
910	—	—	1.89	0.02	8.8	1.87	0.02	3.48	0.07
920	1.96	0.14	1.94	0.02	9.4	1.90	0.02	3.63	0.07
930	—	—	1.99	0.02	10.0	1.95	0.02	3.79	0.08
940	2.06	0.12	2.04	0.02	10.9	1.99	0.02	3.95	0.08
950	—	—	2.10	0.03	11.8	2.04	0.02	4.16	0.09
960	2.20	0.14	2.17	0.04	12.9	2.10	0.02	4.39	0.10
970	—	—	2.26	0.09	13.8	2.16	0.03	4.67	0.12
980	2.36	0.19	2.36	0.09	15.2	2.24	0.03	5.02	0.14
990	—	—	2.45	0.13	16.4	2.34	0.04	5.48	0.18
1000	2.62	0.28	2.56	0.23	18.7	2.48	0.06	6.13	0.29
1010	—	—	2.66	0.29	21.3	2.66	0.11	7.04	0.61
1020	2.87	0.61	2.82	0.40	23.9	2.87	0.25	8.19	1.44
1030	—	—	3.01	0.54	27.1	3.07	0.51	9.18	3.14
1040	3.01	1.13	3.19	0.82	30.0	3.18	0.87	9.36	5.51
1045	—	—	—	—	31.2	3.18	1.06	8.97	6.75
1050	2.98	1.43	3.25	1.19	32.6	3.14	1.25	8.30	7.84
1060	2.83	1.77	3.24	1.59	35.8	3.02	1.56	6.65	9.43
1070	2.62	2.03	2.99	2.05	38.9	2.91	1.87	4.99	10.86
1080	2.40	2.33	2.64	2.43	43.8	2.73	2.30	2.14	12.53
1090	2.12	2.60	2.07	2.71	49.1	2.24	2.73	−2.47	12.23
1095	—	—	—	—	52.7	1.89	2.85	−4.55	10.74
1100	1.50	2.82	1.50	2.77	56.4	1.52	2.86	−5.86	8.66
1105	—	—	—	—	61.1	1.17	2.77	−6.28	6.47
1110	—	—	0.83	2.61	65.8	0.88	2.60	−5.98	4.58
1120	0.52	2.13	0.47	2.26	71.0	0.52	2.17	−4.45	2.28
1124	—	—	—	—	71.5	0.45	2.01	−3.82	1.82
1130	0.41	1.77	0.32	1.90	69.3	0.39	1.78	−3.02	1.42
1140	0.41	1.52	0.31	1.60	62.6	0.38	1.49	−2.07	1.14
1150	—	—	0.36	1.40	53.3	0.41	1.28	−1.46	1.05
1160	0.47	1.18	0.39	1.26	46.3	0.45	1.12	−1.05	1.01
1170	—	—	0.40	1.15	40.0	0.48	1.02	−0.82	0.99
1180	0.51	1.00	0.39	1.05	38.2	0.50	0.95	−0.65	0.95
1190	0.52	0.93	0.39	0.95	35.0	0.50	0.89	−0.53	0.89
1200	0.50	0.87	0.38	0.86	33.0	0.49	0.82	−0.44	0.81
1210	—	—	0.39	0.78	31.6	0.46	0.76	−0.36	0.70
1220	0.45	0.72	0.36	0.70	30.7	0.44	0.68	−0.27	0.59
1230	—	—	0.33	0.60	29.2	0.41	0.59	−0.18	0.48
1240	0.41	0.54	0.32	0.48	27.2	0.40	0.48	−0.07	0.38
1250	0.40	0.42	0.32	0.34	23.2	0.41	0.35	0.04	0.29
1260	0.47	0.29	0.38	0.17	16.8	0.46	0.23	0.16	0.21

Table A1 (continued) Optical Constants and Reflectivity of Vitreous Silica in the 900 to 1200 cm^{-1} Range According to References 79, 169, and 172

Frequency	Optical constants and reflectivity according to sources								
	Ref. 169		Ref. 172		Ref. 79				
(cm^{-1})	$n(\omega)$	$\kappa(\omega)$	$n(\omega)$	$\kappa(\omega)$	$R(\omega)$,%	$n(\omega)$	$\kappa(\omega)$	$\varepsilon'(\omega)$	$\varepsilon''(\omega)$
1270	—	—	0.51	0.08	9.3	0.54	0.14	0.27	0.15
1280	0.62	0.17	0.62	0.06	5.2	0.62	0.09	0.38	0.11
1290	—	—	0.69	0.04	2.9	0.70	0.06	0.48	0.08
1300	0.74	0.06	0.76	0.02	1.7	0.76	0.04	0.58	0.06
1310	—	—	—	—	1.2	0.81	0.03	0.66	0.04
1320	—	—	—	—	0.9	0.86	0.02	0.74	0.04
1330	—	—	—	—	0.6	0.90	0.02	0.80	0.03
1340	—	—	—	—	0.3	0.93	0.01	0.86	0.03
1350	0.95	0.02	—	—	0.2	0.96	0.01	0.92	0.02

Table A2 Optical Constants and Reflectivity of Vitreous Silica in the 200 to 900 cm^{-1} Range According to References 79, 169, and 170

Frequency	Optical constants and reflectivity according to sources								
	Ref. 170		Ref. 169		Ref. 79				
(cm^{-1})	$n(\omega)$	$\kappa(\omega)$	$n(\omega)$	$\kappa(\omega)$	$R(\omega)$,%	$n(\omega)$	$\kappa(\omega)$	$\varepsilon'(\omega)$	$\varepsilon''(\omega)$
200	—	—	2.06	0.03	11.6	2.00	0.02	4.01	0.08
210	—	—	—	—	11.6	2.01	0.02	4.05	0.09
230	—	—	—	—	11.8	2.04	0.03	4.14	0.11
250	—	—	2.12	0.05	12.0	2.06	0.03	4.25	0.14
270	—	—	—	—	12.5	2.09	0.04	4.38	0.18
290	—	—	—	—	13.2	2.13	0.06	4.55	0.24
300	—	—	2.19	0.11	—	2.16	0.06	4.65	0.28
310	—	—	—	—	13.9	2.18	0.07	4.77	0.32
330	—	—	—	—	14.6	2.25	0.10	5.06	0.46
350	—	—	2.33	0.16	16.2	2.35	0.16	5.44	0.70
370	—	—	—	—	18.4	2.45	0.24	5.96	1.17
390	—	—	—	—	20.8	2.59	0.41	6.54	2.11
400	2.53	0.20	2.62	0.41	22.0	2.64	0.54	6.70	2.88
410	—	—	2.71	0.54	23.4	2.67	0.71	6.62	3.80
415	2.65	0.40	—	—	—	2.67	0.80	6.47	4.25
420	—	—	2.77	0.71	25.0	2.66	0.88	6.29	4.68
430	—	—	2.79	0.94	27.0	2.65	1.06	5.89	5.64
433	2.86	0.94	—	—	—	—	—	—	—
440	—	—	2.77	1.18	30.0	2.62	1.31	5.16	6.88
450	2.74	1.83	2.68	1.52	34.0	2.56	1.69	3.76	8.69
455	2.51	2.10	—	—	—	2.43	2.00	1.89	9.74
460	—	—	2.45	1.92	42.0	2.07	2.29	–0.96	9.53
465	1.71	2.31	—	—	47.8	1.56	2.38	–3.24	7.48
470	—	—	1.84	2.30	53.5	1.09	2.23	–3.78	4.86
475	—	—	1.37	2.39	55.6	0.78	1.93	–3.12	2.99
480	—	—	0.98	2.16	53.6	0.62	1.62	–2.24	1.99
481	0.72	1.71	—	—	—	—	—	—	—
490	—	—	0.52	1.39	40.0	0.53	1.12	–0.97	1.19
500	0.67	0.75	0.64	0.87	24.2	0.58	0.75	–0.22	0.86
510	—	—	—	—	9.7	0.68	0.50	0.21	0.68
520	—	—	0.99	0.42	3.1	0.86	0.32	0.65	0.55
529	1.05	0.20	—	—	—	—	—	—	—
540	—	—	1.12	0.32	1.4	1.09	0.19	1.16	0.41
550	—	—	—	—	1.5	1.16	0.16	1.34	0.37
552	1.23	0.08	—	—	—	—	—	—	—
560	—	—	1.30	0.24	1.7	1.23	0.14	1.50	0.34
570	—	—	—	—	2.0	1.28	0.12	1.64	0.31
571	1.35	0.06	—	—	—	—	—	—	—
580	—	—	1.35	0.19	2.2	1.33	0.11	1.75	0.30
588	1.41	0.05	—	—	—	—	—	—	—
590	—	—	—	—	2.5	1.36	0.10	1.85	0.28
600	—	—	1.41	0.14	2.8	1.40	0.10	1.94	0.27
606	1.44	0.05	—	—	—	—	—	—	—
610	—	—	—	—	3.1	1.42	0.09	2.02	0.26
620	—	—	1.47	0.13	3.4	1.45	0.09	2.09	0.25
625	1.46	0.04	—	—	—	1.46	0.08	2.12	0.24
630	—	—	—	—	3.7	1.47	0.08	2.15	0.24

Table A2 (continued) Optical Constants and Reflectivity of Vitreous Silica in the 200 to 900 cm^{-1} Range According to References 79, 169, and 170

Frequency (cm^{-1})	Optical constants and reflectivity according to sources								
	Ref. 170		Ref. 169		Ref. 79				
	$n(\omega)$	$\kappa(\omega)$	$n(\omega)$	$\kappa(\omega)$	$R(\omega)$,%	$n(\omega)$	$\kappa(\omega)$	$\varepsilon'(\omega)$	$\varepsilon''(\omega)$
640	—	—	1.54	0.10	3.8	1.49	0.08	2.21	0.23
645	1.52	0.03	—	—	—	1.50	0.07	2.24	0.22
650	—	—	—	—	4.0	1.51	0.07	2.27	0.21
660	—	—	1.60	0.08	4.1	1.53	0.07	2.33	0.20
667	1.54	0.03	—	—	—	—	—	—	—
680	—	—	1.67	0.08	4.6	1.56	0.06	2.44	0.18
690	1.60	0.03	—	—	4.8	1.58	0.05	2.50	0.16
700	—	—	1.72	0.08	5.1	1.60	0.05	2.57	0.15
714	1.66	0.04	—	—	—	—	—	—	—
720	—	—	1.77	0.08	5.7	1.65	0.04	2.72	0.14
740	—	—	1.86	0.12	6.2	1.71	0.05	2.92	0.17
741	1.70	0.06	—	—	—	—	—	—	—
750	—	—	—	—	6.8	1.74	0.06	3.04	0.22
752	1.72	0.08	—	—	—	—	—	—	—
760	—	—	1.92	0.20	7.1	1.78	0.09	3.15	0.33
775	1.84	0.17	—	—	—	1.80	0.16	3.23	0.57
780	—	—	1.93	0.31	8.5	1.80	0.18	3.22	0.66
790	—	—	—	—	8.5	1.78	0.23	3.12	0.83
800	1.77	0.34	1.82	0.37	7.8	1.74	0.26	2.96	0.92
810	—	—	—	—	7.2	1.68	0.27	2.76	0.91
820	1.64	0.28	1.72	0.32	6.3	1.63	0.24	2.61	0.79
840	—	—	1.71	0.24	5.3	1.61	0.13	2.56	0.43
860	—	—	1.77	0.20	5.6	1.67	0.05	2.79	0.18
862	1.68	0.06	—	—	—	—	—	—	—
870	1.69	0.03	—	—	6.5	1.71	0.03	2.94	0.12
880	—	—	1.84	0.19	7.4	1.75	0.02	3.08	0.09
885	1.71	0.02	—	—	—	1.77	0.02	3.14	0.08
890	—	—	—	—	7.7	1.79	0.02	3.21	0.07

Table A3 Optical Constants and Reflectivities of Some Sodium Silicate Glasses in the 40 to 1260 cm^{-1} Range According to Reference 79 (Glass Compositions in Mol% by Analysis)

	$19.3Na_2O \cdot 80.7SiO_2$			$23.6Na_2O \cdot 76.4SiO_2$			$33.4Na_2O \cdot 66.6SiO_2$		
ω, cm^{-1}	$R(\omega)$,%	$n(\omega)$	$\kappa(\omega)$	$R(\omega)$,%	$n(\omega)$	$\kappa(\omega)$	$R(\omega)$,%	$n(\omega)$	$\kappa(\omega)$
40	19.4	2.53	0.20	21.0	2.65	0.25	20.6	2.58	0.30
60	18.8	2.50	0.33	20.4	2.60	0.39	19.4	2.50	0.42
80	18.2	2.40	0.45	20.0	2.50	0.51	18.5	2.38	0.52
100	17.5	2.26	0.51	19.2	2.36	0.56	16.8	2.24	0.55
120	15.5	2.15	0.50	17.4	2.26	0.56	15.6	2.13	0.53
140	13.8	2.08	0.49	16.3	2.20	0.56	14.5	2.06	0.50
160	13.5	2.02	0.48	16.0	2.13	0.59	13.5	2.02	0.48
180	12.8	1.95	0.46	15.4	2.04	0.61	12.5	1.98	0.47
200	11.2	1.89	0.41	13.5	1.94	0.59	12.4	1.93	0.45
220	10.1	1.85	0.34	11.5	1.85	0.52	11.2	1.90	0.42
240	9.7	1.86	0.28	9.8	1.81	0.42	10.6	1.88	0.38
260	10.3	1.90	0.23	9.9	1.83	0.33	10.6	1.88	0.35
280	10.9	1.95	0.20	10.4	1.89	0.29	10.6	1.89	0.33
300	11.5	2.01	0.21	11.1	1.95	0.29	10.6	1.90	0.32
320	12.2	2.06	0.24	11.7	1.98	0.32	10.7	1.91	0.32
340	13.0	2.09	0.28	12.2	2.00	0.34	10.8	1.93	0.31
360	14.0	2.11	0.31	12.8	2.02	0.34	11.8	1.97	0.31
380	15.5	2.16	0.33	14.2	2.09	0.34	13.2	2.04	0.34
400	17.3	2.28	0.39	16.0	2.21	0.42	14.7	2.13	0.42
410	—	2.36	0.47	—	2.27	0.51	—	2.16	0.50
420	19.6	2.43	0.61	18.2	2.31	0.64	16.5	2.17	0.60
430	—	2.46	0.79	—	2.30	0.79	—	2.15	0.71
440	22.6	2.41	1.02	20.8	2.24	0.96	18.1	2.10	0.82
450	—	2.26	1.24	—	2.12	1.11	—	2.01	0.93
460	26.9	2.03	1.40	23.5	1.94	1.22	19.2	1.89	1.01
470	—	1.74	1.48	—	1.73	1.28	—	1.74	1.06
475	—	1.58	1.47	—	1.62	1.29	—	1.65	1.07
480	28.5	1.43	1.44	24.1	1.50	1.27	18.2	1.57	1.07
490	—	1.16	1.29	—	1.28	1.20	—	1.41	1.02
500	23.4	0.96	1.07	20.1	1.10	1.05	15.9	1.26	0.94
510	—	0.87	0.80	—	0.97	0.85	—	1.14	0.80
520	7.4	0.89	0.55	9.6	0.92	0.63	10.1	1.07	0.64
540	1.1	1.11	0.27	0.9	1.06	0.28	0.5	1.11	0.35
560	2.0	1.28	0.19	2.0	1.26	0.15	2.0	1.25	0.19
580	3.2	1.38	0.18	3.2	1.40	0.13	3.4	1.38	0.14
600	3.8	1.44	0.18	4.0	1.48	0.14	4.1	1.46	0.15
620	4.1	1.47	0.18	4.5	1.53	0.17	4.6	1.48	0.17
640	4.4	1.49	0.16	4.8	1.54	0.17	4.8	1.49	0.14
660	4.8	1.52	0.13	5.2	1.56	0.16	5.2	1.53	0.09
680	5.3	1.56	0.10	5.6	1.59	0.15	5.7	1.60	0.07
700	5.9	1.62	0.08	6.0	1.62	0.14	6.2	1.66	0.09
720	6.6	1.68	0.09	6.6	1.67	0.13	6.8	1.71	0.13
740	7.3	1.74	0.14	7.3	1.72	0.16	7.4	1.72	0.18
750	—	1.76	0.17	—	1.74	0.19	—	1.70	0.20
760	7.8	1.76	0.20	7.8	1.74	0.23	7.3	1.68	0.20
775	—	1.73	0.24	—	1.71	0.27	—	1.67	0.17
780	7.4	1.72	0.25	7.4	1.69	0.26	6.7	1.66	0.16
790	—	1.68	0.25	—	1.66	0.26	—	1.67	0.14

Table A3 (continued) Optical Constants and Reflectivities of Some Sodium Silicate Glasses in the 40 to 1260 cm^{-1} Range According to Reference 79 (Glass Compositions in Mol% by Analysis)

	$19.3Na_2O \cdot 80.7SiO_2$			$23.6Na_2O \cdot 76.4SiO_2$			$33.4Na_2O \cdot 66.6SiO_2$		
ω, cm^{-1}	$R(\omega)$,%	$n(\omega)$	$\kappa(\omega)$	$R(\omega)$,%	$n(\omega)$	$\kappa(\omega)$	$R(\omega)$,%	$n(\omega)$	$\kappa(\omega)$
800	6.3	1.66	0.24	6.1	1.63	0.23	6.4	1.69	0.11
820	5.7	1.65	0.16	5.9	1.64	0.13	7.5	1.76	0.07
840	6.6	1.70	0.10	7.0	1.73	0.08	9.3	1.87	0.06
850	—	1.74	0.08	—	1.78	0.06	—	1.93	0.06
860	8.1	1.79	0.06	8.8	1.84	0.06	11.5	2.01	0.09
880	9.8	1.90	0.06	10.6	1.96	0.08	14.3	2.17	0.20
900	11.9	2.03	0.09	13.1	2.11	0.16	17.2	2.29	0.43
910	—	2.10	0.13	—	2.18	0.22	—	2.30	0.59
920	14.1	2.17	0.19	15.5	2.24	0.31	19.5	2.25	0.75
930	—	2.23	0.27	—	2.28	0.42	—	2.16	0.90
940	16.1	2.28	0.37	17.8	2.30	0.55	20.2	2.01	1.00
950	—	2.30	0.49	—	2.28	0.68	—	1.85	1.04
960	18.0	2.29	0.60	19.4	2.23	0.80	18.6	1.69	1.02
970	—	2.26	0.72	—	2.16	0.90	—	1.58	0.95
980	19.3	2.20	0.80	20.4	2.07	0.97	14.5	1.53	0.85
990	—	2.14	0.87	—	1.98	1.02	—	1.54	0.78
1000	19.9	2.10	0.92	20.4	1.90	1.05	12.9	1.58	0.76
1010	—	2.06	0.97	—	1.84	1.07	—	1.62	0.78
1020	20.8	2.03	1.04	19.6	1.78	1.11	14.2	1.63	0.84
1030	—	1.99	1.13	—	1.73	1.15	—	1.61	0.91
1040	23.3	1.93	1.24	21.6	1.65	1.20	17.2	1.55	0.98
1050	—	1.82	1.34	—	1.55	1.25	—	1.47	1.05
1060	27.4	1.68	1.42	24.8	1.43	1.28	20.0	1.36	1.08
1070	—	1.50	1.46	—	1.30	1.28	—	1.24	1.10
1075	—	1.41	1.47	—	1.23	1.27	—	1.18	1.09
1080	30.4	1.32	1.46	26.3	1.16	1.25	20.9	1.12	1.08
1090	—	1.15	1.41	—	1.04	1.18	—	1.00	1.03
1100	29.8	1.01	1.32	23.6	0.95	1.10	19.9	0.90	0.96
1120	26.6	0.83	1.10	20.6	0.85	0.92	17.4	0.75	0.76
1140	23.1	0.77	0.91	18.0	0.80	0.79	12.7	0.70	0.54
1150	—	0.74	0.85	—	0.78	0.73	—	0.71	0.44
1160	19.4	0.72	0.78	15.1	0.75	0.66	6.5	0.73	0.36
1180	15.6	0.68	0.64	11.5	0.69	0.53	2.0	0.80	0.22
1200	11.8	0.65	0.48	7.3	0.68	0.36	0.5	0.89	0.13
1220	7.6	0.67	0.32	3.4	0.75	0.20	0.2	0.97	0.08
1240	3.4	0.75	0.18	1.0	0.85	0.11	0.3	1.03	0.06
1250	—	0.80	0.14	—	0.89	0.09	—	1.06	0.05
1260	0.8	0.84	0.11	0.2	0.93	0.07	0.5	1.08	0.04
1280	—	0.92	0.07	—	1.00	0.05	—	1.12	0.03
1300	—	0.99	0.04	—	1.05	0.04	—	1.16	0.02

Table A4 Optical Constants and Reflectivities of Some Barium Borate Glasses in the 640 to 1680 cm^{-1} Range According to Reference 79 (Glass Compositions in Mol% by Analysis)

	$21.1BaO \cdot 78.9B_2O_3$			$25.9BaO \cdot 74.1B_2O_3$			$31.1BaO \cdot 68.9B_2O_3$		
ω, cm^{-1}	$R(\omega)$,%	$n(\omega)$	$\kappa(\omega)$	$R(\omega)$,%	$n(\omega)$	$\kappa(\omega)$	$R(\omega)$,%	$n(\omega)$	$\kappa(\omega)$
640	13.0	2.12	0.08	13.6	2.17	0.11	14.0	2.22	0.14
660	13.8	2.16	0.13	14.3	2.21	0.17	14.8	2.24	0.19
670	14.3	2.18	0.17	14.6	2.22	0.21	15.0	2.24	0.23
680	14.6	2.20	0.23	14.8	2.21	0.25	15.2	2.24	0.27
700	14.2	2.12	0.39	14.5	2.14	0.34	14.9	2.19	0.33
710	13.0	2.01	0.41	13.6	2.08	0.35	14.5	2.16	0.35
720	11.1	1.92	0.37	12.6	2.04	0.32	14.0	2.12	0.34
740	10.0	1.89	0.23	12.3	2.02	0.26	13.4	2.08	0.30
760	11.2	1.94	0.16	12.3	2.03	0.23	12.9	2.10	0.24
780	11.1	2.01	0.14	12.7	2.07	0.18	13.3	2.16	0.22
800	12.2	2.06	0.15	13.7	2.16	0.18	15.0	2.24	0.23
820	13.1	2.12	0.18	15.1	2.24	0.23	16.5	2.32	0.29
840	14.1	2.17	0.23	16.2	2.30	0.31	17.7	2.38	0.40
860	15.0	2.21	0.30	17.2	2.32	0.41	18.7	2.39	0.52
880	15.9	2.24	0.38	18.1	2.32	0.51	19.5	2.37	0.64
900	16.5	2.23	0.48	18.7	2.30	0.62	20.2	2.32	0.75
920	16.7	2.19	0.58	19.0	2.25	0.74	20.4	2.25	0.86
940	16.5	2.11	0.66	19.1	2.16	0.84	20.6	2.14	0.96
960	15.8	2.02	0.70	18.6	2.03	0.90	20.5	2.00	1.02
980	15.2	1.94	0.71	17.9	1.90	0.92	20.0	1.86	1.04
1000	14.5	1.87	0.72	16.7	1.78	0.90	18.8	1.72	1.02
1020	13.7	1.81	0.72	15.4	1.69	0.86	16.7	1.60	0.98
1040	13.0	1.74	0.72	13.9	1.62	0.82	14.6	1.50	0.91
1060	11.8	1.66	0.69	12.4	1.54	0.77	12.9	1.41	0.83
1080	10.7	1.59	0.64	10.7	1.47	0.70	11.0	1.33	0.73
1100	9.3	1.55	0.55	9.1	1.42	0.60	8.9	1.29	0.61
1120	7.9	1.55	0.44	7.0	1.41	0.48	6.3	1.30	0.47
1140	6.5	1.61	0.35	5.6	1.46	0.37	4.6	1.36	0.36
1160	6.8	1.71	0.30	5.5	1.56	0.29	4.4	1.45	0.29
1180	9.3	1.84	0.31	7.0	1.69	0.26	5.3	1.56	0.26
1200	13.0	1.97	0.40	10.0	1.84	0.32	7.0	1.66	0.28
1220	15.7	2.02	0.60	12.5	1.92	0.49	8.8	1.74	0.34
1240	15.9	1.92	0.79	13.5	1.86	0.65	10.2	1.78	0.43
1250	15.4	1.82	0.84	13.2	1.79	0.69	—	1.79	0.47
1260	14.6	1.72	0.84	13.0	1.73	0.69	11.0	1.79	0.51
1280	13.1	1.63	0.74	12.5	1.70	0.65	11.7	1.78	0.59
1300	12.8	1.68	0.69	12.5	1.74	0.68	12.9	1.75	0.68
1310	—	1.72	0.72	13.4	1.74	0.73	—	1.73	0.72
1320	14.0	1.74	0.77	14.0	1.73	0.79	14.0	1.69	0.77
1340	15.8	1.70	0.90	15.4	1.64	0.90	14.5	1.58	0.85
1360	17.4	1.57	1.01	16.4	1.50	0.96	14.3	1.44	0.88
1370	—	1.48	1.04	16.5	1.42	0.97	—	1.36	0.08
1380	17.9	1.39	1.05	16.2	1.34	0.96	13.4	1.29	0.85
1400	17.7	1.21	1.00	14.8	1.20	0.90	11.8	1.18	0.78
1420	16.1	1.08	0.90	13.1	1.11	0.81	9.9	1.10	0.68
1440	13.9	1.01	0.79	11.2	1.06	0.72	8.0	1.07	0.59
1460	11.7	0.97	0.70	9.5	1.00	0.66	6.3	1.04	0.52

Table A4 (continued) Optical Constants and Reflectivities of Some Barium Borate Glasses in the 640 to 1680 cm^{-1} Range According to Reference 79 (Glass Compositions in Mol% by Analysis)

	$21.1BaO \cdot 78.9B_2O_3$			$25.9BaO \cdot 74.1B_2O_3$			$31.1BaO \cdot 68.9B_2O_3$		
ω, cm^{-1}	$R(\omega)$,%	$n(\omega)$	$\kappa(\omega)$	$R(\omega)$,%	$n(\omega)$	$\kappa(\omega)$	$R(\omega)$,%	$n(\omega)$	$\kappa(\omega)$
1480	9.9	0.92	0.62	7.9	0.94	0.58	4.7	1.00	0.45
1500	7.7	0.88	0.53	6.0	0.89	0.47	3.0	0.98	0.35
1520	5.5	0.86	0.41	3.8	0.89	0.34	1.4	0.99	0.25
1540	3.4	0.88	0.30	1.9	0.93	0.23	0.3	1.04	0.16
1560	1.4	0.92	0.20	0.1	1.00	0.16	0.2	1.09	0.11
1580	0.3	0.98	0.14	0.3	1.06	0.12	0.4	1.15	0.08
1600	0.2	1.04	0.10	0.3	1.10	0.10	0.7	1.19	0.06
1620	0.1	1.09	0.08	0.4	1.14	0.08	0.9	1.22	0.05
1640	0.4	1.13	0.06	0.7	1.18	0.07	1.1	1.25	0.04
1660	0.6	1.16	0.05	0.8	1.20	0.06	1.3	1.28	0.04
1680	0.8	1.19	0.04	1.0	1.23	0.05	1.5	1.30	0.03

Table A5 Optical Constants and Reflectivity of 36.0BaO · $64.0B_2O_3$ Glass (Mol% by Analysis) in the 20 to 1650 cm^{-1} Range According to Reference 79

ω, cm^{-1}	$R(\omega)$,%	$n(\omega)$	$\kappa(\omega)$	ω, cm^{-1}	$R(\omega)$,%	$n(\omega)$	$\kappa(\omega)$
20	23.5	2.87	0.20	820	16.9	2.30	0.34
30	23.3	2.83	0.26	840	17.9	2.35	0.44
40	23.0	2.80	0.32	860	18.6	2.36	0.55
50	22.9	2.77	0.38	880	19.4	2.34	0.67
60	23.0	2.75	0.43	900	20.0	2.30	0.78
70	23.3	2.74	0.48	920	20.5	2.22	0.89
80	23.6	2.74	0.56	940	20.9	2.11	0.99
90	24.0	2.72	0.67	955	21.0	2.02	1.04
100	24.3	2.68	0.80	970	20.8	1.91	1.07
110	24.6	2.59	0.95	980	20.6	1.83	1.08
120	24.8	2.43	1.08	990	19.9	1.76	1.07
130	24.8	2.23	1.17	1000	19.3	1.68	1.06
140	23.1	1.99	1.19	1020	17.0	1.54	1.00
150	19.0	1.75	1.12	1040	14.8	1.45	0.92
160	17.0	1.54	0.98	1060	13.2	1.37	0.82
170	11.0	1.41	0.77	1080	11.0	1.33	0.73
180	8.0	1.39	0.55	1100	8.6	1.30	0.63
200	6.0	1.40	0.43	1120	6.3	1.30	0.53
220	5.8	1.56	0.14	1140	4.9	1.33	0.44
240	6.0	1.67	0.09	1160	4.4	1.38	0.36
260	7.0	1.74	0.06	1180	4.9	1.45	0.30
280	8.0	1.79	0.05	1200	5.8	1.54	0.28
300	8.7	1.83	0.04	1210	6.4	1.58	0.28
320	9.0	1.86	0.03	1220	6.8	1.62	0.29
340	9.5	1.89	0.03	1230	7.4	1.66	0.31
360	9.8	1.92	0.02	1240	8.0	1.70	0.34
370	10.0	1.92	0.02	1250	8.7	1.73	0.38
380	10.2	1.93	0.02	1260	9.4	1.75	0.42
400	10.5	1.95	0.03	1270	10.2	1.77	0.48
420	10.7	1.97	0.03	1280	10.8	1.78	0.55
440	10.9	1.99	0.03	1290	11.4	1.76	0.59
460	11.2	2.01	0.04	1300	12.0	1.73	0.64
480	11.7	2.04	0.04	1310	12.4	1.70	0.69
500	12.2	2.06	0.06	1320	12.6	1.66	0.74
510	12.5	2.07	0.06	1330	12.8	1.60	0.77
520	12.5	2.08	0.07	1340	12.8	1.54	0.80
540	12.3	2.09	0.10	1350	12.8	1.47	0.81
560	12.5	2.09	0.11	1360	12.6	1.40	0.81
580	12.8	2.10	0.10	1370	12.2	1.34	0.80
600	13.3	2.13	0.11	1380	11.7	1.28	0.78
620	13.5	2.16	0.12	1390	11.0	1.22	0.74
640	14.1	2.20	0.15	1400	10.4	1.17	0.71
660	14.7	2.23	0.20	1420	8.5	1.11	0.62
680	15.4	2.25	0.27	1440	5.6	1.07	0.53
690	15.5	2.24	0.31	1460	4.9	1.04	0.45
700	15.3	2.22	0.35	1480	3.1	1.03	0.37
720	14.4	2.14	0.39	1490	2.2	1.03	0.33
730	13.8	2.10	0.38	1510	0.8	1.04	0.24

Table A5 (continued) Optical Constants and Reflectivity of $36.0BaO \cdot 64.0B_2O_3$ Glass (Mol% by Analysis) in the 20 to 1650 cm^{-1} Range According to Reference 79

ω, cm^{-1}	$R(\omega)$,%	$n(\omega)$	$\kappa(\omega)$	ω, cm^{-1}	$R(\omega)$,%	$n(\omega)$	$\kappa(\omega)$
740	13.3	2.08	0.35	1530	0.1	1.07	0.17
750	13.3	2.07	0.32	1550	0.4	1.11	0.11
770	13.2	2.11	0.27	1570	0.8	1.16	0.08
780	13.6	2.15	0.26	1590	1.1	1.20	0.05
790	14.4	2.19	0.26	1620	1.4	1.26	0.04
800	15.3	2.23	0.28	1650	1.6	1.31	0.02

Table A6 Optical Constants of Mixed Borate Glasses in the 640 to 1680 cm^{-1} Range According to Reference 79 (Glass Compositions in Mol% by Analysis[a])

	$15.0(Me_2O + MeO) \cdot 85.0B_2O_3$		$25.1(Me_2O + MeO) \cdot 74.9B_2O_3$		$37.3(Me_2O + MeO) \cdot 62.7B_2O_3$	
ω, cm^{-1}	$n(\omega)$	$\kappa(\omega)$	$n(\omega)$	$\kappa(\omega)$	$n(\omega)$	$\kappa(\omega)$
640	2.08	0.09	2.13	0.10	2.10	0.07
660	2.13	0.14	2.17	0.15	2.13	0.09
680	2.17	0.24	2.18	0.25	2.17	0.14
690	2.16	0.33	2.16	0.31	2.18	0.17
700	2.11	0.43	2.12	0.36	2.19	0.22
720	1.86	0.46	1.96	0.35	2.14	0.31
730	1.78	0.38	1.93	0.30	2.08	0.32
740	1.76	0.29	1.92	0.25	2.04	0.30
750	1.78	0.23	1.94	0.21	2.02	0.26
770	1.84	0.16	2.00	0.16	2.05	0.20
780	1.87	0.14	2.03	0.16	2.08	0.19
790	1.90	0.13	2.07	0.16	2.12	0.20
800	1.93	0.13	2.10	0.16	2.15	0.20
820	1.98	0.15	2.17	0.20	2.21	0.25
840	2.02	0.17	2.23	0.26	2.26	0.32
860	2.05	0.21	2.27	0.35	2.30	0.41
880	2.08	0.26	2.29	0.45	2.31	0.51
900	2.09	0.32	2.28	0.57	2.29	0.64
920	2.07	0.39	2.22	0.69	2.23	0.76
940	2.03	0.43	2.13	0.78	2.14	0.86
955	2.01	0.44	2.05	0.82	2.04	0.92
970	1.99	0.45	1.97	0.84	1.94	0.96
980	1.99	0.45	1.92	0.84	1.87	0.97
990	1.99	0.46	1.87	0.84	1.80	0.97
1000	1.99	0.48	1.83	0.84	1.74	0.96
1020	1.98	0.52	1.75	0.83	1.62	0.92
1040	1.94	0.58	1.67	0.82	1.52	0.87
1060	1.87	0.61	1.58	0.79	1.44	0.80
1080	1.79	0.60	1.50	0.74	1.37	0.72
1100	1.72	0.54	1.43	0.64	1.33	0.63
1120	1.70	0.45	1.41	0.52	1.32	0.52
1140	1.74	0.35	1.45	0.41	1.34	0.43
1160	1.86	0.28	1.54	0.33	1.39	0.35
1180	2.02	0.29	1.66	0.29	1.46	0.31
1200	2.20	0.42	1.80	0.33	1.54	0.30
1210	2.26	0.53	1.86	0.40	1.57	0.31
1220	2.29	0.67	1.90	0.48	1.60	0.32
1230	2.26	0.82	1.90	0.58	1.62	0.34
1240	2.18	0.95	1.85	0.66	1.64	0.36
1250	2.07	1.05	1.78	0.71	1.65	0.38
1260	1.94	1.09	1.71	0.71	1.66	0.41
1270	1.83	1.08	1.68	0.69	1.68	0.43
1280	1.75	1.04	1.67	0.67	1.68	0.46
1290	1.72	1.00	1.69	0.67	1.69	0.50
1300	1.72	0.98	1.71	0.70	1.69	0.53

Table A6 (continued) Optical Constants of Mixed Borate Glasses in the 640 to 1680 cm^{-1} Range According to Reference 79 (Glass Compositions in Mol% by Analysis[a])

	$15.0(Me_2O + MeO) \cdot 85.0B_2O_3$		$25.1(Me_2O + MeO) \cdot 74.9B_2O_3$		$37.3(Me_2O + MeO) \cdot 62.7B_2O_3$	
ω, cm^{-1}	$n(\omega)$	$\kappa(\omega)$	$n(\omega)$	$\kappa(\omega)$	$n(\omega)$	$\kappa(\omega)$
1310	1.73	0.99	1.71	0.74	1.68	0.57
1320	1.72	1.03	1.70	0.79	1.66	0.62
1330	1.70	1.08	1.67	0.85	1.64	0.66
1340	1.65	1.14	1.62	0.90	1.60	0.70
1350	1.58	1.18	1.57	0.94	1.56	0.73
1360	1.49	1.21	1.50	0.97	1.50	0.76
1370	1.40	1.22	1.42	0.98	1.44	0.78
1380	1.31	1.22	1.34	0.98	1.37	0.78
1390	1.22	1.19	1.26	0.97	1.30	0.78
1400	1.14	1.15	1.19	0.94	1.24	0.75
1420	1.01	1.05	1.08	0.85	1.13	0.67
1440	0.93	0.95	1.01	0.76	1.07	0.57
1460	0.86	0.85	0.96	0.68	1.04	0.50
1480	0.81	0.75	0.90	0.61	0.99	0.43
1500	0.76	0.65	0.84	0.49	0.96	0.30
1510	0.74	0.58	0.83	0.42	0.98	0.24
1520	0.72	0.52	0.84	0.35	1.01	0.19
1540	0.73	0.38	0.88	0.24	1.07	0.12
1560	0.78	0.27	0.95	0.16	1.12	0.09
1580	0.85	0.19	1.01	0.12	1.17	0.07
1600	0.91	0.14	1.06	0.10	1.20	0.06
1620	0.97	0.10	1.10	0.08	1.23	0.04
1640	1.02	0.08	1.14	0.06	1.26	0.04
1660	1.06	0.07	1.16	0.06	1.28	0.03
1680	1.09	0.06	1.19	0.05	1.30	0.03

[a] Ionic oxide content of glasses is $2.9K_2O \cdot 9.0CaO \cdot 3.1BaO$, $4.7K_2O \cdot 15.4CaO \cdot 5.0BaO$, and $6.2K_2O \cdot 23.5CaO \cdot 7.6BaO$, respectively.

Table A7 Optical Constants and Reflectivity of Vitreous Germania in the 420 to 1100 cm^{-1} Range According to Reference 79

ω, cm^{-1}	$R(\omega)$,%	$n(\omega)$	$\kappa(\omega)$	ω, cm^{-1}	$R(\omega)$,%	$n(\omega)$	$\kappa(\omega)$
400	4.0	1.35	0.32	750	10.4	1.95	0.03
410	4.5	1.42	0.33	760	11.4	2.00	0.03
420	5.1	1.48	0.34	770	12.4	2.07	0.04
430	5.9	1.52	0.36	780	13.6	2.15	0.04
440	6.4	1.55	0.36	790	15.3	2.25	0.06
450	7.2	1.58	0.37	800	16.9	2.38	0.08
460	7.4	1.60	0.38	810	19.2	2.54	0.15
470	8.2	1.62	0.38	820	22.7	2.74	0.28
480	8.4	1.65	0.38	830	25.4	2.92	0.53
490	8.6	1.67	0.39	840	28.1	3.02	0.90
500	9.1	1.70	0.40	850	31.2	2.94	1.33
510	9.5	1.72	0.43	860	33.2	2.69	1.69
520	10.0	1.74	0.46	870	35.8	2.35	1.87
530	10.3	1.74	0.51	880	38.8	2.11	1.95
540	10.7	1.72	0.55	890	42.6	1.87	2.16
550	10.9	1.68	0.59	900	48.5	1.34	2.32
560	10.6	1.63	0.62	910	57.0	0.76	2.09
570	10.1	1.56	0.63	920	64.1	0.47	1.64
580	9.2	1.49	0.62	930	50.7	0.42	1.25
590	8.0	1.41	0.58	940	35.2	0.48	0.96
600	6.6	1.35	0.51	950	25.2	0.57	0.77
610	5.2	1.32	0.42	960	19.2	0.67	0.65
620	3.8	1.32	0.32	970	14.3	0.75	0.58
630	3.4	1.34	0.24	980	11.5	0.80	0.54
640	2.5	1.39	0.17	990	8.8	0.84	0.50
650	3.0	1.45	0.12	1000	6.4	0.86	0.47
660	4.1	1.51	0.08	1010	4.5	0.88	0.44
670	4.8	1.56	0.06	1020	3.1	0.89	0.41
680	5.3	1.61	0.05	1030	1.9	0.89	0.37
690	6.2	1.66	0.04	1040	1.4	0.90	0.33
700	7.3	1.71	0.03	1050	0.8	0.91	0.29
710	7.9	1.75	0.03	1060	0.3	0.92	0.25
720	8.7	1.80	0.03	1070	0.3	0.94	0.21
730	9.0	1.84	0.03	1080	0.4	0.96	0.17
740	9.7	1.89	0.03	1090	0.4	0.99	0.14

Table A8 Optical Constants of Some Sodium Germanate Glasses in the IR Range According to Reference 79 (Glass Compositions in Mol% by Analysis)

	$2.0Na_2O \cdot 98.0GeO_2$			$4.9Na_2O \cdot 95.1GeO_2$			$32.5Na_2O \cdot 67.5GeO_2$[a]		
ω, cm^{-1}	$R(\omega)$,%	$n(\omega)$	$\kappa(\omega)$	$R(\omega)$,%	$n(\omega)$	$\kappa(\omega)$	$R(\omega)$,%	$n(\omega)$	$\kappa(\omega)$
20	—	—	—	—	—	—	24.0	2.91	0.20
40	—	—	—	—	—	—	23.4	2.83	0.30
60	—	—	—	—	—	—	22.7	2.76	0.39
80	—	—	—	—	—	—	22.0	2.68	0.46
100	—	—	—	—	—	—	21.2	2.59	0.51
120	—	—	—	—	—	—	20.4	2.51	0.55
140	—	—	—	—	—	—	19.5	2.44	0.56
160	—	—	—	—	—	—	18.6	2.37	0.55
180	—	—	—	—	—	—	18.2	2.33	0.54
200	20.0	2.60	0.24	21.4	2.70	0.30	18.0	2.32	0.52
220	21.3	2.68	0.40	22.5	2.75	0.47	18.0	2.33	0.51
240	22.8	2.70	0.63	23.8	2.73	0.70	18.3	2.36	0.53
260	23.8	2.59	0.91	25.0	2.60	0.95	20.2	2.42	0.61
280	24.2	2.33	1.13	25.3	2.35	1.14	21.4	2.44	0.75
300	23.4	1.97	1.21	23.4	2.02	1.19	22.5	2.37	0.95
320	20.2	1.61	1.08	19.4	1.69	1.08	23.1	2.18	1.12
340	12.2	1.39	0.80	13.4	1.48	0.83	22.2	1.88	1.18
360	5.4	1.38	0.48	7.8	1.45	0.54	18.9	1.59	1.06
380	5.2	1.51	0.29	6.5	1.55	0.34	13.1	1.42	0.81
400	6.8	1.65	0.21	7.0	1.68	0.24	7.4	1.42	0.56
420	8.4	1.78	0.20	9.0	1.81	0.22	5.9	1.53	0.39
440	9.5	1.85	0.25	10.2	1.91	0.26	8.0	1.69	0.31
460	10.3	1.88	0.31	11.0	1.94	0.32	11.0	1.84	0.35
480	10.7	1.88	0.34	11.4	1.95	0.37	12.6	1.91	0.49
500	11.0	1.89	0.36	11.8	1.96	0.41	12.7	1.84	0.63
520	11.6	1.91	0.41	12.2	1.96	0.48	12.0	1.71	0.66
540	12.0	1.90	0.49	12.4	1.90	0.55	10.7	1.62	0.62
560	11.8	1.82	0.58	12.0	1.80	0.59	9.4	1.54	0.60
580	10.8	1.67	0.60	11.2	1.69	0.56	7.8	1.44	0.51
600	8.4	1.53	0.51	9.2	1.60	0.48	6.0	1.42	0.34
620	5.2	1.47	0.35	6.1	1.57	0.35	4.6	1.50	0.20
640	4.5	1.52	0.19	4.9	1.60	0.23	5.3	1.63	0.13
660	5.8	1.62	0.10	5.7	1.68	0.15	7.7	1.76	0.11
680	7.4	1.73	0.07	7.7	1.78	0.10	10.3	1.92	0.13
700	9.1	1.83	0.06	9.8	1.89	0.09	13.8	2.11	0.23
720	10.0	1.94	0.07	11.6	2.02	0.10	17.4	2.27	0.50
740	12.2	2.06	0.10	14.0	2.15	0.15	20.3	2.20	0.88
760	14.2	2.20	0.18	16.2	2.30	0.26	21.4	1.88	1.14
780	16.4	2.32	0.32	18.8	2.43	0.45	19.3	1.51	1.10
800	19.3	2.43	0.49	21.8	2.51	0.74	15.4	1.36	0.93
820	23.2	2.55	0.78	25.1	2.43	1.12	14.7	1.31	0.89
840	27.5	2.51	1.24	28.1	2.14	1.47	15.1	1.16	0.88
860	32.1	2.12	1.70	31.4	1.67	1.63	13.0	0.96	0.77
880	37.2	1.48	1.85	33.7	1.19	1.52	8.5	0.84	0.53
900	43.2	0.89	1.60	30.1	0.87	1.20	3.3	0.88	0.29
920	36.0	0.61	1.12	19.9	0.79	0.87	0.8	1.00	0.15
940	20.0	0.66	0.74	12.6	0.81	0.66	0.3	1.10	0.09
960	12.6	0.74	0.58	8.0	0.81	0.50	0.7	1.17	0.06

Table A8 (continued) Optical Constants of Some Sodium Germanate Glasses in the IR Range According to Reference 79 (Glass Compositions in Mol% by Analysis)

	$2.0Na_2O \cdot 98.0GeO_2$			$4.9Na_2O \cdot 95.1GeO_2$			$32.5Na_2O \cdot 67.5GeO_2$[a]		
ω, cm^{-1}	$R(\omega)$,%	$n(\omega)$	$\kappa(\omega)$	$R(\omega)$,%	$n(\omega)$	$\kappa(\omega)$	$R(\omega)$,%	$n(\omega)$	$\kappa(\omega)$
980	7.7	0.73	0.44	4.7	0.83	0.32	0.9	1.23	0.05
1000	3.8	0.77	0.24	2.3	0.93	0.18	1.3	1.27	0.04
1020	1.2	0.90	0.10	0.9	1.03	0.10	—	—	—
1040	0.5	1.02	0.05	0.4	1.12	0.06	—	—	—
1060	0.3	1.10	0.03	0.6	1.18	0.04	—	—	—
1080	0.7	1.16	0.02	1.0	1.24	0.03	—	—	—
1100	0.9	1.21	0.02	1.3	1.28	0.03	—	—	—

[a] Glass composition by synthesis.

Table A9 Optical Constants of Some Sodium Tellurite Glasses in the 20 to 900 cm^{-1} Range According to Reference 79 (Glass Compositions in Mol% by Synthesis)

	$12Na_2O \cdot 88TeO_2$			$20Na_2O \cdot 80TeO_2$			$35Na_2O \cdot 65TeO_2$		
ω, cm^{-1}	$R(\omega)$,%	$n(\omega)$	$\kappa(\omega)$	$R(\omega)$,%	$n(\omega)$	$\kappa(\omega)$	$R(\omega)$,%	$n(\omega)$	$\kappa(\omega)$
20	49.2	5.57	0.82	48.3	5.34	1.22	38.1	4.18	0.43
30	49.0	5.43	1.06	48.0	5.08	1.29	38.0	4.13	0.61
40	48.8	5.27	1.32	47.0	4.86	1.38	37.8	4.05	0.81
50	48.2	5.06	1.56	46.0	4.67	1.47	37.6	3.92	1.00
60	47.5	4.79	1.76	44.9	4.49	1.54	37.0	3.75	1.17
70	46.8	4.49	1.90	43.8	4.31	1.60	36.4	3.53	1.30
80	45.2	4.18	1.95	42.9	4.14	1.65	35.2	3.30	1.35
90	44.2	3.90	1.91	42.0	3.99	1.67	33.6	3.08	1.33
100	41.5	3.68	1.82	41.2	3.84	1.68	31.6	2.91	1.25
120	38.5	3.46	1.59	39.6	3.60	1.67	27.0	2.76	1.05
140	37.0	3.42	1.49	38.8	3.42	1.65	27.0	2.77	0.97
160	37.0	3.38	1.50	38.0	3.28	1.63	27.4	2.76	1.01
180	37.4	3.31	1.54	37.1	3.17	1.64	27.3	2.69	1.08
200	37.5	3.23	1.60	36.5	3.07	1.66	26.5	2.58	1.11
220	36.8	3.15	1.67	36.0	2.97	1.71	25.4	2.48	1.10
240	36.8	3.04	1.77	36.2	2.84	1.78	25.0	2.41	1.09
260	37.8	2.89	1.88	36.8	2.68	1.85	25.0	2.38	1.10
280	39.0	2.67	1.97	37.1	2.46	1.89	25.0	2.33	1.15
300	38.3	2.40	1.99	35.9	2.23	1.87	25.1	2.24	1.23
320	36.4	2.13	1.92	34.0	2.00	1.78	25.1	2.10	1.29
340	34.0	1.90	1.78	31.7	1.81	1.63	24.8	1.91	1.30
360	30.6	1.74	1.58	28.1	1.69	1.45	23.4	1.71	1.24
380	26.3	1.66	1.39	23.6	1.62	1.28	20.3	1.54	1.12
400	22.3	1.63	1.22	20.1	1.59	1.13	15.2	1.42	0.94
420	18.6	1.63	1.07	17.1	1.58	0.99	10.8	1.39	0.74
440	16.3	1.64	0.93	14.6	1.60	0.86	8.5	1.42	0.57
460	14.5	1.68	0.80	13.0	1.64	0.73	7.5	1.50	0.44
480	13.1	1.76	0.68	12.0	1.72	0.61	7.2	1.59	0.36
500	12.8	1.89	0.58	11.8	1.85	0.53	7.9	1.70	0.31
520	13.8	2.08	0.55	12.8	2.02	0.51	8.8	1.81	0.30
540	18.0	2.30	0.64	16.4	2.20	0.59	10.7	1.93	0.34
560	23.4	2.46	0.88	21.2	2.34	0.80	13.2	2.02	0.45
580	27.4	2.45	1.23	24.7	2.33	1.09	15.5	2.04	0.61
600	30.4	2.22	1.55	27.3	2.14	1.36	15.9	1.95	0.76
620	32.2	1.83	1.69	27.9	1.80	1.48	15.2	1.78	0.83
640	31.8	1.44	1.60	26.6	1.48	1.39	13.3	1.62	0.78
660	29.1	1.20	1.38	23.4	1.30	1.20	11.1	1.56	0.66
680	23.6	1.08	1.15	19.2	1.20	1.05	9.7	1.58	0.59
700	17.6	1.04	0.95	15.4	1.13	0.89	10.2	1.62	0.60
720	13.1	1.05	0.80	11.1	1.12	0.74	10.5	1.61	0.66
740	10.6	1.04	0.70	8.5	1.13	0.65	11.5	1.51	0.73
760	7.7	0.99	0.59	7.0	1.10	0.56	10.6	1.36	0.72
780	4.3	0.97	0.40	4.5	1.07	0.42	8.5	1.20	0.62
800	1.6	1.06	0.22	2.0	1.11	0.27	4.8	1.12	0.44
820	0.6	1.19	0.13	0.8	1.20	0.17	1.6	1.15	0.28
840	1.4	1.29	0.10	1.3	1.29	0.11	1.2	1.23	0.17
860	2.3	1.36	0.08	2.4	1.36	0.08	1.9	1.30	0.12
880	3.2	1.42	0.07	3.0	1.42	0.07	2.4	1.36	0.08
900	3.8	1.46	0.06	3.8	1.46	0.06	2.9	1.40	0.07

Table A10 Optical Constants of Vitreous Silica in the VUV Range According to References 159 and 289

	Optical Constants According to Sources			
	Reference 159 dispersion analysis based on Model (50)		Reference 289 Kramers-Kronig transform	
Wavelength, Å	$n(\lambda)$	$\kappa(\lambda)$	$n(\lambda)$	$\kappa(\lambda)$
580	0.67	0.32	—	—
600	0.70	0.40	—	—
620	0.74	0.46	—	—
640	0.78	0.51	—	—
660	0.81	0.55	—	—
680	0.86	0.60	—	—
700	0.91	0.65	—	—
720	0.98	0.67	—	—
740	1.05	0.67	—	—
760	1.10	0.66	—	—
780	1.13	0.64	—	—
800	1.14	0.62	—	—
820	1.13	0.63	—	—
840	1.14	0.66	—	—
860	1.17	0.69	—	—
880	1.22	0.72	—	—
900	1.28	0.73	—	—
920	1.33	0.71	—	—
940	1.36	0.68	—	—
960	1.36	0.67	—	—
980	1.36	0.68	—	—
1000	1.38	0.70	1.39	0.82
1020	1.41	0.73	1.42	0.85
1040	1.46	0.75	—	—
1050	1.49	0.76	1.51	0.89
1060	1.53	0.77	—	—
1075	—	—	1.63	0.87
1080	1.62	0.76	—	—
1100	1.70	0.68	1.72	0.78
1120	1.70	0.57	1.69	0.67
1140	1.62	0.52	—	—
1150	1.57	0.53	1.58	0.65
1160	1.52	0.57	1.56	0.67
1180	1.50	0.76	1.55	0.83
1190	1.58	0.87	1.58	0.88
1200	1.73	0.95	1.67	1.00
1210	1.91	0.94	1.82	1.06
1220	2.07	0.85	1.98	1.02
1240	2.20	0.57	—	—
1250	2.18	0.45	2.23	0.67
1260	2.15	0.36	—	—
1275	—	—	2.21	0.53
1280	2.07	0.25	—	—
1300	2.00	0.18	2.17	0.40
1350	1.88	0.11	2.07	0.25
1400	1.80	0.08	1.98	0.14
1450	1.74	0.06	1.89	0.08
1500	1.70	0.05	1.83	0.04

REFERENCES

1. **Korff, S. A. and Breit, G.,** Optical dispersion, *Rev. Mod. Phys.*, 4, 471, 1932.
2. **Seitz, F.,** *The Modern Theory of Solids,* McGraw-Hill, New York, 1940.
3. **Moss, T. S.,** *Optical Properties of Semiconductors,* Butterworths, London, 1959.
4. **Born, M. and Wolf, E.,** *Principles of Optics,* 3rd ed., Pergamon Press, Oxford, 1965.
5. **Garbuny, M.,** *Optical Physics,* Academic Press, New York, 1965.
6. **King, F. W.,** Sum rules for the optical constants, *J. Math. Phys.*, 17, 1509, 1976.
7. **Efimov, A. M. and Makarova, E. G.,** Precision interpolation and extrapolation of glass refractive index in the transparency range, *Glass Phys. Chem.*, 19, 6, 1993.
8. **Pai, K. F., Parker, T. J., Tornberg, N. E., Lowndes, R. P., and Chambers, W. G.,** Determination of the complex refractive indices of solids in the far-infrared by dispersive Fourier transform spectroscopy. I. Alkali halides, *Infrared Phys.*, 18, 199, 1978.
9. **Pantelides, S. T.,** Universal valence bands for rocksalt-type compounds and their connection with those of tetrahedral crystals, *Phys. Rev. B,* 11, 5082, 1975.
10. **Chelikowsky, J. R. and Cohen, M. L.,** Nonlocal pseudopotential calculations for the electronic structure of eleven diamond and zinc-blende semiconductors, *Phys. Rev. B,* 4, 556, 1976.
11. **Vinogradov, V. S.,** Applying the Green functions method to the calculation of infrared absorption in ionic crystals, *Fiz. Tverd. Tela* (Sov. Solid State Phys.), 4, 712, 1962.
12. **Wallis, R. F., Ipatova, I. P., and Maradudin, A. A.,** On the temperature dependence of linewidth for the main lattice absorption in ionic crystals, *Fiz. Tverd. Tela* (Sov. Solid State Phys.), 8, 1064, 1966.
13. **Spitzer, W. G. and Kleinman, D. A.,** Infrared lattice bands of quartz, *Phys. Rev.*, 121, 1324, 1961.
14. **Barker, A. S. and Hopfield, J. J.,** Coupled optical phonon mode theory of the infrared dispersion in $BaTiO_3$, $SrTiO_3$, and $KTaO_3$, *Phys. Rev.*, 135, 1732, 1964.
15. **Martin, B. G. and Wallis, R. F.,** The dielectric constant associated with direct interband transitions in germanium, *Solid State Commun.*, 21, 385, 1977.
16. **Tauc, J.,** Optical properties of semiconductors in the visible and ultra-violet ranges, in *Progress in Semiconductors,* Vol. 9, Heywood Book Temple Press Books, London, 1965, 87.
17. **Nozieres, P. and Pines, D.,** Electron interaction in solids. Collective approach to the dielectric constant, *Phys. Rev.*, 109, 762, 1958.
18. **Burckhardt, W., Feltz, A., and Bussemer, P.,** The influence of the structure of different arsenic chalcogenide glasses on the dispersion parameters of a Sellmeier and a Lorentz-Lorenz relation, *J. Non-Cryst. Solids,* 89, 273, 1987.
19. **Bakhshiev, N. G.,** *Spektroskopiya Mezhmolekulyarnykh Vzaimodeistvii* (Spectroscopy of Intermolecular Interactions), Nauka, Leningrad, 1972.
20. *Kolebaniya Okisnykh Reshetok* (Vibrations of Oxide Lattices), Lazarev, A. N., Ed., Nauka, Leningrad, 1980.
21. **Libov, V. S. and Arzheukhova, N. B.,** The analysis of vibrational spectra of glass with resonance interactions taken into account, *Fiz. Khim. Stekla* (Sov. J. Glass Phys. Chem.), 16, 458, 1990.
22. **Efimov, A. M.,** Use of the dispersion formulas in describing the frequency dependence of refractive index. Results and prospects, *Fiz. Khim. Stekla* (Sov. J. Glass Phys. Chem.), 3, 434, 1977.
23. **Batsanov, S. S.,** *Strukturnaya Refraktometriya* (Structural Refractometry), 2nd ed., Vysshaya Shkola, Moscow, 1976.
24. **Bauer, N. and Fajans, K.,** Molar dispersion and refraction of free bonded ions, *J. Am. Chem. Soc.*, 64, 3023, 1942.
25. **Efimov, A. M. and Mazurina, E. K.,** Refractivity of glass network and the structure of silicate, borate, and germanate glasses, *Fiz. Khim. Stekla* (Sov. J. Glass Phys. Chem.), 3, 1, 1977.

26. **Kitamura, N., Toguchi, Y., Funo, S., Yamashita, H., and Kinoshita, M.,** Refractive index of densified silica glass, *J. Non-Cryst. Solids,* 159, 241, 1993.
27. **Wemple, S. H. and DiDomenico, M.,** Behavior of the electronic dielectric constant in covalent and ionic materials, *Phys. Rev. B,* 3, 1338, 1971.
28. **DiDomenico, M.,** Material dispersion in optical fiber waveguides, *Appl. Opt.,* 11, 652, 1972.
29. **Wemple, S. H.,** Refractive index behavior of amorphous semiconductors and glasses, *Phys. Rev. B,* 7, 3767, 1973.
30. **Wemple, S. H.,** Optical oscillator strengths and excitation energies in solids, liquids, and molecules, *J. Chem. Phys.,* 67, 2151, 1977.
31. **Wemple, S. H.,** Material dispersion in optical fibers, *Appl. Opt.,* 18, 31, 1979.
32. **Penn, D. R.,** Wave-number-dependent dielectric function of semiconductors, *Phys. Rev.,* 128, 2093, 1962.
33. **Van Vechten, J. A.,** Quantum dielectric theory of electronegativity in covalent systems. I. Electronic dielectric constant, *Phys. Rev.,* 182, 891, 1969.
34. **Berreman, D. W. and Unterwald, F. D.,** Adjusting poles and zeros of dielectric dispersion to fit reststrahlen of $PrCl_3$ and $LaCl_3$, *Phys. Rev.,* 174, 791, 1968.
35. **Lyddane, R. H., Sachs, R. G., and Teller, E.,** On the polar vibrations of alkali halides, *Phys. Rev.,* 59, 673, 1941.
36. **Gervais, F. and Piriou, B.,** Anharmonicity in several-polar-mode crystals: adjusting phonon self-energy of LO and TO modes in Al_2O_3 and TiO_2 to fit infrared reflectivity, *J. Phys. C,* 7, 2374, 1974.
37. **Kopilevich, Yu. I. and Makarova, E. G.,** On the models of dielectric constant in the dispersion analysis of reflection spectra, *Opt. Spektrosk.* (Sov. Optics and Spectroscopy), 63, 147, 1987.
38. **Kachare, A., Andermann, G., and Soriaga, M. P.,** Temperature dependence of the infrared optical properties of LiF, *J. Phys. Chem. Solids,* 36, 83, 1975.
39. **Zernik, W.,** Theory of infrared optical properties of fluorite crystals, *Rev. Mod. Phys.,* 39, 432, 1967.
40. **Pavinich, V. F. and Lazarev, A. N.,** On the interpretation of complicated contours of IR reflection bands, *Opt. Spektrosk.* (Sov. Optics and Spectroscopy), 43, 584, 1977.
41. **Born, M. and Huang, K.,** *Dynamical Theory of Crystal Lattices,* Clarendon Press, Oxford, 1954.
42. **Phillips, J. C.,** Optical spectra of solids, in *Solid State Physics,* Ehrenreich, H., Zeits, F., and Turnbull, D., Eds., Academic Press, New York, 1966.
43. **Gaskell, P. H. and Johnson, D. W.,** The optical constants of quartz, vitreous silica, and neutron-irradiated vitreous silica, *J. Non-Cryst. Solids,* 20, 153, 1976.
44. **Bonch-Bruevich, V. L., Zvyagin, I. R., Kaiper, R., Mironov, A. G., Enderline, R., and Esser, B.,** *Electronic Theory of Disordered Semiconductors,* Nauka, Moscow, 1981, Chap. 1.
45. **Zakis, Yu. R. and Klyava, Ya. R.,** Spectral manifestations of discrete and continuous disorder in non-metallic glasses, in *Stekloobraznoe Sostoyanie* (Vitreous State), Proc. 8th All-Union Conf., Porai-Koshits, E. A., Ed., Nauka, Leningrad, 1988, 70.
46. **Wright, A. C.,** Diffraction studies of glass structure, in *Glass '89, Survey Papers of the 15th Int. Congr. Glass,* Mazurin, O. V., Ed., Nauka, Leningrad, 1988, 30.
47. **Galeener, F. L.,** Spectroscopic investigations of glass structure, in *Glass '89, Survey Papers of the 15th Int. Congr. Glass,* Mazurin, O. V., Ed., Nauka, Leningrad, 1988, 103.
48. **Gaskell, P. H.,** Novel aspects of the structure of glass, in *Proc. 16th Int. Congr. Glass,* Vol. 1, Madrid, 1992, Boletin Soc. Esp. Ceram. Vid., 31-C, 1992 (3), 25.
49. **Mozzi, R. L. and Warren, B. E.,** The structure of vitreous silica, *J. Appl. Cryst.,* 2, 164, 1969.
50. **Uhlmann, D. R. and Wicks, G. G.,** Structural analysis using fluorescence excitation, *Wiss. Z. Jena Univ.,* Math.-Naturwiss. Reihe, 28, 231, 1979.
51. **Gladden, L. F.,** Structure and dynamics of 4-2 coordinated glasses, in *The Physics of Non-Crystalline Solids,* Pye, L. D., LaCourse, W. C., and Stevens, H. J., Eds., Taylor and Francis, London, 1992, 91.

52. **Stoneham, A. M.,** Shapes of inhomogeneously broadened resonance lines in solids, *Rev. Mod. Phys.,* 41, 82, 1969.
53. **Ageeva, L. E., Brachkovskaya, N. V., Grubin, A. A., Lunter, S. G., Raaben, E. L., and Przhevuskii, A. K.,** Determination of Stark structure of absorption and luminescence bands in glass spectra with a computer, in *Spektrosk. Kristallov* (Spectroscopy of Crystals), Kaminskii, A. A., Morgenshtern, Z. L., and Sviridov, D. T., Eds., Moscow, 1975, 339.
54. **Bakhshiev, N. G. and Zelikman, P. I.,** On the mechanisms of band broadening in the vibrational spectra of liquid systems, *Dokl. Akad. Nauk SSSR* (Sov. Acad. Sci. Reports), 250, 128, 1980.
55. **Seshadri, K. S. and Jones, R. N.,** The shapes and intensities of infrared absorption bands. A review, *Spectrochim. Acta,* 19, 1013, 1963.
56. **Armstrong, B. H.,** Spectrum line profiles: the Voigt function, *J. Quant. Spectros. Radiat. Transfer,* 7, 61, 1967.
57. **Painter, P. C., Coleman, M. M., and Koenig, J. L.,** *The Theory of Vibrational Spectroscopy and Its Applications to Polymeric Materials,* John Wiley & Sons, New York, 1982, Chap. 15.
58. **Galeener, F. L., Leadbetter, A. J., and Stringfellow, M. W.,** Comparison of the neutron, Raman, and infrared vibrational spectra of vitreous SiO_2, GeO_2, and BeF_2, *Phys. Rev. B,* 27, 1052, 1983.
59. **Bell, R. J. and Dean, P.,** Atomic vibrations in vitreous silica, *Nature,* 212, 1356, 1966.
60. **Bell, R. J., Bird, N. F., and Dean, P.,** The vibrational spectra of vitreous silica, germania, and beryllium fluoride, *J. Phys. C,* 1, 299, 1968.
61. **Bell, R. J. and Dean, P.,** Atomic vibrations in vitreous silica, *Disc. Faraday Soc.,* 50, 55, 1970.
62. **Bell, R. J., Dean, P., and Hibbins-Butler, D. C.,** Localization of normal modes in vitreous silica, germania, and beryllium fluoride, *J. Phys. C,* 3, 2111, 1970.
63. **Bell, R. J., Dean, P., and Hibbins-Butler, D. C.,** Normal mode assignments in vitreous silica, germania, and beryllium fluoride, *J. Phys. C,* 4, 1214, 1971.
64. **Galeener, F. L. and Sen, P. N.,** Theory for the first-order vibrational spectra of disordered solids, *Phys. Rev. B,* 17, 1928, 1978.
65. **Bell, R. J., Carnevale, A., Kurkjian, C. R., and Peterson, G. E.,** Structure and phonon spectra of vitreous SiO_2, B_2O_3, and mixed SiO_2-B_2O_3 glasses, *J. Non-Cryst. Solids,* 35-36, 1185, 1980.
66. **Hass, M.,** Raman spectra of vitreous silica, germania, and sodium silicate glasses, *J. Phys. Chem. Solids,* 31, 323, 1970.
67. **Bell, R. J., Dean, P., and Hibbins-Butler, D. C.,** Infrared activity of normal modes in vitreous silica, germania, and beryllium fluoride, *J. Phys. C,* 9, 1171, 1976.
68. **Silin, A. R., Bray, P. J., and Mikkelsen, J. C.,** High-temperature intrinsic defects in fused silica, *J. Non-Cryst. Solids,* 64, 185, 1984.
69. **Galeener, F. L. and Mikkelsen, J. C.,** Vibrational dynamics in ^{18}O substituted vitreous silica, *Phys. Rev. B,* 23, 5527, 1981.
70. **Revesz, A. G. and Walrafen, G. E.,** Structural interpretations for some Raman lines from vitreous SiO_2, *J. Non-Cryst. Solids,* 54, 323, 1983.
71. **Leeuw, S. W. de, He, H., and Thorpe, M. F.,** Vibrational spectra of rings in vitreous silica, *Solid State Commun.,* 56, 343, 1985.
72. **Mysen, B. O., Finger, L. W., Virgo, D., and Seifert, F. A.,** Curve-fitting of Raman spectra of silicate glasses, *Am. Miner.,* 67, 686, 1982.
73. **Mitchell, D. L., Bishop, S. G., and Taylor, P. C.,** Infrared conductivity of chalcogenide glasses, *J. Non-Cryst. Solids,* 8-10, 231, 1972.
74. **Taylor, P. C., Bishop, S. G., and Mitchell, D. L.,** Far infrared and microwave optical properties of vitreous As_2Se_3, *Solid State Commun.,* 8, 1783, 1970.
75. **Efimov, A. M. and Makarova, E. G.,** Vitreous state and dispersion theory, in *Stekloobraznoe Sostoyanie* (Vitreous State), Proc. 7th All-Union Conf., Porai-Koshits, E. A., Ed., Nauka, Leningrad, 1983, 165.

76. **Efimov, A. M. and Makarova, E. G.,** Dispersion equation for the complex dielectric constant of vitreous solids and dispersion analysis of their reflection spectra, *Fiz. Khim. Stekla* (Sov. J. Glass Phys. Chem.), 11, 385, 1985.
77. **Efimov, A. M.,** A model for the IR spectra of silicate glasses, *Fiz. Khim. Stekla* (Sov. J. Glass Phys. Chem.), 15, 366, 1989.
78. **Efimov, A. M. and Makarova, E. G.,** Fundamental spectra simulation as a method of investigation of glass structure and its optical properties, in *Proc. 15th Int. Congr. Glass,* Vol. 1a, Mazurin, O. V., Ed., Nauka, Leningrad, 1989, 108.
79. **Efimov, A. M.,** Dispersion of optical constants of vitreous solids, D. Sc. Thesis, Vavilov State Optical Institute, Leningrad, 1990.
80. **Efimov, A. M. and Makarova, E. G.,** The problems of the dispersion analysis of the IR spectra of glasses, in *Sbornik Statei, Posvyazhshennykh 90-letiyu so Dnya Rozhdeniya Akademika A. A. Lebedeva* (Collection of Papers Devoted to the 90th Anniversary of Academician A. A. Lebedev), Miroshnikov, M. M., Ed., Leningrad, State Optical Institute, 1985, Part 2, p. 238.
81. **Laughlin, R. B.,** Optical absorption edge of SiO_2, *Phys. Rev. B,* 22, 3021, 1980.
82. **Benneth, A. J. and Roth, L. M.,** Calculation of optical properties of amorphous SiO_x materials, *Phys. Rev. B,* 4, 2686, 1971.
83. **Griscom, D. L.,** The electronic structure of SiO_2: a review of recent spectroscopic and theoretical advances, *J. Non-Cryst. Solids,* 24, 155, 1977.
84. **Schluter, M. and Chelikowsky, J. R.,** Electron states in α-quartz, *Solid State Commun.,* 21, 381, 1977.
85. **Pantelides, S. T.,** The optical absorption spectrum of SiO_2, in *The Physics of SiO_2 and Its Interfaces,* Pantelides, S. T. (Ed.), Pergamon Press, New York, 1978, 80.
86. **Philipp, H. R.,** Optical transitions in crystalline and fused quartz, *Solid State Commun.,* 4, 73, 1966.
87. **Rossinelli, M. and Bosh, M. A.,** Reflectance spectrum of crystalline and vitreous SiO_2 at low temperatures, *Phys. Rev. B,* 25, 6482, 1982.
88. **Silin, A. R. and Trukhin, A. N.,** *Point Defects and Elementary Excitations in Crystalline and Vitreous SiO_2,* Zinatne, Riga, 1985.
89. **Platzoder, K.,** Temperature effects on the vacuum ultraviolet reflectance of α-quartz, *Phys. Status Solidi,* 29, K63, 1968.
90. **Sumi, H. and Toyozawa, Y.,** Urbach-Martiensen rule and exciton trapped momentarily by lattice vibration, *J. Phys. Soc. Jpn.,* 31, 342, 1971.
91. **Cho, K.,** Antiresonance (Fano effect) with phonon contribution, *Phys. Status Solidi B,* 54, 583, 1972.
92. **Ellis, E., Johnson, D. W., Breeze, A., Magee, P. M., and Perkins, P. G.,** The electronic band structure and optical properties of oxide glasses. I. SiO_2, Na_2O:SiO_2, and Na_2O:CaO:SiO_2, *Philos. Mag. B,* 40, 105, 1979.
93. **Ellis, E., Johnson, D. W., Breeze, A., Magee, P. M., and Perkins, P. G.,** The electronic band structure and optical properties of oxide glasses. II. Lead silicates, *Philos. Mag. B,* 40, 125, 1979.
94. **Pajasova, L.,** Optical properties of GeO_2 in the ultraviolet region, *Czech. J. Phys. B,* 19, 1265, 1969.
95. **Sigel, G. H.,** Vacuum ultraviolet absorption in alkali doped fused silica and silicate glasses, *J. Phys. Chem. Solids,* 32, 2373, 1971.
96. **Bourdillon, A. J., Khumalo, F., and Bordas, J.,** The reflection spectra of lead glass, *Philos. Mag. B,* 37, 731, 1978.
97. **Efimov, A. M., Sidorin, V. K., and Sidorin, K. K.,** Reflection spectra of binary sodium silicate glasses in the photon energy range from 7 to 21 eV, *Fiz. Khim. Stekla* (Sov. J. Glass Phys. Chem.), 12, 245, 1986.
98. **Savel'ev, V. L., D'yachkova, Yu. V., and Pajasova, L.,** Influence of surface characteristics of silicate glasses on their optical spectra, in *Electronic Processes and the Structure of Defects in Glassforming Systems,* Zakis, Yu. R., Ed., Zinatne, Riga, 1982, 116.

99. **Izumitani, T., Hirota, S., Tanaka, K., Onuki, H., and Saitoh, T.,** Reflection spectra of fluoride and oxide glasses in the extreme ultraviolet region, *J. Non-Cryst. Solids,* 87, 108, 1986.
100. **Risen, W. M.,** Optical spectra of glasses, *J. Non-Cryst. Solids,* 76, 97, 1985.
101. **Onari, S., Saegusa, H., Mori, T., and Arai, T.,** One and two-phonon vibrational spectra of amorphous As- and Sb-chalcogenide systems, *J. Non-Cryst. Solids,* 59-60, 871, 1983.
102. **Ljundahl, G., Fellman, J., and Ribbing, C. G.,** Infrared optical constants of mineral wool raw materials, *J. Non-Cryst. Solids,* 136, 137, 1991.
103. **Gervais, F., Blin, A., Massiot, D., Coutures, J. P., Chopinet, M. H., and Naudin, F.,** Infrared reflectivity spectroscopy of silicate glasses, *J. Non-Cryst. Solids,* 89, 384, 1987.
104. **Parot-Rajaona, T., Vaills, Y., Massiot, D., and Gervais, F.,** Analysis of lithium aluminosilicate glasses by Raman scattering, infrared reflectivity, and ^{29}Si MAS-NMR spectroscopy, in *Proc. 16th Int. Congr. Glass,* Vol. 3, Madrid, 1992, Boletin Soc. Esp. Ceram. Vid., 31-C, 1992 (3), 383.
105. **Brendel, R. and Bormann, D.,** An IR dielectric function model for amorphous solids, *J. Appl. Phys.,* 71, 1, 1992.
106. **Morozov, V. N., Shakaryan, E. S., Molochnikov, B. I., Leikin, M. V., and Zolotarev, V. M.,** Measurement of optical constants in the polarized light at single incidence angle, *Opt. Mekh. Prom.,* (Optical and Mechanical Industry), 8, 14, 1979.
107. **Morozov, V. N., Shakaryan, E. S., Molochnikov, B. I., Shmulevich, M. I., and Vasil'eva, I. S.,** Measurement of optical constants at two magnitudes of incidence angle, *Opt. Mekh. Prom.* (Optical and Mechanical Industry), 9, 8, 1979.
108. **Parker, T. J., Ford, J. E., and Chambers, W. G.,** The optical constants of pure fused quartz in the far-infrared, *Infrared Phys.,* 18, 215, 1978.
109. **Vartiainen, E. M., Peiponen, K.-E., and Asakura, T.,** Maximum entropy model in reflection spectrum analysis, *Opt. Commun.,* 89, 37, 1992.
110. **Peiponen, K.-E. and Vartiainen, E. M.,** Kramers-Kronig relations in optical data inversion, *Phys. Rev. B,* 44, 8301, 1991.
111. **Harrick, N. J.,** *Internal Reflection Spectroscopy,* John Wiley, New York, 1967.
112. **Robinson, T. S.,** Optical constants by reflection, *Proc. Phys. Soc. B (London),* 65, 910, 1952.
113. **Young, R. H.,** Validity of the Kramers-Kronig transformation used in reflection spectroscopy, *J. Opt. Soc. Am.,* 67, 520, 1977.
114. **Makarova, E. G.,** Methods for calculating the optical constants of condensed media by reflection spectra, Ph.D. Thesis, Vavilov State Optical Institute, Leningrad, 1985.
115. **Philipp, H. R. and Ehrenreich, H.,** Optical properties in the range of fundamental absorption, in *Semiconductors and Semimetals. Vol. 3. Optical Properties of III-V Compounds,* Willardson, R. K. and Beer, A. C., Eds., Academic Press, New York, 1967.
116. **Velicky, B.,** The use of the Kramers-Kronig relations in determining optical constants, *Czech. J. Phys.,* 11B, 787, 1961.
117. **Al'perovich, L. I.,** *Method of Dispersion Relations and its use for Determining the Optical Constants,* Irfon, Dushanbe, 1973.
118. **Arkatova, T. G., Gopshtein, N. M., Makarova, E. G., and Mikhailov, B. A.,** Determining the optical constants of a material with Kramers-Kronig transform, *Opt. Mekh. Prom.* (Optical and Mechanical Industry), 9, 44, 1981.
119. **Kopilevich, Yu. I. and Makarova, E. G.,** On the correctness of dispersion analysis of reflection spectra, *Opt. Spektrosk.* (Sov. Optics and Spectroscopy), 63, 356, 1987.
120. **Belousov, M. V. and Pogarev, D. E.,** Kramers-Kronig analysis of reflection spectra with preliminary dispersion analysis, *Opt. Spektrosk.* (Sov. Optics and Spectroscopy), 43, 379, 1977.
121. **Efimov, A. M.,** On the nature of interconnection between the relative partial dispersion and the Abbe number of optical materials. Normal line and abnormal glasses, *Opt. Spektrosk.* (Sov. Optics and Spectroscopy), 67, 1127, 1989.

122. **Nagano, N., Saito, M., Miyagi, M., Baba, N., and Sawanobori, N.,** Refractive indices of SiO_2- and GeO_2-based glasses near the infrared absorption peaks, *J. Non-Cryst. Solids,* 135, 114, 1991.
123. **Fock, M. V.,** On separating complicated spectra into individual constituents, *Zh. Prikl. Spektr.* (Sov. J. Appl. Spectr.), 11, 926, 1969.
124. **Spitzer, W. G., Kleinman, D. A., and Walsh, D.,** Infrared properties of hexagonal silicon carbide, *Phys. Rev.,* 113, 127, 1959; **Spitzer, W. G., Kleinman, D. A., and Frosch, C. J.,** Infrared properties of cubic silicon carbide films, *Phys. Rev.,* 113, 133, 1959.
125. **Kleinman, D. A. and Spitzer, W. G.,** Infrared lattice absorption of GaP, *Phys. Rev.,* 118, 110, 1960.
126. **Russell, E. E. and Bell, E. E.,** Measurement of the optical constants of crystal quartz in the far infrared with the asymmetric Fourier transform method, *J. Opt. Soc. Am.,* 57, 341, 1967.
127. **Roessler, D. M. and Albers, W. A.,** Infrared reflectance of single crystal tetragonal GeO_2, *J. Phys. Chem. Solids,* 33, 293, 1972.
128. **Spitzer, W. G., Miller, R. C., Kleinman, D. A., and Howarth, L. E.,** Far infrared dielectric dispersion in $BaTiO_3$, $SrTiO_3$, and TiO_2, *Phys. Rev.,* 126, 1710, 1962.
129. **Barker, A. S.,** Infrared lattice vibrations and dielectric dispersion in corundum, *Phys. Rev.,* 132, 1475, 1963.
130. **Korn, D. M., Pine, A. S., and Dresselhaus, G.,** Infrared reflectivity of paratellurite, TeO_2, *Phys. Rev. B,* 8, 768, 1973.
131. **Collins, R. J. and Kleinman, D. A.,** Infrared reflectivity of zinc oxide, *J. Phys. Chem. Solids,* 33, 293, 1972.
132. **Jasperse, J. E., Kahan, A., Plendl, J. M., and Mitra, S. S.,** Temperature dependence of infrared dispersion in ionic crystals LiF and MgO, *Phys. Rev.,* 146, 526, 1966.
133. **Kachare, A., Andermann, G., and Brantley, L. R.,** Reliability of classical dispersion of LiF and MgO reflectance data, *J. Phys. Chem. Solids,* 33, 467, 1972.
134. **Hass, M.,** Temperature dependence of the infrared reflection of sodium chloride, *Phys. Rev.,* 117, 1497, 1960.
135. **Barker, A. S.,** Transverse and longitudinal modes study in MgF_2 and ZnF_2, *Phys. Rev.,* 136, 1290, 1964.
136. **Kaizer, W., Spitzer, W. G., Kaizer, R. H., and Howarth, L. E.,** Infrared properties of CaF_2, SrF_2, and BaF_2, *Phys. Rev.,* 127, 1950, 1962.
137. **Sieskind, M., Ayadi, M., and Zachmann, G.,** Infrared lattice vibration, dielectric dispersion and lattice dynamics of BaFCl, *Phys. Status Solidi B,* 136, 489, 1986.
138. **Belousov, M. V., Pogarev, D. E., and Shultin, A. A.,** Intense bands of out-of-plane vibration, v_2, of CO_3^{2-} and NO_3^{2-} in crystals with the structure of calcite and aragonite, *Fiz. Tverd. Tela* (Sov. Solid State Phys.), 11, 2697, 1969.
139. **Gervais, F., Piriou, B., and Cabannes, F.,** Anharmonicity of infrared vibration modes in beryl, *Phys. Status Solidi B,* 51, 701, 1972.
140. **Gervais, F., Piriou, B., and Cabannes, F.,** Anharmonicity in silicate crystals: temperature dependence of A_u type vibration modes in $ZrSiO_4$ and $LiAlSi_2O_6$, *J. Phys. Chem. Solids,* 34, 1785, 1973.
141. **Pavinich, V. F. and Belousov, M. V.,** Dispersion analysis of reflection spectra of monoclinic crystals, *Opt. Spektrosk.* (Sov. Optics and Spectroscopy), 5, 1114, 1978.
142. **Kravits, L. C., Kingsley, J. D., and Elkin, E. L.,** Raman and infrared studies of coupled PO_4^{3-} vibrations, *J. Chem. Phys.,* 49, 4600, 1968.
143. **Armbruster, A.,** Infrared reflection studies on the phosphates, arsenates, and vanadates of lutetium and yttrium, *J. Phys. Chem. Solids,* 37, 321, 1976.
144. **Lazarev, A. N., Mazhenov, N. A., and Mirgorodskii, A. P.,** Optical vibrations of YPO_4 crystal and its analogs; resonance splitting of internal vibrations of complex anions, *Izv. AN SSSR, Neorgan. Mater.* (Acad. Sci. USSR, Inorganic Materials), 14, 2107, 1978.
145. **Barker, A. S.,** Dielectric properties and optical phonons in $LiNbO_3$, *Phys. Rev.,* 158, 433, 1967.
146. **Barker, A. S.,** Infrared lattice vibrations in calcium tungstate and calcium molybdate, *Phys. Rev.,* 135, 742, 1964.

147. **Neufeld, J. D., Brantley, L. R., Sakamoto, P., and Andermann, G.,** Classical dispersion analysis of multiband system in the infrared region, *Appl. Spectros.*, 27, 198, 1973.
148. **Verleur, H. W.,** Determination of optical constants from reflectance or transmittance measurements on bulk crystals or thin films, *J. Opt. Soc. Am.*, 58, 1356, 1968.
149. **Gervais, F. and Piriou, B.,** Temperature dependence of transverse and longitudinal modes in TiO_2 (rutile), *Phys. Rev. B*, 10, 1642, 1974.
150. **Gervais, F. and Piriou, B.,** Temperature dependence of transverse and longitudinal modes in the α and β phases of quartz, *Phys. Rev. B*, 11, 3944, 1975.
151. **Belousov, M. V. and Pogarev, D. E.,** Dispersion analysis of complicated spectra, *Opt. Spektrosk.* (Sov. Optics and Spectroscopy), 38, 1018, 1975.
152. **Efimov, A. M., Makarova, E. G., Evtushenko, I. V., and Arkatova, T. G.,** Interpretation of IR spectra of sodium germanate glasses as follows from the data of dispersion analysis, in *Proc. 6th All-Union Conf. "Optical and Spectroscopic Properties of Glasses,"* Zakis, Yu. R., Ed., Zinatne, Riga, 1986, 82.
153. **Shevlyakova, N. M., Makarova, E. G., Efimov, A. M., and Volkova, V. V.,** Dispersion analysis of reflection spectrum in the frequency range of internal vibrations of tellurite polyanion, in *Proc. 12th All-Union Conf. "Application of Vibrational Spectra to Studying the Inorganic and Coordination Compounds,"* Komyak, A. I., Ed., Minsk, 1989, 9.
154. **Mikhailov, B. A., Arkatova, T. G., and Efimov, A. M.,** Quantitative registration of reflectivity and calculation of optical constants of borate glasses in the IR range, *Fiz. Khim. Stekla* (Sov. J. Glass Phys. Chem.), 5, 681, 1979.
155. **Mazurin, O. V., Streltsina, M. V., and Shvaiko-Shvaikovskaya, T. V.,** *Handbook on Glass Data. Part A. Silica Glass and Binary Silicate Glasses*, Elsevier, New York, 1983, 145-146.
156. **Von Hippel, A. R., Ed.,** *Dielectric Materials and Application*, Technology Press, New York, 1954.
157. **Efimov, A. M., Khar'uzov, V. A., and Proskuryakov, M. V.,** Dielectric properties of borate and germanate glasses, *Fiz. Khim. Stekla* (Sov. Phys. Chem. Glass), 6, 737, 1980.
158. **Vlasov, A. G. and Florinskaya, V. A., Eds.,** *Struktura i Fiziko Khimicheskie Svoistva Neorganicheskikh Stekol* (Structure, Physical and Chemical Properties of Inorganic Glasses), Khimiya, Leningrad, 1974, 320.
159. **Efimov, A. M. and Makarova, E. G.,** Dispersion analysis of reflection spectrum of vitreous SiO_2 in the vacuum ultraviolet, *Opt. Spektrosk.* (Sov. Optics and Spectroscopy), 67, 659, 1989.
160. **Wong, J. and Angell, C. A.,** *Glass Structure by Spectroscopy*, Marcel Dekker, New York, 1976.
161. **Lazarev, A. N.,** *Vibrational Spectra and Structure of Silicates*, Consultants Bureau, New York, 1972.
162. **Lazarev, A. N., Mirgorodskii, A. P., and Ignat'ev, I. S.,** *Kolebatel'nye Spektry Slozhnykh Okislov* (Vibrational Spectra of Complex Oxides), Nauka, Leningrad, 1975, Chaps. 2-5.
163. **Efimov, A. M. and Makarova, E. G.,** Developing the quantitative methods of data treatment in the IR and VUV spectroscopy of glass and using the methods for studying glass structure, in *Stekloobraznoe Sostoyanie* (Vitreous State), Proc. 8th All-Union Conf., Porai-Koshits, E. A., Ed., Nauka, Leningrad, 1988, 52.
164. **Simon, I. and McMahon, H. O.,** Study of the structure of quartz, cristobalite, and vitreous silica by reflection in infrared, *J. Chem. Phys.*, 21, 23, 1953.
165. **Neuroth, N.,** Uber die Bestimmung der optischen Konstanten n, κ aus Reflexionsmessungen, *Zs. Phys.*, 144, 85, 1956.
166. **Miler, M.,** Infrared absorption of glassy silicon dioxide, *Czech. J. Phys. B*, 18, 354, 1968.
167. **Girin, O. P., Kondrat'ev, Yu. N., and Raaben, E. L.,** Optical constants and spectral microcharacteristics of Na_2O-SiO_2 glasses in the infrared range, *Opt. Spektrosk.* (Sov. Optics and Spectroscopy), 29, 745, 1970.
168. **Zolotarev, V. M.,** Optical constants of amorphous oxides SiO_2 and GeO_2 in the range of the stretching vibration, *Opt. Spektrosk.* (Sov. Optics and Spectroscopy), 29, 66, 1970.

169. **Popova, S. I., Tolstykh, T. S., and Vorob'ev, V. G.,** Optical characteristics of amorphous quartz in the 1400-200 cm^{-1} range, *Opt. Spektrosk.* (Sov. Optics and Spectroscopy), 33, 801, 1972.
170. **Steyer, T. R., Day, K. L., and Huffman, D. R.,** Infrared absorption by small amorphous quartz spheres, *Appl. Opt.,* 13, 1586, 1974.
171. **Saidov, G. V. and Bernstein, E. B.,** Optical constants of surface layer of fused quartz in the 900-1300 cm^{-1} range, *Fiz. Khim. Stekla* (Sov. J. Glass Phys. Chem.), 8, 75, 1982.
172. **Mamedov, R. K., Mansurov, G. M., and Dubovikov, N. I.,** Optical constants of quartz glass in the IR range, *Opt. Mekh. Prom.* (Optical and Mechanical Industry), 4, 56, 1982.
173. **Zolotarev, V. M., Morozov, V. N., and Smirnova, E. V.,** *Optical Constants of Natural and Commercial Media,* Bakhshiev, N. G., Ed., Khimiya, Leningrad, 1984, Chap. 6 and Appendix 1.
174. **Mansurov, G. M., Mamedov, R. K., Sudarushkin, A. S., Sidorin, V. K., Sidorin, K. K., Pshenitsyn, V. I., and Zolotarev, V. M.,** Studying the nature of polished surface of silica glass with ellipsometry and spectroscopy, *Opt. Spektrosk.* (Sov. Optics and Spectroscopy), 52, 852, 1982.
175. **Mansurov, G. M., Rozanov, N. N., Zolotarev, V. M., and Sutovskii, S. M.,** Determining the optical characteristics of materials in the bulk and surface layer by the internal reflection spectra, *Opt. Spektrosk.* (Sov. Optics and Spectroscopy), 53, 301, 1982.
176. **Crozier, D. and Douglas, R. W.,** Study of sodium silicate glasses in the infra-red by means of thin films, *Phys. Chem. Glasses,* 6, 240, 1965.
177. **Sen, P. N. and Thorpe, M. F.,** Phonons in AX_2 glasses: from molecular to band-like modes, *Phys. Rev. B,* 15, 4030, 1977.
178. **Lazarev, A. N., Mirgorodskii, A. P., and Smirnov, M. B.,** *Kolebatel'nye Spektry i Dinamika Inonno-Kovalentnykh Kristallov* (Vibrational Spectra and Dynamics of Ionic and Covalent Crystals), Nauka, Leningrad, 1985, Chap. 3.
179. **Vlasov, A. G. and Florinskaya, V. A., Eds.,** *Infrakrasnye Spektry Neorganicheskikh Stekol i Kristallov* (Infrared Spectra of Inorganic Glasses and Crystals), Khimiya, Leningrad, 1972, Part 1.
180. **Sanders, D. M., Person, W. B., and Hench, L. L.,** Quantitative analysis of glass structure with the use of infrared reflection spectra, *Appl. Spectros.,* 28, 247, 1974.
181. **Ghivelder, L. and Phillips, W. A.,** Far infrared absorption in disordered solids, *J. Non-Cryst. Solids,* 109, 280, 1989.
182. **Bagdade, W. and Stolen, R.,** *J. Phys. Chem. Solids,* 29, 2001, 1968.
183. **Wong, P. T. T. and Whalley, E.,** Infrared and Raman spectra of glasses. Part 2. Far infrared spectrum of vitreous silica in the range 100-15 cm^{-1}, *Disc. Faraday Soc.,* 50, 94, 1970.
184. **Stolen, R. H., Krause, J. T., and Kurkjian, C. R.,** Raman scattering and far infrared absorption in neutron-compacted silica, *Disc. Faraday Soc.,* 50, 103, 1970.
185. **Etchepare, J., Merian, M., and Smetankine, L.,** Vibrational normal modes of SiO_2, *J. Chem. Phys.,* 60, 1873, 1974.
186. **Sweet, J. R. and White, W. B.,** Study of sodium silicate glasses and liquids by infrared reflectance spectra, *Phys. Chem. Glasses,* 10, 246, 1969.
187. **Domine, F. and Piriou, B.,** Study of sodium silicate melt and glass by infrared reflectance spectroscopy, *J. Non-Cryst. Solids,* 1983, 55, 125.
188. **Smirnova, E. V., Makarova, E. G., and Morozov, V. M.,** Optical constants of lead glasses, *Zh. Prikl. Spektrosk.* (Sov. J. Appl. Spectr.), 21, 940, 1974.
189. **Vlasov, A. G. and Florinskaya, V. A., Eds.,** *Struktura i Fiziko-Khimicheskie Svoistva Neorganicheskikh Stekol* (Structure, Physical and Chemical Properties of Inorganic Glasses), Khimiya, Leningrad, 1974.
190. **Worrell, C. A. and Henshall, T.,** Vibrational spectroscopic studies of some lead silicate glasses, *J. Non-Cryst. Solids,* 29, 283, 1978.
191. **Bagley, B. G., Vogel, E. M., French, W. G., Pasteur, G. A., Gan, J. N., and Tauc, J.,** The optical properties of a soda-lime-silica glass in the region from 0.006 to 22 eV, *J. Non-Cryst. Solids,* 22, 423, 1976.

192. **Hsieh, C. K. and Su, K. C.,** Thermal radiative properties of glass from 0.32 to 206 μm, *Solar Energy,* 22, 37, 1979.
193. **Vlasov, A. G. and Florinskaya, V. A., Eds.,** *Infrakrasnye Spektry Zhshelochnykh Silikatov* (Infrared Spectra of Alkali Silicates), Khimiya, Leningrad, 1970.
194. **Ferraro, J. R. and Manghnani, M. H.,** Infrared absorption spectra of sodium silicate glasses at high pressures, *J. Appl. Phys.,* 43, 1595, 1972.
195. **Kolesova, V. A.,** Investigation of structure of alkalialuminosilicate glasses by their infrared absorption spectrum, in *Stekloobraznoe Sostoyanie* (Vitreous State), Proc. 3rd All-Union Conf., Porai-Koshits, E. A., Ed., Nauka, Leningrad, 1960, 203.
196. **Ocana, M., Fornes, V., and Serna, C. J.,** The variability of the infrared powder spectrum of amorphous SiO_2, *J. Non-Cryst. Solids,* 107, 187, 1989.
197. **Exarhos, G. J., Miller, P. J., and Risen, W. M.,** Interionic vibrations and glass transitions in ionic oxide metaphosphate glasses, *J. Chem. Phys.,* 60, 4145, 1974.
198. **Exarhos, G. J., Miller, P. J., and Risen, W. M.,** Calculation of ionic conductivity activation energies in ionic oxide glasses from spectroscopic data, *Solid State Commun.,* 17, 29, 1975.
199. **Kamitsos, E. I. and Risen, W. M.,** Vibrational spectra of single and mixed alkali pentasilicate glasses, *J. Non-Cryst. Solids,* 65, 333, 1984.
200. **Hanna, R.,** The structure of sodium silicate glasses and their far infrared absorption spectra, *J. Phys. Chem.,* 69, 3846, 1965.
201. **Smirnova, E. V. and Vuks, V. M.,** Optical constants of flint glasses in the 1300-300 cm^{-1} range, *Zh. Prikl. Spektrosk.* (Sov. J. Appl. Spectr.), 20, 731, 1974.
202. **Jellison, J. E. and Bray, P. J.,** A structural interpretation of ^{10}B and ^{11}B NMR spectra in sodium borate glasses, *J. Non-Cryst. Solids,* 29, 187, 1978.
203. **Bray, P. J., Geissberger A. E., Bucholtz, F., and Harris, J. A.,** Glass structure, *J. Non-Cryst. Solids,* 52, 45, 1982.
204. **Jellyman, P. E. and Procter, J. P.,** Infrared reflection spectra of glasses, *J. Soc. Glass. Technol.,* 39, 173t, 1955.
205. **Parsons, J. L. and Millberg, M. E.,** Vibrational spectra of vitreous $B_2O_3 \cdot xH_2O$, *J. Am. Ceram. Soc.,* 43, 326, 1960.
206. **Borelli, N. F., McSwain, B. D., and Su, G.-J.,** The infrared spectra of vitreous boron dioxide and sodium borate glasses, *Phys. Chem. Glasses,* 4, 11, 1963.
207. **Krogh-Moe, J.,** Interpretation of the infrared spectra of boron oxide and alkali borate glasses, *Phys. Chem. Glasses,* 6, 46, 1965.
208. **Konijendijk, W. L.,** The structure of borosilicate glasses, *Philips Res. Rep. Suppl.,* 1, 1, 1975.
209. **Chryssikos, G. D., Turkotte, D. E., Mulkern, R. V., Bray, P. J., and Risen, V. M.,** Infrared study of cadmium borogermanate glasses, *J. Non-Cryst. Solids,* 85, 54, 1986.
210. **Kolesova, V. A.,** Vibrational spectra and the structure of alkali borate glasses, *Fiz. Khim. Stekla* (Sov. J. Glass Phys. Chem.), 12, 4, 1986.
211. **Tenney, A. S. and Wong, J.,** Vibrational spectra of vapor-deposited binary borosilicate glasses, *J. Chem. Phys.,* 56, 5516, 1972.
212. **Efimov, A. M., Polushkin, Yu. I., Mikhailov, B. A., Fomina, T. N., and Arkatova, T. G.,** Low-frequency IR reflection spectra and optical constants of some borate glasses, *Fiz. Khim. Stekla* (Sov. J. Glass Phys. Chem.), 6, 706, 1980.
213. **Sinclair, R. N., Desa, J. A., Etherington, G., Johnson, P. A., and Wright, A. C.,** Neutron diffraction studies of amorphous solids, *J. Non-Cryst. Solids,* 42, 107, 1980.
214. **Weir, C. E. and Schroeder, R. A.,** Infrared spectra of the crystalline inorganic borates, *J. Res. Nat. Bur. Stand.,* 68A, 465, 1964.
215. **Bray, P. J.,** Investigation of glass structure with the nuclear magnetic resonance, in *Stekloobraznoe Sostoyanie* (Vitreous State), Proc. 4th All-Union Conf., Porai-Koshits, E. A., Ed., Nauka, Moscow-Leningrad, 1965, 237.
216. **Murthy, M. K. and Kirby, E. M.,** Infrared spectra of alkali germanate glasses, *Phys. Chem. Glasses,* 5, 144, 1964.

217. **Venediktov, A. A., Morozov, V. N., and Polykhin, V. N.,** Reflection spectra of the crystalline and vitreous modifications of SiO_2 and GeO_2 in the 5-50 micron range, Zh. Prikl. Spektrosk. (Sov. J. Appl. Spectr.), 10, 969, 1969.
218. **Chen, B. T. K. and Su, G.-J.,** Infrared spectra of vitreous germanium dioxide and their interpretation, *Phys. Chem. Glasses,* 12, 33, 1971.
219. **Zolotarev, V. M. and Morozov, V. N.,** Optical constants of oxides SiO_2 and GeO_2 in the 4000-20 cm^{-1} range, *Opt. Spektrosk.* (Sov. Optics and Spectroscopy), 34, 319, 1973.
220. **Kolesova, V. A.,** Glasses and solid solutions in the GeO_2-SiO_2 system, *Fiz. Khim. Stekla* (Sov. J. Glass Phys. Chem.), 1, 70, 1975.
221. **Riebling, E. F.,** Infrared study of polymerization in silver, thallium, and thallium aluminogermanate glasses, *J. Mater. Sci.,* 7, 40, 1972.
222. **Efimov, A. M. and Rogova, N. P.,** Reflection spectra of La_2O_3-B_2O_3-GeO_2 glasses, *Fiz. Khim. Stekla* (Sov. J. Glass Phys. Chem.), 2, 21, 1976.
223. **Ingri, N. and Lundgren, G.,** The crystal structure of $Na_4Ge_9O_{20}$, *Acta Chem. Scand.,* 17, 617, 1963.
224. **Roessler, D. M. and Albers, W. A.,** Infrared reflectance spectrum of single crystal tetragonal GeO_2, *J. Phys. Chem. Solids,* 33, 293, 1972.
225. **Scott, J. F.,** Raman spectra of GeO_2, *Phys. Rev.,* 3488, 1970.
226. **Dimitriev, Y., Dimitrov, V., and Arnaudov, M.,** IR spectra and structures of tellurite glasses, *J. Mater. Sci.,* 18, 1353, 1983.
227. **Ovcharenko, N. V., Volkova, V. V., Arkatova, T. G., and Fomina, T. N.,** Investigation of structure of tellurite glasses with the IR spectroscopy, in *Proc. 7th All-Union Symp. "Optical and Spectroscopic Properties of Glasses,"* State Optical Institute, Leningrad, 1989, 187.
228. **Heo, J., Lam, D., Sigel, G. H., Mendoza, E. A., and Hensley, D. A.,** Spectroscopic analysis of the structure and properties of alkali tellurite glasses, *J. Am. Ceram. Soc.,* 75, 277, 1992.
229. **Korn, D. M., Pine, A. S., Dresselhaus, G., and Reed, T. B.,** Infrared reflectivity of paratellurite, TeO_2, *Phys. Rev. B,* 8, 768, 1973.
230. **Aurault, B., Decamps, E.-A., Abba, F., Marqueton, Y., and Durand, M.,** Lattice dynamics of paratellurite, TeO_2, *Solid State Commun.,* 11, 639, 1972.
231. **Libov, V. S., Perova, T. S., and Efimov, A. M.,** Comparison between the Raman spectra of alkali tellurite glasses and of paratellurite crystal, in *Proc. 7th All-Union Symp. "Optical and Spectroscopic Properties of Glasses,"* State Optical Institute, Leningrad, 1989, 187.
232. **Sekiya, T., Mochida, N., Ohtsuka, A., and Tonokawa, M.,** Raman spectra of $MO_{1/2}$-TeO_2 (M = Li, Na, K, Rb, Cs, and Tl) glasses, *J. Non-Cryst. Solids,* 144, 128, 1992.
233. **Kolesova, V. A.,** Vibrational spectra of some crystalline and vitreous phosphates, *Opt. Spektrosk.* (Sov. Optics and Spectroscopy), 2, 165, 1957.
234. **Shih, C. K. and Su, G.-J.,** Infrared spectra of binary metaphosphate glasses, *Proc. 7th Int. Congr. Glass,* Brussels, 1965, paper 48.
235. **Bartolomew, R. F.,** Structure and properties of silver phosphate glasses — infrared and visible spectra, *J. Non-Cryst. Solids,* 7, 221, 1972.
236. **Wong, J.,** Vibrational spectra of vapor-deposited binary phosphosilicate glasses, *J. Non-Cryst. Solids,* 20, 83, 1976.
237. **Hogarth, C. A. and Hosseini, A. A.,** Infrared absorption and structural investigation of some vanadate glasses, *J. Mater. Sci. Lett.,* 3, 359, 1984.
238. **Cherches, G. Kh., Pechkovskii, V. V., Kuz'menkov, M. I., and Barannikova, T. I.,** IR spectra ov vitreous polyphosphates of alkali metals, *Fiz. Khim. Stekla* (Sov. J. Glass Phys. Chem.), 4, 233, 1978.
239. **Miroshnichenko, O. Ya. and Mombelli, V. V.,** IR spectra and the structure of BaO-WO_3-P_2O_5 glasses, *Fiz. Khim. Stekla* (Sov. J. Glass Phys. Chem.), 5, 30, 1979.
240. **Miroshnichenko, O. Ya. and Mombelli, V. V.,** IR spectra and the structure of Na_2O-V_2O_5-P_2O_5 glasses, *Fiz. Khim. Stekla* (Sov. J. Glass Phys. Chem.), 9, 521, 1983.

241. **Miroshnichenko, O. Ya. and Mombelli, V. V.,** IR spectra and the structure of molybdenophosphate glasses, *Fiz. Khim. Stekla* (Sov. J. Glass Phys. Chem.), 13, 914, 1987.
242. **Gray, P. and Klein, L. C.,** Optical spectra of sodium phosphate glasses, *J. Non-Cryst. Solids,* 68, 75, 1984.
243. **Nelson, B. N. and Exarhos, G. J.,** Vibrational spectroscopy of cation-site interactions in phosphate glasses, *J. Chem. Phys.,* 71, 2739, 1979.
244. **Bues, W. and Gehrke, H.-W.,** Schwingungsspektren Gechmolzener, Glasiger und Kristallisierter Hochpolymerer Phosphate, *Z. Anorg. Allg. Chem.,* 288, 307, 1956.
245. **Corbridge, D. E. C.,** The infrared spectra of phosphorus compounds, in *Topics in Phosphorus Chemistry,* Grayson, M. and Griffith, E. J., Eds., Interscience, New York, 1969, Vol. 6, p. 235.
246. **Corl, E. A., Silverman, S. L., and Kim, Y. S.,** Non-destructive detection of phosphorus oxide layers on semiconductor wafers, *Solid State Electr.,* 9, 1009, 1966.
247. **Murthy, M. K.,** Phosphate-halide systems: II, Infrared spectra of glasses in the system $NaPO_3$-NaF, *J. Am. Ceram. Soc.,* 46, 558, 1963.
248. **Tarlakov, Yu. P., Khalilev, V. D., and Pogosyan, M. A.,** IR spectroscopic study of ($0.3LiF \cdot 0.7NaPO_3$)-MeF_x (Me being Mg, Ca, Al) glasses, *Fiz. Khim. Stekla* (Sov. J. Glass Phys. Chem.), 4, 175, 1978.
249. **Pronkin, A. A., Tarlakov, Yu. P., Igitkhanyan, Yu. G., Dzhavuktsyan, S. G., and Khalilev, V. D.,** Electrical conductivity and the IR spectra of $Ba(PO_3)_2$.AlF_3-RF_2 (R = Mg, Ca) glasses, *Fiz. Khim. Stekla* (Sov. J. Glass Phys. Chem.), 5, 252, 1979.
250. **Smirnova, E. V. and Urusovskaya, L. N.,** Infrared spectra of alkali fluoroaluminophosphate glasses, *Zh. Prikl. Spektrosk.* (Sov. J. Appl. Spectr.), 30, 1026, 1979.
251. **Galant, V. E., Smirnova, E. V., and Urusovskaya, L. N.,** Infrared spectra of $Al(PO_3)_3$-BaF_2 glasses, *Fiz. Khim. Stekla* (Sov. J. Glass Phys. Chem.), 6, 321, 1980.
252. **Gan Fuxi,** *Optical and Spectroscopic Properties of Glass,* Springer-Verlag, Berlin, Shanghai, 1992.
253. **Almeida, R. M. and Mackenzie, J. D.,** Infrared absorption and structure of chlorophosphate glasses, *J. Non-Cryst. Solids,* 40, 535, 1980.
254. **Baldwin, C. M., Almeida, R. M., and Mackenzie, J. D.,** Halide glasses, *J. Non-Cryst. Solids,* 43, 309, 1981.
255. **Almeida, R. M.,** Vibrational spectroscopy of glasses, *J. Non-Cryst. Solids,* 106, 347, 1988.
256. **Almeida, R. M.,** Fluoride glasses, in *Handbook on the Physics and Chemistry of Rare Earths,* Vol. 15, Gschneider, K. A. and Eyring, L., Eds., Elsevier, Amsterdam, 1991, Chap. 101.
257. **Bendow, B., Drexhage, M. G., Banerjee, P. K., Goltman, J., Mitra, S. S., and Moynihan, C. T.,** Vibrational spectra of fluorohafnate glasses, *Solid State Commun.,* 37, 485, 1981.
258. **Bendow, B., Banerjee, P. K., Drexhage, M. G., El-Bayoumi, O. H., Mitra, S. S., Moynihan, C. T., Gavin, D., Fonteneau, G., Lucas, J., and Poulain, M.,** Effect of aluminium content on infrared characteristics of BaF_2/ThF_4-containing glasses, *J. Am. Ceram. Soc.,* 66, C-64, 1983.
259. **Almeida, R. M.,** Short and intermediate range structures in fluoride glasses by vibrational spectroscopy, *J. Non-Cryst. Solids,* 140, 92, 1992.
260. **Almeida, R. M., Pereira, J. C., and Alcala, R.,** Vibrational spectra and structure of glasses based on ZnF_2-CdF_2, *Phys. Chem. Glasses,* 33, 187, 1992.
261. **Almeida, R. M., Pereira, J. C., Messaddeq, Y., and Aegerter, M. A.,** Vibrational spectra and structure of fluoroindate glasses, *J. Non-Cryst. Solids,* 161, 105, 1993.
262. **Hilton, A. R., Jones, C. E., and Brau, M.,** Non-oxide IVA-VA-VIA chalcogenide glasses. Part 3. Structural studies, *Phys. Chem. Glasses,* 7, 116, 1966.
263. **Zlatkin, L. B. and Markov, Yu. F.,** On the nature of the optical vibrations in As_2S_3 and As_2Se_3 glasses, *Phys. Status Solidi (a),* 4, 391, 1971.
264. **Lucovsky, G. and Galeener, F. L.,** Infrared studies of amorphous semiconductors, in *Structure and Properties of Non-Crystalline Semiconductors,* Proc. 6th Int. Conf. Amorphous and Liquid Semiconductors, Kolomiets, Leningrad, 1975, B. T., Ed., 207.

265. **Lucovsky, G., Keezer, R. C., Six, H. A., and Geils, R. H.,** IR active modes in the amorphous alloy systems: GeS_2-$GeSe_2$, As_2S_3-GeS_2 and As_2Se_3-$GeSe_2$, in *Structure and Properties of Non-Crystalline Semiconductors,* Proc. 6th Int. Conf. Amorphous and Liquid Semiconductors, Kolomiets, Leningrad, 1975, B. T., Ed., 296.
266. **Galeener, F. L. and Lucovsky, G.,** Second order vibrational spectra of vitreous GeS_2 and As_2S_3, in *Structure and Properties of Non-Crystalline Semiconductors,* Proc. 6th Int. Conf. on Amorphous and Liquid Semiconductors, Kolomiets, Leningrad, 1975, B. T., Ed., 305.
267. **Khiminets, V. V., Gerasimenko, V. S., Khiminets, O. V., Turyanitsa, I. D., and Chepur, D. V.,** Vibrational spectra of binary As-S glasses, *Fiz. Khim. Stekla* (Sov. J. Glass Phys. Chem.), 4, 648, 1978.
268. **Kato, M., Onari, S., and Arai, T.,** Far infrared and Raman spectra in $(As_2S_3)_{1-x}(Sb_2S_3)_x$ glasses, *Jpn. J. Appl. Phys.,* 22, 1382, 1983.
269. **Kato, M., Onari, S., and Arai, T.,** Far infrared and Raman spectra in $(As_2Se_3)_{1-x}(Sb_2Se_3)_x$ glasses, *Jpn. J. Appl. Phys.,* 20, 2013, 1981.
270. **Lucovsky, G., Keezer, R. C., Geils, R. H., and Six, H. A.,** Structural interpretation of the infrared and Raman spectra of glasses in the alloy system $Ge_{1-x}S_x$, *Phys. Rev. B,* 10, 5134, 1974.
271. **Cooper, W. C., Ed.,** *Physics of Selenium and Tellurium,* Pergamon Press, New York, 1969, 255.
272. **Felty, E. J., Lucovsky, G., and Myers, M. B.,** Optical properties of the mixed amorphous system $As_2S_xSe_{3-x}$, *Solid State Commun.,* 5, 555, 1967.
273. **Moynihan, C. T., Macedo, P. B., Maklad, M. S., Mohr, R. K., and Howard, R. E.,** Intrinsic and impurity infrared absorption in As_2Se_3 glass, *J. Non-Cryst. Solids,* 17, 369, 1975.
274. **Mollot, F., Cernogora, J., Ribeyron, A., and Laurant, J. M.,** Variation of far infrared absorption after optical excitation in amorphous As_2Se_3 at 4.2 K, in *Structure and Properties of Non-Crystalline Semiconductors,* Proc. 6th Int. Conf. on Amorphous and Liquid Semiconductors, Kolomiets, Leningrad, 1975, B. T., Ed., 237.
275. **Onari, S., Matsuishi, K., and Arai, T.,** Far-infrared absorption spectra and the spatial fluctuation of charges on amorphous As-S and As-Se systems, *J. Non-Crystalline Solids,* 86, 22, 1986.
276. **Ohsaka, T.,** IR spectra of glassy Se containing small amounts of S, Te, As, or Ge, *J. Non-Cryst. Solids,* 17, 121, 1975.
277. **Zirke, J. and Tausend, A.,** Disorder-induced effects on the infrared spectra of amorphous selenium, in *Structure and Properties of Non-Crystalline Semiconductors,* Proc. 6th Int. Conf. on Amorphous and Liquid Semiconductors, Kolomiets, B. T., Ed., Leningrad, 1975, 310.
278. **Lucovsky, G., Mooradian, A., Taylor, W., Wright, G. B., and Keezer, R. C.,** Identification of the fundamental vibrational modes of trigonal, α-monoclinic, and amorphous selenium, *Solid State Commun.,* 5, 113, 1967.
279. **Lucovsky, G. and Martin, R. M.,** A molecular model for the vibrational modes in chalcogenide glasses, *J. Non-Cryst. Solids,* 8-10, 185, 1972.
280. **Efimov, A. M. and Khar'yusov, V. A.,** Dielectric properties and structure of chalcogenide glasses, in *Structure and Properties of Non-Crystalline Semiconductors,* Proc. 6th Int. Conf. on Amorphous and Liquid Semiconductors, Kolomiets, Leningrad, 1975, B. T., Ed., 113.
281. **Savage, J. A., Webber, P. J., and Pitt, A. N.,** Infrared optical glasses for applications in 8-12-mm thermal imaging systems, *Appl. Opt.,* 16, 2938, 1977.
282. **Savage, J. A., Webber, P. J., and Pitt, A. N.,** The potential of Ge-As-Se-Te glasses as 3-5 and 8-12-μm infrared optical materials, *Infrared Phys.,* 20, 313, 1980.
283. **Gerasimenko, V. S., Khiminets, O. V., Khiminets, V. V., and Turyanitsa, I. D.,** Vibrational spectra of As-S-Br glasses, *Fiz. Khim. Stekla* (Sov. J. Glass Phys. Chem.), 4, 112, 1978.
284. **Sichka, M. Yu., Migolinets, I. M., and Gerasimenko, V. S.,** Glass formation in the Ge-Se-Te-I system and the basic physical properties of glasses, *Fiz. Khim. Stekla* (Sov. J. Glass Phys. Chem.), 15, 176, 1989.

285. **Harfmann, E., Videau, J. J., Portier, J., and Tanguy, B.,** New IR transmitting 40HgS-40PbBr$_2$-20PbF$_2$ glasses, *J. Non-Cryst. Solids,* 161, 309, 1993.
286. **Philipp, H. R.,** Optical properties of non-crystalline Si, SiO, SiO$_x$, and SiO$_2$, *J. Phys. Chem. Solids,* 32, 1935, 1971.
287. **Bosio, C., Czaja, W., and Mertins, H. C.,** VUV reflectivity of crystalline α-SiO$_2$ and of amorphous SiO$_2$, *Europhys. Lett.,* 18, 319, 1992.
288. **Osantovsky, J. F.,** Reflectance and optical constants for Cer-Vit from 250 to 1050 Å, *J. Opt. Soc. Am.,* 64, 834, 1974.
289. **Lamy, P. L.,** Optical constants of crystalline and fused quartz in the far ultraviolet, *Appl. Opt.,* 16, 2212, 1977.
290. **Efimov, A. M., Sidorin, V. K., and Sidorin, K. K.,** Reflection spectra of borate glasses in the vacuum ultraviolet, *Fiz. Khim. Stekla* (Sov. J. Glass Phys. Chem.), 7, 647, 1981.
291. **Izumitani, T., Hirota, S., Tanaka, K., and Onuki, H.,** Dispersion and reflection spectra of fluoride and oxide glasses in the extreme ultraviolet region, *J. Non-Cryst. Solids,* 86, 361, 1986.
292. **Belle, M. L., Kolomiets, B. T., and Pavlov, B. V.,** Comparative study of optical properties of arsenic chalcogenides in crystalline and vitreous state, *Fiz. Tekh. Poluprovodn.* (Sov. J. Phys. Techn. Semicond.), 2, 1448, 1968.
293. **Zallen, R., Drews, R. E., Emerald, R. L., and Slade, M. L.,** Electronic structure of crystalline and amorphous As$_2$S$_3$ and As$_2$Se$_3$, *Phys. Rev. Lett.,* 26, 1564, 1971.
294. **Drews, R. E., Emerald, R. L., Slade, M. L., and Zallen, R.,** Interband spectra of As$_2$S$_3$ and As$_2$Se$_3$ crystals and glasses, *Solid State Commun.,* 10, 293, 1972.
295. **Velicky, B., Zavetova, M., and Pajasova, L.,** Interband optical transitions in glassy arsenic telluride, in *Structure and Properties of Non-Crystalline Semiconductors,* Proc. 6th Int. Conf. on Amorphous and Liquid Semiconductors, Kolomiets, Leningrad, 1975, B. T., Ed., 273.
296. **Aspnes, D. E., Phillips, J. C., Tai, K. L., and Bridenbaugh, P. M.,** Optical spectra and electronic structure of crystalline and glassy Ge(S, Se)$_2$, *Phys. Rev. B,* 23, 816, 1981.
297. **Bendow, B., Drexhage, M. G., and Lipson, H. G.,** Infrared absorption in highly transparent fluorozirconate glasses, *J. Appl. Phys.,* 52, 1460, 1981.
298. **Payne, D. N. and Gambling, W. A.,** Zero material dispersion in optical fibres, *Electron Lett.,* 11, 176, 1975.
299. **Fleming, J. W.,** Material and mode dispersion in GeO$_2$ · B$_2$O$_3$ · SiO$_2$ glasses, *J. Am. Ceram. Soc.,* 59, 503, 1976.
300. **Fleming, J. W.,** Material dispersion in lightguide glasses, *Electron Lett.,* 14, 326, 1978.
301. **Fonteneau, G., Chiaruttini, I., Mitachi, S., and Lucas, J.,** Optical dispersion of an indium-based fluoride glass, *J. Non-Cryst. Solids,* 140, 340, 1992.
302. **Feltz, A., Burckhardt, W., Voigt, B., and Linke, D.,** Optical glasses for IR transmittance, *J. Non-Cryst. Solids,* 129, 31, 1991.
303. **Petrovskii, G. T.,** Properties and structure of glasses based on beryllium fluoride, in *Opticheskoe Steklo* (Optical Glass), Petrovskii, G. T., Ed., Mashinostroenie, Leningrad, 1972, 58.
304. **Ovcharenko, N. V. and Yakhind, A. K.,** New high-refracting optical glasses based on tellurium dioxide, *Opt. Mekh. Prom.* (Optical and Mechanical Industry), 3, 47, 1968.
305. **Aio, L. G., Efimov, A. M., and Kokorina, V. F.,** Refractive index of chalcogenide glasses over a wide range of compositions, *J. Non-Cryst. Solids,* 27, 299, 1978.
306. **Heo, J. and Mackenzie, J. D.,** Chalcohalide glasses, *J. Non-Cryst. Solids,* 111, 29, 1989.
307. **Ma, H. L., Zhang, X. H., Lucas, J., Iizuka, R., Yamagishi, T., and Yamashita, T.,** Refractive index measurement of the TeX glasses, *J. Non-Cryst. Solids,* 161, 198, 1993.
308. **Herzberger, M.,** The dispersion of optical glass, *J. Opt. Soc. Am.,* 32, 70, 1942.
309. **Malitson, J. H.,** Interspecimen comparison of the refractive index of fused silica, *J. Opt. Soc. Am.,* 55, 1205, 1965.
310. **Hirota, S. and Izumitani, T.,** Influence of oscillator strength on the refractive index of optical glasses, *Yoguo Kyokai-Shi* (J. Ceram. Assoc. Japan), 84, 435, 1976.

311. **Hirota, S. and Izumitani, T.,** Effect of cations on the inherent absorption wavelength and the oscillator strength of ultraviolet absorption in borate glasses, *J. Non-Cryst. Solids,* 29, 109, 1978.
312. **Bohl, M.,** Electronic structure, UV absorption, and dispersion of optical glasses, *Phys. Status Solidi A,* 63, 529, 1981.
313. **Stozharov, A. I. and Frolov, V. I.,** Calculation of the dispersion of optical glasses, *Opt. Mekh. Prom.* (Optical and Mechanical Industry), 7, 44, 1974.
314. **Agafonova, K. A. and Efimov, A. M.,** The calculation of parameters of the dispersion formula for describing the frequency dependence of glass refractive index in the transparency range, *Fiz. Khim. Stekla* (Sov. J. Glass Phys. Chem.), 4, 707, 1978.
315. **Bagley, B. G., Aspnes, D. E., and Fleming, J. W.,** Refractive index dispersion in oxide glasses, *J. Non-Cryst. Solids,* 40, 623, 1980.
316. **Engl, J.,** Zur Dispersion des kristallisierten und glasigen Quarzes, *Ann. Phys.,* 25, 600, 1936.
317. **Florinskaya, V. A.,** On the variation of the optical glass dispersion after heat treatment, *Zh. Tekh. Fiz.* (Sov. J. Appl. Phys.), 15, 681, 1945.
318. **Young, J. C. and Finn, A. N.,** Effect of composition and other factors on the specific refraction and dispersion of glasses, *J. Res. Nat. Bur. Stand.,* 25, 759, 1940.
319. **Huggins, M. L.,** The dispersion of silicate glasses as a function of composition, *J. Opt. Soc. Am.,* 30, 514, 1940.
320. **Huggins, M. L., Sun, K.-H., and Davis, D. O.,** The dispersion of silicate glasses as a function of composition, *J. Opt. Soc. Am.,* 32, 635, 1942.
321. **Ramachandran, G. N.,** Thermo-optic behavior of solids. VI. Optical glasses, *Proc. Ind. Acad. Sci.,* A25, 498, 1947.
322. **Polukhin, V. N. and Krymova, S. V.,** On the dependence of the specific refractivity and specific dispersion of optical glasses on frequency, *Fiz. Khim. Stekla* (Sov. J. Glass Phys. Chem.), 7, 374, 1981.
323. **Mitachi, S. and Miyashita, T.,** Refractive-index dispersion for BaF_2-GdF_3-ZrF_4-AlF_3 glasses, *Appl. Opt.,* 22, 2419, 1983.
324. **Sun, K.-H., Safford, H. W., and Silverman, A.,** Review of the relation of density and refractive index to the composition of glass: II, *J. Am. Ceram. Soc.,* 23, 343, 1940.
325. **Appen, A. A.,** *Khimiya Stekla,* Leningrad, Khimiya, 1974, Chap. 8.
326. **Demkina, L. I.,** A system for calculating glass properties via their chemical composition, in *Fiz. Khim. Silikatov* (Physics and Chemistry of Silicates), Shultz, M. M. and Grebenzhchikov, R. G., Eds., Leningrad, Nauka, 1987, p. 132.
327. **Appen, A. A.,** Calculation of the optical properties, density, and expansion coefficient of optical glasses via their composition, *Dokl. Akad. Nauk SSSR* (USSR Acad. Sci. Reports), 69, 841, 1949.
328. **Demkina, L. I.,** On the additivity of the optical properties of glass, *Dokl. Akad. Nauk SSSR* (USSR Acad. Sci. Reports), 58, 807, 1947.
329. **Demkina, L. I. and Loskutova, S. A.,** A system for calculating density, the refractive index, and the main dispersion of lanthanum borate glasses via their chemical composition, *Fiz. Khim. Stekla* (Sov. J. Glass Phys. Chem.), 15, 717, 1989.
330. **Gan Fuxi,** New system for calculating some physical properties of silicate glasses, *Sci. Sinica,* 12, 1365, 1963.

INDEX

D

E

F

S

T

U

V

W